Biodiversity in India
Volume 8

The Editors

Prof. T. Pullaiah obtained his M.Sc. (1973) and Ph.D. (1976) degrees in Botany from Andhra University. He was a Post Doctoral Fellow at Moscow State University, Russia during 1976-78. He traveled widely in Europe and USA and visited Universities and Botanical Gardens in about 17 countries. Professor Pullaiah joined Sri Krishnadevaraya University as Lecturer in 1979 and became Professor in 1993. He held several positions in the University, which include Dean, Faculty of Life Sciences, Head of the Department of Botany, Chairman, BOS in Botany, Head of the Department of Sericulture, Coordinator and Chairman, BOS in Biotechnology, Vice Principal and Principal, S.K. University College. He retired from active service on 31st May 2011. He was selected by UGC as UGC-BSR Faculty Fellow and is working in Sri Krishnadevaraya University. Prof. Pullaiah has published 62 books, 305 research papers and 35 popular articles. He is Principal Investigator of 20 major Research Projects totaling more than a Crore of Rupees funded by DBT, DST, CSIR, UGC, BSI, WWF, GCC etc. Under his guidance 53 students obtained their Ph.D. degrees and 35 students their M.Phil. degrees. He is recipient of Best Teacher Award from Government of Andhra Pradesh, Prof. P. Maheswari Gold Medal and Prof. G. Panigrahi Memorial Lecture of Indian Botanical Society and Prof. Y.D. Tyagi Gold medal of Indian Association for Angiosperm Taxonomy. He was Past President of Indian Association for Angiosperm Taxonomy and presently he is President of Indian Botanical Society. He was Member of Species Survival Commission of International Union for Conservation of Nature and Natural Resources (IUCN).

Dr. S. Sandhya Rani did her Master degree (1991) in Botany with first rank from Sri Venkatewara University, Tirupati and thereafter obtained her Ph.D. degree (1998) from Sri Krishnadevaraya University, Anantapur, Andhra Pradesh, India. At present she is working as Assistant Professor in the Department of Botany, Sri Krishnadevaraya University, Anantapur. She has 15 years of research experience in plant taxonomy and 5 years of teaching experience. Her publications include 52 research papers and 5 books. She was Principal Investigator of a Research Project funded by Department of Atomic Energy, Government of India and life member of IAAT, IBS and NESA.

Biodiversity in India
Volume 8

— Editors —
T. Pullaiah
S. Sandhya Rani

Department of Botany
Sri Krishnadevaraya University
Anantapur – 515 003
Andhra Pradesh

2016
Regency Publications
A Division of
Astral International Pvt. Ltd.
New Delhi – 110 002

Cataloging in Publication Data--DK
Courtesy: D.K. Agencies (P) Ltd. <docinfo@dkagencies.com>

Biodiversity in India / editors, by T. Pullaiah, S. Sandhya Rani.
  volumes cm
  Includes bibliographical references and index.
  ISBN 978-93-5130-952-9 (International Edition)

  1. Biodiversity--India.   2. Biology--India--Classification.
I. Pullaiah, T., editor. II. Sandhya Rani, S., editor.
QH183.B56 2016   DDC 578.70954   2

Published by	:	**Daya Publishing House®**
		A Division of
		Astral International Pvt. Ltd.
		– ISO 9001:2008 Certified Company –
		4760-61/23, Ansari Road, Darya Ganj
		New Delhi-110 002
		Ph. 011-43549197, 23278134
		E-mail: info@astralint.com
		Website: www.astralint.com
Laser Typesetting	:	**Classic Computer Services,** Delhi - 110 035
Printed at	:	**Thomson Press India Limited**

Preface

Species extinction is one of the main problem of the present century. Hence most of the Governments have made Biodiversity as the centre for all their developmental activities. Public is more concerned about the loss of Biodiversity and its resultant effect on food and medicine. One of the mandate of Convention of Biological Diversity is Documentation of Biological diversity of the country. With a view to document this Biodiversity in our country we continued with the present volume 8. We are elated to know that volume 1 has gone out of print and got reprinted in 2015. We wish to express our grateful thanks to all the authors who contributed their research/review articles for the present volume. We thank for their cooperation and erudition. We thank the readers for their appreciation and request them to continue their support. We also request them to give their suggestions and comments.

As said earlier we would like to continue the series and welcome researchers to send their articles for publication in the next volume.

T.Pullaiah

S.Sandhya Rani

Contents

Chapter 1

Floristic Diversity in Uttara Kannada District, Karnataka

T.V. Ramachandra, M.D. Subash Chandran,
G.R. Rao, Vishnu Mukri and N.V. Joshi

Energy and Wetlands Research Group, Centre for Ecological Sciences,
Indian Institute of Science, Bangalore – 560 012, Karnataka
E-mail: cestvr@ces.iisc.ernet.in; energy@ces.iisc.ernet.in

ABSTRACT

The forests are valuable resources on innumerable counts *viz.*, as sources of various useful products to humans, for their environmental and ecosystem services (soil and water conservation, regulation of water flow, carbon sequestration, nutrient cycling, etc.) and as centres of biodiversity. Out of the total 329 million ha land area of India, 43 per cent is under cropping and 23 per cent classified as forests. The total area of forest cover in India, as per the latest assessment is about 692,027 km^2 or 21.05 per cent of the total geographical area.

The Western Ghats range of hills, running close and parallel to the Arabian Sea along the western Peninsular India for about 1600 km from the south of Gujarat to Kanyakumari, covers an area of about 1,60,000 sq.km. This region harbours very rich flora and fauna and there are records of over 4,000 species of flowering plants (38 per cent endemics). Western Ghats is among the 34 global biodiversity hotspots on account of exceptional plant endemism and serious levels of habitat degradation. The complex geography, wide variations in annual rainfall from 1000-6000 mm, and altitudinal decrease in temperature, coupled with anthropogenic factors, have produced a variety of vegetation types in the Western Ghats. While tropical evergreen forest is the natural climax vegetation of the more humid western slopes, along the rain-shadow region eastwards vegetation changes rapidly from semi-evergreen to moist and dry deciduous forests, the last one being characteristic of the semi-arid Deccan region as well. Lower temperature, especially in altitudes exceeding 1500 m, has produced a unique mosaic of montane 'shola' evergreen forests alternating with rolling grasslands, mainly in the Nilgiris and the Anamalais. All these types of natural vegetation are prone to or have already undergone degradations due to human impacts.

Uttara Kannada district with 76 per cent of its 10,291 sq.km area covered with forests has the distinction of having highest forest area. This is the northernmost coastal district of Karnataka State (13.9220° N to 15.5252° N and 74.0852° E to 75.0999° E) has a geographical area of 10, 291 km². Topographically the district can be divided into three zones – the narrow and relatively flat to low hilly coastal along the west of Karwar, Ankola, Kumta, Honavar and Bhatkal taluks; the precipitously rising main range of Western Ghats towards the eastern interior of these taluks, the crestline zone composed of Sirsi, Siddapur Supa and Yellapur taluks and Haliyal and Mundgod taluks towards the north-east flattening and merging with the Deccan Plateau. The district can be divided broadly into five vegetation zones namely: Coastal, Northern evergreen, Southern evergreen, Moist deciduous and Dry deciduous. The evergreen to semi-evergreen forests form major portion of the district especially towards the more rainy western parts. Towards the eastern rain-shadow portion, the forests change rapidly into moist and dry deciduous types.

Whereas substantial areas of natural forests, through forestry practices over a period of more than one century, have been converted into monoculture tree plantations of teak, eucalypts (in the past) and into Acacia plantations in recent decades, there also remained in many places blocks of ancient patches of evergreens, known as *kans*, which were or still are sacred to the local people being the seats of village deities. These are relatively less impacted areas of climax evergreen forests, being sacred groves protected by the people through generations. Being preserved forests from ancient times these *kans* or their remains still might harbor rare species of plants, with high degree of Western Ghats endemism, and also endemic faunal elements. *e.g.*, Asollikan (Ankola), Kathalekan (Siddapur), Karikan (Honavar) etc.

Slash and burn cultivation that prevailed almost till close of 19th century, especially in the heavy rainfall zone created considerable areas of secondary forests that replaced primary evergreen. Wherever clear felling has taken place in the past in the heavy rainfall belt, for shifting cultivation or under forestry operations, very sensitive evergreens and those without coppicing character tend to vanish. Old growth forests in stages of late secondary almost resemble the primary forests. But conspicuously absent in them are climax evergreen forest trees like *Dipterocarpus indicus, Vateria indica, Palaquium ellipticum*, and species confined to *Myristica* swamps like *Myristica fatua (M. magnifica)*, etc. The forests bearing centuries of history is a grant mosaic of evergreen and semi-evergreen to secondary moist deciduous (in the high rainfall areas) to deciduous types. These are intermingled in many places with degraded stages like savannah, and scrub or entirely changed into grassy blanks used for cattle grazing, which within forest zone also have crucial role of supporting wild herbivores.

Karnataka has five National Parks and 21 Wildlife Sanctuaries. Uttara Kannada has mainly two important protected areas namely Anshi National Park and Dandeli Wildlife Sanctuary. These two PA's are brought together under

Dandeli-Anshi Tiger Reserve with focus on tiger conservation. The DATR presently covers an area of 1365 sq.km. in the taluks Joida, Karwar and Haliyal. We could not carry out forest studies within the DATR due to want of permission from the Wildlife wing of Forest Department. However, prior to the imposition of restrictions on studies within Tiger Reserves we had carried out a study on the grassland resources within the Reserve. Recently (in 2011) Attivery Bird Sanctuary was declared in Mundgod taluk covering 2.23 sq.km area, mainly composed of a reservoir and its peripheral areas.

Conservation Reserves are a new concept within the framework of PAs under the Wildlife (Protection) Amendment Act of 2002. They seek to protect habitats that are under private ownership also, through active stakeholder participation. They are typically buffer zones or connectors and migration corridors between established national parks, wildlife sanctuaries and other RFs. They are designated as conservation reserves if they are uninhabited and completely owned by the government but used for subsistence by communities, and community reserves if part of the lands are privately owned. Administration of such reserves would be through joint participation of forest officials and local bodies like gram sabhas and gram panchayats. They do not involve any displacement and protect user rights of communities. In Uttara Kannada, four such Conservation Reserves were set up by the Government of Karnataka:

1. **Aghanashini LTM Conservation Reserve** (299.52 sq.km), to protect Lion tailed macaque and Myristica Swamps.

2. **Bedthi Conservation Reserve** (59.07 sq.km) as Hornbill habitats and for medicinal plant species like *Coscinium fenestratum.*

3. **Shalmala Riparian Eco-system Conservation Reserve** (4.89 sq.km) for conservation flora and fauna of a riverine ecosystem and

4. **Hornbill Conservation Reserve** (52.5 sq.km) covering part of Kali River basin for specifically hornbill conservation.

The current study investigates *floristic diversity associated with different forests and computes basal area, biomass and carbon sequestration in forests. Apart from this inventorying and mapping of endemic tree species has been done to find out areas of high endemism and congregations of threatened species. A set of criteria for holistic conservation of forest ecosystems, particularly of high endemism of Western Ghats has been prepared based on field investigation, interaction with stakeholders (researchers working in this region, forest officials, local people, subject experts).*

Forests of all major kinds were studied using transect cum quadrat methods (altogether 116 transects, each transect with five quadrats of 400 sq.m each for tree vegetation, 10 sub-quadrats each of 25 sq.m for shrubs and tree saplings and 20 subquadrats of one sq.m for herb layer diversity. Out of 116 transects 8 were studied using point-centre quarter method). Altogether for tree vegetation 540 quadrats, each of 400 sq.m were studied. Necessary permission was, however, not granted for forest studies within the Dandeli-Anshi Tiger Reserve areas.

Altogether 1068 species of flowering plants were inventorised during the study period, through sample surveys and opportunistic surveys outside the transect zones. These species represented 138 families. Of these 278 were trees species (from 59 families), 285 shrubs species (73 families) and 505 herb species (55 families). Moraceae, the family of figs (*Ficus* spp.), keystone resources for animals, had maximum tree sp (18), followed by Euphorbiaceae (16 sp.), Leguminosae (15 sp.), Lauraceae (14 sp.), Anacardiaceae (13 sp.) and Rubiaceae (13 sp.). Shrub species richness was pronounced in Leguminosae (32 sp.), Rubiaceae (24 sp.) and Euphorbiaceae (24 sp.). Among herbs grasses (Poaceae) were most specious (77 sp.), followed by sedges (Cyperaceae) with 67 sp. Orchids (Orchidaceae) were in good number.

Tropical forests are major reservoirs of carbon in the terrestrial areas of the planet which is confronted with the prospects of imminent climatic change. World over all countries need to be alert to this major catastrophe. Apart from regulating pollution levels from various sources carbon sequestration in biomass has to be increased considerably. Our estimates on carbon sequestration based on tree biomass estimates from 116 forest samples show that the average carbon sequestration per hectare of forest (barren areas, scrub and grasslands excluded from sampling) was 154.251 ha.

It is a significant find that the sacred *kan* forests of pre-colonial era, despite their merger with state reserved forests, and subjection of most to timber extraction pressures in the post-independence era, continue to lead the chart of sites having some of the highest carbon sequestration per unit area. Thus the *kan* forest adjoining the Karikanamman temple in Honavar taluk had the highest carbon sequestration at 363.07 t/ha in the tree biomass alone. This is followed by Tarkunde-Birgadde in Yellapur (357.67 t/ha), and some of the swamp-stream forest samples in Kathalekan (299.66 t/ha, 275.18 t/ha, 259.21 t/ha etc.). Likewise Kanmaski-Vanalli in Sirsi had 242.43 t/ha of carbon.

The lowest carbon sequestered was found to be in the savannized forests, for obvious reasons of low to very low number of trees in them. These savannas whether they be in high evergreen forest belt (in Siddapur or Joida for instance) or be in drier zone of Haliyal or Mundgod have carbon storage of <50 t/ha in the tree biomass. Savannization was a necessity in the past for agricultural occupation of humans in the Western Ghats, for cattle grazing and slash and burn cultivation. Today the process is repeating to some extent still as forest encroachments have happened rampantly in all taluks increasing the porosity of otherwise in tact forests. Most bettalands allotted to arecanut orchard owners for exercising the privilege of leaf manure collection are in poor state of biomass and carbon sequestration (*e.g.*, 14.19 t/ha in Gondsur-Sampekattu betta in Sirsi, Talekere betta in Siddapur 41.47 t/ha).

The study highlights the importance of conservation of riparian forests occurring along streams and swamps, not only from high species endemism but also for higher carbon sequestration. A very detailed study in Kathalekan involving nine samples of such forests versus nine samples away from such water courses reveal that the average carbon sequestration in the former was 225.506 t/ha against 165.541 t/ha in the latter. This is despite the fact both types occur within what is traditionally designated as a *kan* forest. We therefore recommend that forests adjoining or covering streams, swamps and riverbanks of the Western Ghats be considered sacrosanct and as critical areas for hydrology not only of the coast but of the entire Indian peninsula.

As regards trees are concerned, in principle, there are close associations between areas of rich tree endemism and occurrence of RET tree species. Forests with high tree endemism also tend to shelter endemic/RET non-tree species and fauna- especially fishes and amphibians- which are indicators of other such organisms as well. Tree species in danger of local or total extinction mainly exist in and closer to the Myristica swamps. These include *Syzygium travancoricum* (Critically Endangered), *Myristica fatua* (*M. magnifica*) (Endangered), *Gymnacranthera canarica* (Vulnerable), *Semecarpus kathalekanensis* (newly discovered), *Mastixia arborea* (rare endemic) etc. *Madhuca bourdillonii*, a Critically Endangered tree, was not in our samples, but occurred very sparingly close to some Myristica swamps. The Kathalekan swamp forest sheltered at least 35 species of amphibians, most of them within a range of few hundred meters. While 26 species (74 per cent) of them were Western Ghat endemics, one species *Philautus ponmudi* is Critically Endangered and five species each were Endangered and Vulnerable. Scores of *Myristica* dominated forest swamps would have perished in the Western Ghats in past centuries having given way to human impacts, notably due to reclamation of primeval forest clad valleys for making rice fields and arecanut-spice orchards. The last remains are also under threat, mainly being looked upon for areca orchards by encroachers. Swamps being excellent sources of perennial streams we recommend tracing out all such swamps and potential swamps (of degraded vegetation or waters diverted for agriculture) for hydrological needs. The swamps along with their catchments, even if they have secondary forests, need to be safeguarded as prime areas of hydrological significance and as the last refugia of rain forests in the central Western Ghats.

Keywords: Floristic diversity, Biodiversity, Western Ghats.

Introduction

The forests are valuable resources on innumerable counts *viz.*, as sources of various useful products to humans, for their environmental and ecosystem services (soil and water conservation, regulation of water flow, carbon sequestration, nutrient cylcing etc.) and as centres of biodiversity (Ramachandra, 2007). A wide range of benefits to mankind particularly comes from the tropical forests. Estimating the 'total economic value' of forests has become a popular topic and research discussion in the conservation community (Lele *et al.*, 2000). Out of the total 329 million ha land area of India, 43 per cent was under cropping and 23 per cent classified as forests (Ministry of Environment and Forests, 1999). The total area of forest cover in India, as per the latest assessment by Forest Survey of India (2011), has been put at 692,027 km^2, or merely 21.05 per cent of the total geographical area. Indian forests were classified by Champion and Seth (1968) into four major groups, namely, tropical, sub-tropical, temperate, and alpine. These were further divided into 16 type groups: Tropical (wet evergreen, semi-evergreen, moist deciduous, littoral and swamp, dry deciduous, thorn, dry evergreen), Sub-tropical (broad leaved hill forests, pine, and dry evergreen), Temperate (montane wet, Himalayan moist temperate, Himalayan dry temperate), and Alpine (sub-alpine, moist alpine and dry alpine scrub). Of these 16 types, tropical dry deciduous constitute the major percentage of the forest cover in India.

The Western Ghats range of hills, running close and parallel to the Arabian Sea along the western Peninsular India for about 1600 km from the south of Gujarat to Kanyakumari, covers an area of about 1,60,000 sq.km. (Ramachandra, 2007). It is among the 34 global biodiversity hotspots on account of exceptional plant endemism and serious levels of habitat degradation. The complex geography, wide variations in annual rainfall from 1000-6000 mm, and altitudinal decrease in temperature, coupled with anthropogenic factors, have produced a variety of vegetation types in the Western Ghats. While tropical evergreen forest is the natural climax vegetation of the more humid western slopes, along the rain-shadow region eastwards vegetation changes rapidly from semi-evergreen to moist and dry deciduous forests, the last one being characteristic of the semi-arid Deccan region as well. Lower temperature, especially in altitudes exceeding 1500 m, has produced a unique mosaic of montane 'shola' evergreen forests alternating with rolling grasslands, mainly in the Nilgiris and the Anamalais. All these types of natural vegetation are prone to or have already undergone degradations due to human impacts (Pascal, 1986; 1988).

The Western Ghats harbours very rich flora and fauna and there are records of over 4,000 species of flowering plants (38 per cent endemics), 330 butterflies (11 per cent endemics), 156 reptiles (62 per cent endemics), 508 birds (4 per cent endemics), 120 mammals (12 per cent endemics), 135 amphibians (75 per cent endemics) and 289 fishes (41 per cent endemics) (Daniels, 2003; Gururaja, 2004; Sreekantha _et al._, 2007; Ramachandra _et al._, 2007).

The major forests and associated vegetation types of the Western Ghats are the following:

1. Tropical wet evergreen forests – natural climax of high rainfall areas (>200 cm/year)
2. Semi evergreen forests – natural climax of moderate rainfall areas (150-200 cm/year) and caused also by disturbances to evergreens
3. Moist deciduous forests – natural to 100-150 cm rainfall areas and anthropogenic factors, especially fire in higher rainfall zones
4. Dry deciduous forests – in areas with less than 100 cm/year rainfall
5. Shola forests – stunted evergreens in the wind protected high altitude valleys (>1500 m)
6. Shola grasslands and grassy blanks – former in high altitude exposed slopes and tops and latter anywhere else due to human impacts
7. Savannas – manmade alterations in natural forests; composed grasslands with distantly placed trees
8. Scrubs and thickets in highly disturbed areas.

Profile of Uttara Kannada Forests

The Uttara Kannada the northernmost coastal district of Karnataka State (13.9220° N to 15.5252° N lat. and 74.0852° E to 75.0999° E long.) has a geographical area of 10, 291 km². Topographically the district can be divided into three zones – the narrow and relatively flat to low hilly coastal along the west of Karwar, Ankola, Kumta, Honavar and Bhatkal taluks; the precipitously rising main range of Western Ghats towards the eastern interior of these taluks, the crestline zone composed of Sirsi, Siddapur Supa and Yellapur taluks and Haliyal and Mundgod taluks towards the north-east flattening and merging with the Deccan Plateau. Forest Survey of India (2011) reveals 76 per cent of the district's area as covered by forests. Akbar Shah

Figure 1.1: Evergeen to Semi-evergreen Forest Clad Hills in Honavar Taluk.

(1988) treated 1388.89 km^2 of the district's forests as partially open, 1646.16 km^2 as of medium density and only 714.55 km^2 as closed forest. Daniels *et al.* (1989; 1993) divided Uttara Kannada vegetation into 5 broad zones namely – Coastal, Northern evergreen, Southern evergreen, Moist deciduous and Dry deciduous zones. The evergreen to semi-evergreen forests (Figure 1.1) form major portion of the district especially towards the west which experiences heavy rainfall. With the decline of rainfall towards the eastern portion, the forests change rapidly from moist deciduous to dry deciduous types. Most of the forests towards the western are considered to be of secondary nature owing mainly to the slash and burn cultivation practices which were widely prevalent up to the mid of 19th century and thereafter in an attenuate form until the close of the century. These forests today are in different stages of secondary succession, and in many places old growth forests would appear like the primary forest itself (Chandran 1997, 1998; Chokkalingam *et al.*, 2000).

The earlier accounts of vegetation studies from this district include details of botanical excursions by Santapau (1955) followed by Puri (1960) who gave a general account of the forests of this district. Champion and Seth (1968) classified the vegetation of western and crestal Uttara Kannada as west coast evergreen/ semi-evergreen forest while Pascal (1982, 1984; Pascal *et al.*, 1982) in his vegetation maps (on 1: 250000 scale) classified the forests in the high to moderate rainfall areas of Uttara Kannada as mainly *Persea macrantha-Diospyros* spp.-*Holigarna* spp. type along with fragments of *Dipterocarpus indicus-Diospyros candolleana-Diospyros oocarpa* types towards the southern portions, mainly in the Siddapur hills. Chandran (1995) pointed out that before the entry of agricultural and pastoral people, over three millennia ago, primary forests would have covered the district, and quotes palynological evidences based on Caratini *et al.* (1991), which highlighted the decline in forest pollens and increase in savannah pollens from marine core studies from Karwar coast towards middle of fourth millennium BP. The forest clearances could have been for shifting cultivation and human settlements. Daniels *et al.* (1995) attributes the reasons for the disappearance of several evergreen species to their inability to coppicing following industrial usages and several other human disturbances. There is a need to monitor the history of fire episodes for different forest types in order to interpret the changes in the semi-evergreen forests. Most of the remnants of good forests could have been converted to secondary forests due to large scale human activities. Based on regeneration patterns of forests in Uttara Kannada and afforestation of almost 6500 ha of land in the post1980's, Bhat *et al.* (2000; 2001) gave good prospects for secondary forests. Based on land cover and land use analysis of remote sensing data Ramachandra (2007) observed that high anthropogenic pressures in the district are causing slow transformation of evergreen forests to semi-evergreen and other human impacted landscape elements.

Uttara Kannada possessed many forests traditionally designated by locals as '*kans*' which were relatively less impacted areas of climax evergreen forests, being sacred groves protected by the people through generations. *Kans* were referred to first, in the travel account of Francis Buchanan, a British officer designated by Lord Wellington to study the newly conquered British territories along south-west India (Buchanan, 1870; Chandran and Gadgil, 1998). Buchanan found these *kans* as lofty evergreen forests preserved for growing pepper amidst relative barrenness of the coastal hills. Near Karwar the locals claimed the sacredness of forests to them (referring obviously to the *kans*) and the need for permission from the village headman, also the priest to village deities, for any inevitable tree cutting which was otherwise considered a taboo. Wingate (1888), the then forest settlement officer of the district noted that the *kans* were of "great economic and climatic importance and favored the existence of springs, perennial streams and generally indicated the proximity of valuable spice gardens, which derive from them both shade and moisture."

The *kans* were large sacred groves, each covering few to few hundred hectares in their original state. There were many such sacred *kans* in the rest of the evergreen forest belt of the district as well as in other Malnadu districts of Karnataka (Chandran and Gadgil, 1993; 1998). About 4,000 ha of *kans* from Sirsi and Siddapur taluks were included in a forest working plan for extraction of industrial timbers (Shanmukhappa, 1966). Siddapur taluk of Uttara Kannada, in the Bombay Presidency under the British regime, had 113 *kans* according to Village Forest Registers (Gokhale, 2001). Some *Kans* of southern Uttara Kannada, harbour fragments of *Myristica* swamps, an endangered and ancient habitat of high watershed value (Chandran and Mesta, 2001). Gadgil and Chandran (1989) noted that *Dipterocarpus indicus* of Western Ghats had its northern limits in Uttara Kannada with some of the *kans* being the only refuge for it. Chandran *et al.* (2010) studied the detailed ecology including species diversity, basal area, biomass and carbon sequestration of the swampy relic forests of Kathalekan in Uttara Kannada. The study revealed that this forest harbored relic trees such as *Dipterocarpus indicus* (Endangered) and *Palaquium ellipticum* and has a network of perennial streams and swamps sheltering endemic and rare population in the world of the tree *Semecarpus kathalekanensis* (new discovery by Dasappa and Swaminath, 2000), *Syzygium travancoricum* (Critically Endangered), *Myristica magnifica* (Endangered), *Gymnacranthera canarica* (Vulnerable).

The notable tree species found in the different forests are:

Evergreen Forests

Dipterocarpus indicus (in few locations only), *Ficus nervosa, Palaquium ellipticum, Syzygium gardenerii, Holigarna grahamii, Artocarpus hirsutum, Dysoxylum malabaricum, Lophopetalum wightianum, Diospyros sylvatica* etc. are found as emergent species. In the second strata that make an unbroken canopy are medium height trees like *Actinodaphne agustifolia, Cinnamomum* spp., *Diospyros candolleana, Hopea ponga, Myristica* spp., *Garcinia* spp., *Knema attenuata,* etc. In the lowermost woody strata occur smaller trees and shrubs such as *Aglaia anamalayana, Diospyros saldanhae, Syzygium laetum,* and some palms like *Arenga wightii* and *Pinanga dicksonii.* Lianas are many as well as canes. Most ancient patches of climax forests alone have *Myristica* swamps and *Dipterocarpus indicus.*

Semi evergreen Forests

Forests in heavy to moderate rainfall areas subjected to human pressures currently, or in the past due to shifting cultivation etc. tend to be semi-evergreen in nature. These share most evergreen tree species with the evergreen climax forests, but are distinguished by the presence of certain deciduous trees such as *Lagerstroemia microcarpa, Terminalia* spp. *Schleichera oleosa, Stereospermum personatum, Tetrameles nudiflora, Vitex altissima* etc. In addition some evergreen species like *Alstonia scholaris, Holigarna arnottiana, Mammea suriga, Carallia integerrima* etc. are more characteristic of the latter. Climbers and lianas and canes are abundant.

Moist Deciduous Forests

Natural moist deciduous forests occur along the rainshadow region where annual rainfall is less than 1500 mm., especially in the taluks of Yellapur east, Haliyal and western parts of Mundgod. Secondary moist deciduous forests could occur anywhere as regeneration in fire affected areas (Mesta, 2008). The species that frequently occur here are *Adina cordifolia, Aporosa lindleyana, Careya arborea, Dillenia pentagyna, Lagerstroemia microcarpa, Spondias* spp., *Strychnos nux-vomica, Terminalia paniculata, T. tomentosa, Xylia xylocarpa* etc. Teak, *Tectona grandis,* occurs naturally along with bamboos, especially *Bambusa arundinacea* and *Dendrocalamus strictus.* Teak plantations are plentiful in this zone. Climbers and lianas are present in these forests but lesser in diversity compared to the evergreen-semievergreen forests.

Dry Deciduous Forests

In plains and undulating terrain in Mundgod and eastern Haliyal, where rainfall drastically reduces to less than 1000 mm, these forests constitute the natural climax. Trees mainly found here are *Albizzia* spp., *Anogeissus* spp., *Careya arborea, Bauhinia racemosa, Bombax ceiba, Bridelia retusa, Dalbergia latifolia, Diospyros melanoxylon,*

D. montana, Lagerstroemia parviflora, Tectona grandis, Terminalia paniculata, Terminalia tomentosa, etc. Teak trees here do not attain large girths and these forests were formerly also known as 'teak pole' forests.

Scrub-Savannas

These are formations owing their origin to severe human impacts of past and present. They occur anywhere in the district. The scrub is characterized by shrubby plants, often thorny and spinous ones, with stunted trees. The latter include *Careya arborea, Lannea coromandelica, Phyllanthus emblica, Sapium insigne, Sterculia urens, Strychnos nux-vomica, Terminalia chebula, Zanthoxylum rhetsa* etc. Of the shrubby plants are *Embelia tsjeriam-cottam, Grewia microcos, Helectris isora, Ixora coccinia, Leea macrophylla, Tephrosia purpurea* etc. The thorny/spinous shrubs and climbers include *Capparis spinosa, Zizyphus rugosa, Plectronia parviflora* etc.

Forest Plantations

Over a period of more than a century monoculture forest plantations have been raised in the district. The earliest plantations were of teak and *Casuarina,* the latter along the sandy coastal stretches, especially in Karwar and Kasarkod (in Honavar taluk). Teak plantations became widespread throughout the district in an effort to meet the rising timber needs and for raising revenues. In the 1960's began plantations of eucalypts and *Acacia auriculiformis* a couple of decades later. Mixed plantations of utility trees are being raised in the degraded forests in the vicinity of villages through the involvement of village forest committees.

Objectives of the Study

☆ Floristic diversity associated with different forests

☆ Basal area, biomass and carbon sequestration in forests

☆ Presence and quantification of endemic tree species and find out areas of high endemism and congregations of threatened species

☆ Presence of rare habitats

☆ Prepare a set of criteria for holistic conservation of forest ecosystems, particularly of high endemism of Western Ghats.

Literature Review of Vegetation Studies in different Forest Habitats of Uttara Kannada

The earliest record of the plants from Uttara Kannada was provided by Buchanan (1870) wherein he gave a brief general account of the plants he encountered in the course of his journey through the district. The district's flora was covered in the works of Hooker (1872-1897), Cooke (1901-1908), Talbot (1909) and Saldanha (1984). Account of status and working prescriptions for the forests were available from numerous forest working plans for the district, through a period of about the last 110 years. In addition, numerous works such as monographs, research papers etc. carried valuable information about the districts' plants and forest ecology. Bhat *et al.* (1985) carried out plant diversity studies in the forests of Uttara Kannada covering some reserved forest and minor forest areas. Shastri *et al.* (2002) in their study recorded a total of 144 species of trees from the Sirsimakki village ecosystem in Uttara Kannada district out of which 93 tree species were found in agro-ecosystem area (including home gardens, paddy and areca garden boundary) and 104 species were recorded from non-agricultural lands such as *soppina betta*, minor and reserve forest. Rao *et al.* (2008) documented the floristic diversity of 29 different wetlands in 9 taluks of Uttara Kannada district wherein 167 plant species belonging to 32 families were recorded. Chandran *et al.* (2008) rediscovered the occurrence of two Critically Endangered tree species *Madhuca bourdillonii* and *Syzygium travancoricum* from some relic forests of Uttara Kannada, almost 700 km north of their recorded home range in southern Kerala. Ali *et al.* (2010) reported the occurrence of *Burmannia championii* Thw., a saprophytic herb, for the first time in Karnataka from a *Myristica* swamp in Uttara Kannada district, thereby extending its northern limit of distribution in Western Ghats. During a botanical exploration in Anshi National Park, Uttara Kannada district, Punekar and Lakshminarasimhan (2010) observed and described a new species *Stylidium darwinii* Punekar & Lakshman belonging to family Stylidaceae.

Effects of Mega Projects on Forest Ecosystems

Ever since the arrival of early agriculturists in Uttara Kannada, over three millennia ago, forest have undergone substantial changes because of shifting cultivation and clearance of valley forests for garden cultivation and rice fields. The savannisation of coastal lateritic hills for shifting cultivation, savannization for cattle grazing had beginnings in the pre-historical Uttara Kannada (Chandran, 1997, 1998). However, the early settlers came into equilibrium with nature and Uttara Kannada remained till modern days as one of the most forested districts in the country. Another wave of serious alterations in forests began with the British arrival when timber became a major commodity for sale. The early depletion of natural teak by the close of 19th century was followed by widespread deforestation for raising teak monoculture plantations, not only in its natural deciduous forest zone but also in the heavy rainfall western parts of the district, upsetting ecological conditions substantially.

The post-independence era saw arrival of forest based megaprojects which due to non-sustainable use of timber and bamboo caused considerable forest impoverishment. These industries were given raw materials at abysmally low rates prompting non-sustainable and exhaustive harvesting of resources. The earliest megaproject, Indian Plywood Company started in 1940's in Dandeli. When the prime timbers from deciduous forests were exhausted the factory was given leases for timber from evergreen forest zone. The system of selection felling in the climax evergreen forests had devastating effects on the forest ecosystems. The factory was eventually closed when it could not get adequate raw materials. The West Coast Paper Mills set up at Dandeli in 1958 was given bamboo at very low rates. As lakhs of tons of bamboo were harvested bamboo resources got depleted, forcing the factory to get bamboo from elsewhere and later switching onto pulpwood. Matchwood companies and packing case units were also given concessional timber. All this went on until late 1980's when the Government prohibited all kinds of green fellings in the forests (Gadgil and Chandran, 1989). Considerable areas of land, including good share of forests, were released for various developmental projects (Table 1.1).

Table 1.1: The Extent of Forest Areas Released for other Purposes from 1956

Sl.No.	Particulars	Area in Ha.
1.	The forest area released for cultivation by 3 member committee from 1964 to 1969	6042.500
2.	Forest area released as per special G.O.No.AFD.116 of 16/4/69	11593.342
3.	Forest area released as per G.O.No.AFD-282-FGL74 of 17/19-12-1974	3399.400
4.	Forest area released for long lease	162.100
5.	Hangami Lagan in Notified area	8034.450
6.	Extension of Gouthana	390.400
7.	Forest area released for township	1096.900
8	Mining area leased and area actually in operation	1591.250
9.	Released to House sites to Houseless (1972-1979)	366.000
10.	Rehabilitation of Tibetans, displaced Ryots of Sharavathi Ghataprabha and Malaprabha, Gowli families etc.	4548.170
11.	Area under submersion and other Project	
	1. Kali Hydro Project	14602.000
	2. Bedti Project (for colony)	300.000
	3. Other irrigation tanks etc.	303.365
12.	Released to KSFIC for Napier Hybrid grass cultivation (Sirsi Division)	441.450
13.	Released to KAMCO (Dairy and fruit processing Unit)	153.993
14.	Released to KSFIC for Pineapple cultivation	163.320
15.	Karnataka State Veneers Ltd.	24.000
16.	Power transmission lines	677.979
17.	For establishment of Industries	95.000
18.	Area released to Horticulture department (1969-70)	71.847
19.	Released to Agricultural University, Dharwad	214.000
20.	Sharavathi Tail Race	700.000
21.	Kaiga Atomic Power Project	732.000
22.	Sea Bird Naval Base Project	2259.000
23.	Rehabilitation of Sea Bird out seas	643.720
24	Area released for non-agriculture and other purposes	394.870
25.	Konkan Railway	272.140
26.	Area released for improvement and widening of Ankola-Hubli Road	49.431
27.	Area released for rehabilitation of displaced persons of KHEP and Kaiga Project	316.410
28.	Area released to regularise the encroachments, which have taken place before 27-04-1978	2845.446
29.	Area released to construction of 400 KVDC alternate transmission line between Kaiga NPP and 200 KV sub-station at Narendra in favour of M/s. P.G.C.I.L, Karnataka	330.00
	TOTAL	**62814.483**

*Source: Forest working plan of Kanara circle (year 2009-10).

Methods and Study Area

The forest vegetation was sampled using transect-based quadrats, a method found useful especially in surveying forested landscaped of central Western Ghats (Chandran and Mesta, 2001; Ramachandra *et al.*, 2006; Ali *et al.*, 2007). Remote sensing imageries were used and ground surveys made to select sample plots (Figure 1.2 for locations). Along a transect of 180 m, 5 quadrats each of 20x20 m were laid alternatively on the right and left, for tree study (minimum girth of 30 cm at GBH or 130 cm height from the ground), keeping intervals of 20 m length between successive quadrats. Within each tree quadrat, at two diagonal corners, two sub-quadrats of 5 m × 5 m were laid for shrubs and tree saplings (< 30 cm girth). Within each of these 2 herb layer quadrats, 1 sq.m area each, were also laid down for herbs and tree seedlings (Figure 1.3). Climbers and other associated species were noted. A rapid assessment was made to track vegetational changes from the densely populated coast through the rugged mountainous terrain to the undulating and drier eastern lands using point-centred quarter method along line transects in Ankola (coastal) and Yellapur (hilly to undulating) taluks. Sampling efforts were higher in high endemism areas (*e.g.*, Kathalekan in Siddapur). Details of formulas used for various calculations are given in the Table 1.2.

The data collected was analyzed to calculate the species diversity using Shanon-Weiner's diversity index, Simpson dominance index, IVI, regeneration status, basal area, biomass and carbon sequestration potential. The Pearson Correlation Matrix with probabilities was calculated by using the 'R' software. A Principal Component Analysis (PCA) was carried out using the package PAST version 2.16. This ordination technique was used to analyze the relationship between the samples and to understand the main factors influencing the forest vegetation in Uttara Kannada (Table 1.2). In addition, the loading score for each variable was calculated; the values were then converted into conservation scoring system and added to the respective variable (next section).

The above ground standing biomass of trees is referred to the weight of the trees above ground, in a given area, if harvested at a given time. The change in standing biomass over a period of time is called productivity. The standing biomass helps to estimate the productivity of an area and also gives information on the carrying capacity of land (Ramachandra and Kamakshi, 2005). It also helps in estimating the biomass that can be continuously extracted. Carbon storage in forests is estimated by taking 50 per cent of the biomass as carbon. The mathematical equations for biomass estimations of trees have been developed and used by many researchers for different biogeography regions. For the current study, the standing above-ground biomass was calculated using the basal area equation and indirect estimation was done for calculating below ground biomass (Brown, 1997; Ravindranath *et al.*, 1997; Murali *et al.*, 2005; Ravindranath and Ostwald, 2008). The resultant total biomass was multiplied by 0.5 for estimating the carbon storage.

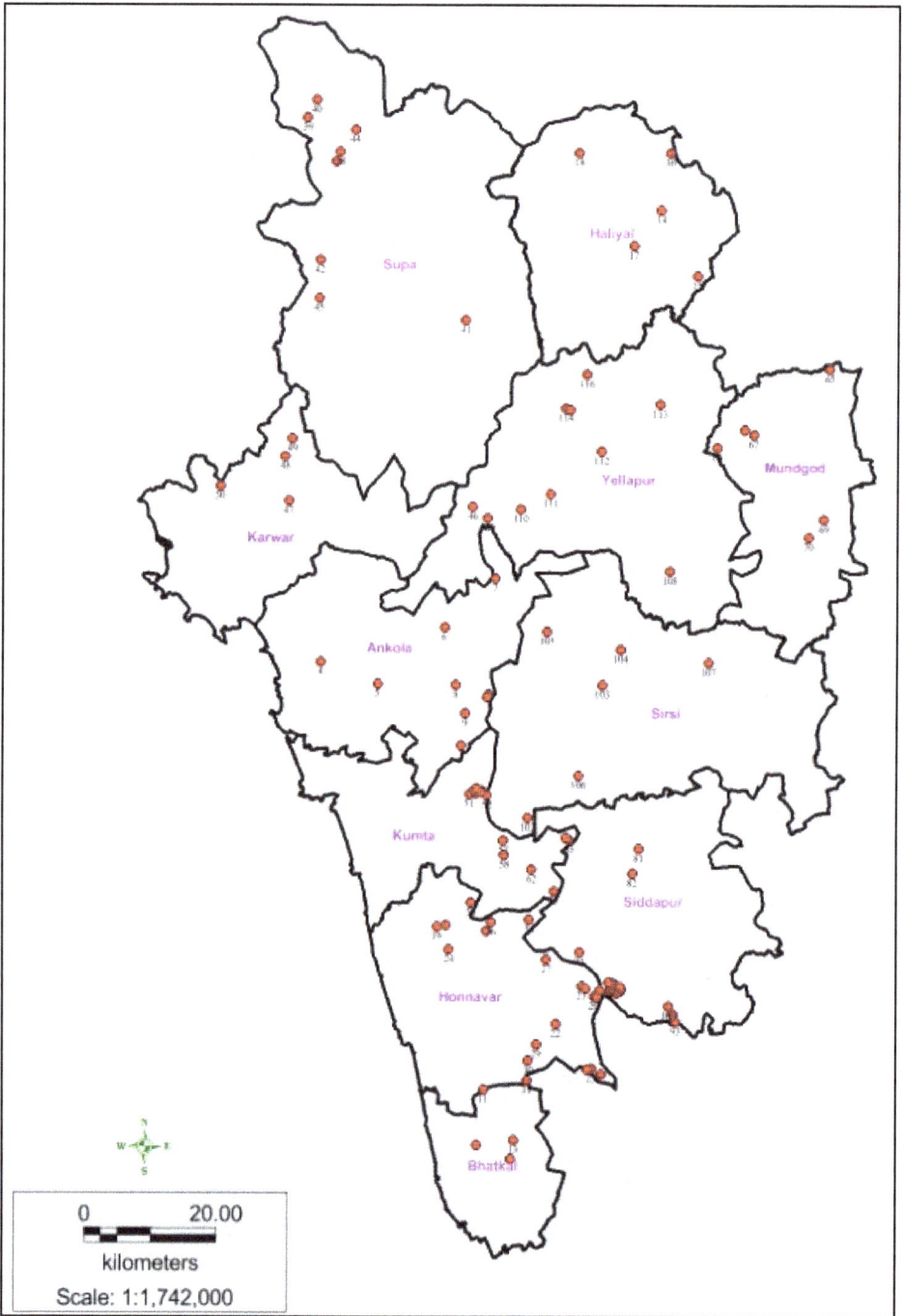

Figure 1.2: Distribution of Sample Study Sites in Uttara Kannada Forests.

Table 1.2: Details of Indices and Calculations and Softwares Used

Index	Equation	Notes
Per cent Evergreenness (trees)	$\dfrac{\text{No. of evergreen trees} \times 100}{\text{Total no. of trees}}$	To estimate how evergreen a forest is.
Per cent Endemism (trees)	$\dfrac{\text{No. of endemic trees} \times 100}{\text{Total no. of trees}}$	Percentage endemism of a forest patch
Basal area (m²)	$(GBH)^2/4\pi$	
Important Value Index	R. density + R. frequency + R. basal area	To know dominant and co-dominant species
Density	No. Species/Total no. of trees	Provides information on the compactness with which a species exists in an area.
Relative Density	$\dfrac{\text{Density of Species A} \times 100}{\text{Total density of all species}}$	
Frequency	$\dfrac{\text{No. points with Species A}}{\text{Total No. points Sampled}}$	Provides information on the repeated occurrence of a species
Relative Frequency	$\dfrac{\text{Frequency of Species A} \times 100}{\text{Total Frequency of all Species}}$	
Relative basal area	$\dfrac{\text{Basal area (m}^2\text{) of Species A} \times 100}{\text{Total basal area of all species}}$	
Shannon Weiner's diversity index	$H' \quad \displaystyle\sum_{i=1}^{s} p_i \ln p_i$	The value of Shannon's diversity index is usually found to fall between 1.5 and 3.5 and only rarely surpasses 4.5.

Contd...

Table 1.2–*Contd...*

Index	Equation	Notes
Simpson's dominance index	$SIDI = 1 - \sum_{i=1}^{N} p_i \times p_i$	
Above Ground Biomass (AGB) (t/ha)	-2.81 + 6.78 (BA)	
Below Ground Biomass (BGB) (t/ha)	0.26 * AGB	
Carbon storage	(AGB + BGB) * 0.5	
Pearson correlation matrix	R software	To estimate correlations between the parameters under consideration
Principal component analysis	Package PAST version 2.16	to analyze the relationship between the samples; to understand the main factors influencing the forest vegetation
Composite conservation index	Ranking of sites based on values assigned to key parameters of trees. Valuation system adopted for based on principal component analysis with loading score	PCA was used to analyze relationship between samples and to understand main factors influencing forest vegetation. Loading score obtained was added to a relative valuation score of key parameters decisive for conservation. Degree of correlation between key parametes was arrived at through Pearson correlation matrix. As evergreenness and endemism are strongly correlated, only one parameter (per cent endemism) was assigned a conservation value. The rest- height, basal area and diversity index having relatively lesser importance in endemism were given lesser values. Presence in a site of IUCN Red Listed tree species given higher valuation related to degree of threat.

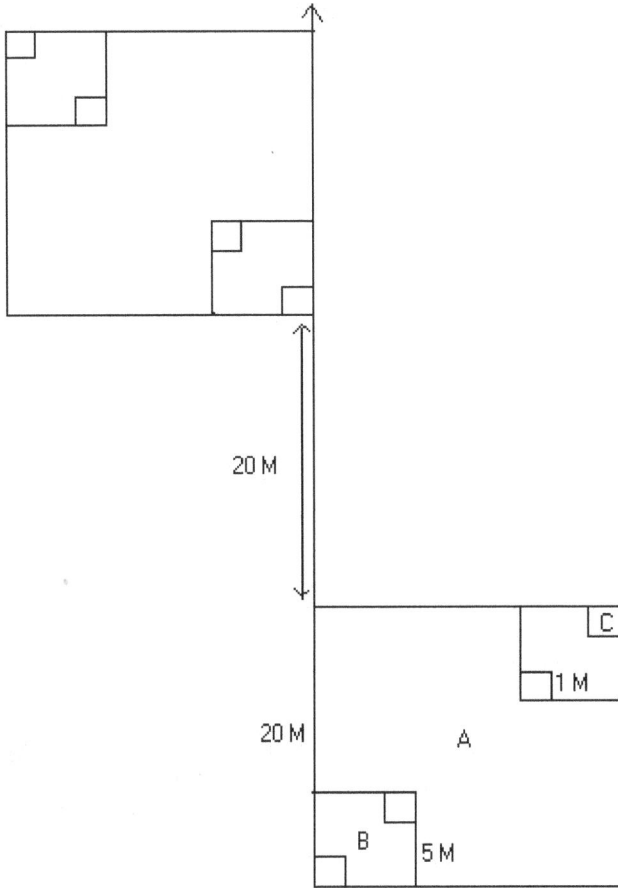

**Figure 1.3: Design of Transect cum Quadrats
(2 of 5 quadrats of 20x20 m only shown).**

Results and Discussion

Floristic Richness

A total of 116 transects studied along with opportunistic surveys yielded a list of 1068 species of flowering plants (about 25 per cent of Western Ghat species) from 138 families. Of these 278 were trees species (from 59 families), 285 shrubs species (73 families) and 505 herb species (55 families) (Figure 1.4). Among trees Moraceae had maximum representation (18 sp.), followed by Euphorbiaceae (16 sp.), Leguminosae (15 sp.), Lauraceae (14 sp.), Anacardiaceae (13 sp.) and Rubiaceae (13 sp.) and so on (Figure 1.5). The genus *Ficus*, members of which are considered as keystone resources for large number of birds and mammals, was the most well represented of Moraceae.

In Shrubs Leguminosae (32 sp.), Rubiaceae (24 sp.), Euphorbiaceae (24 sp.) were the leading families in species richness (Figure 1.6). Grasses (Poaceae) were most speciose (77 sp.) among herbs, followed by sedges –Cyperaceae- (67 sp.) and orchids –Orchidaceae- (35 sp.) (Figure 1.7). Grasses occur everywhere, except underneath the

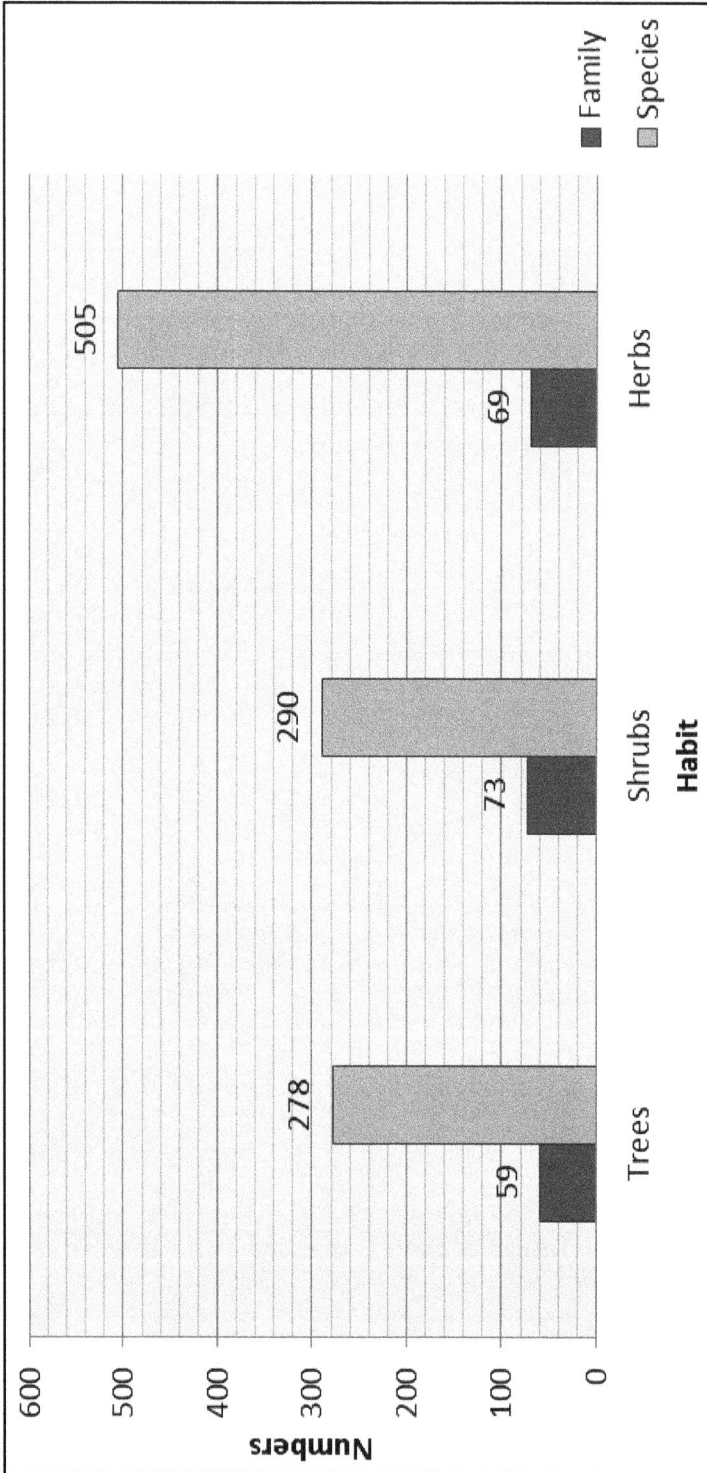

Figure 1.4: Family and Species Number for Trees, Shrubs and Herbs.

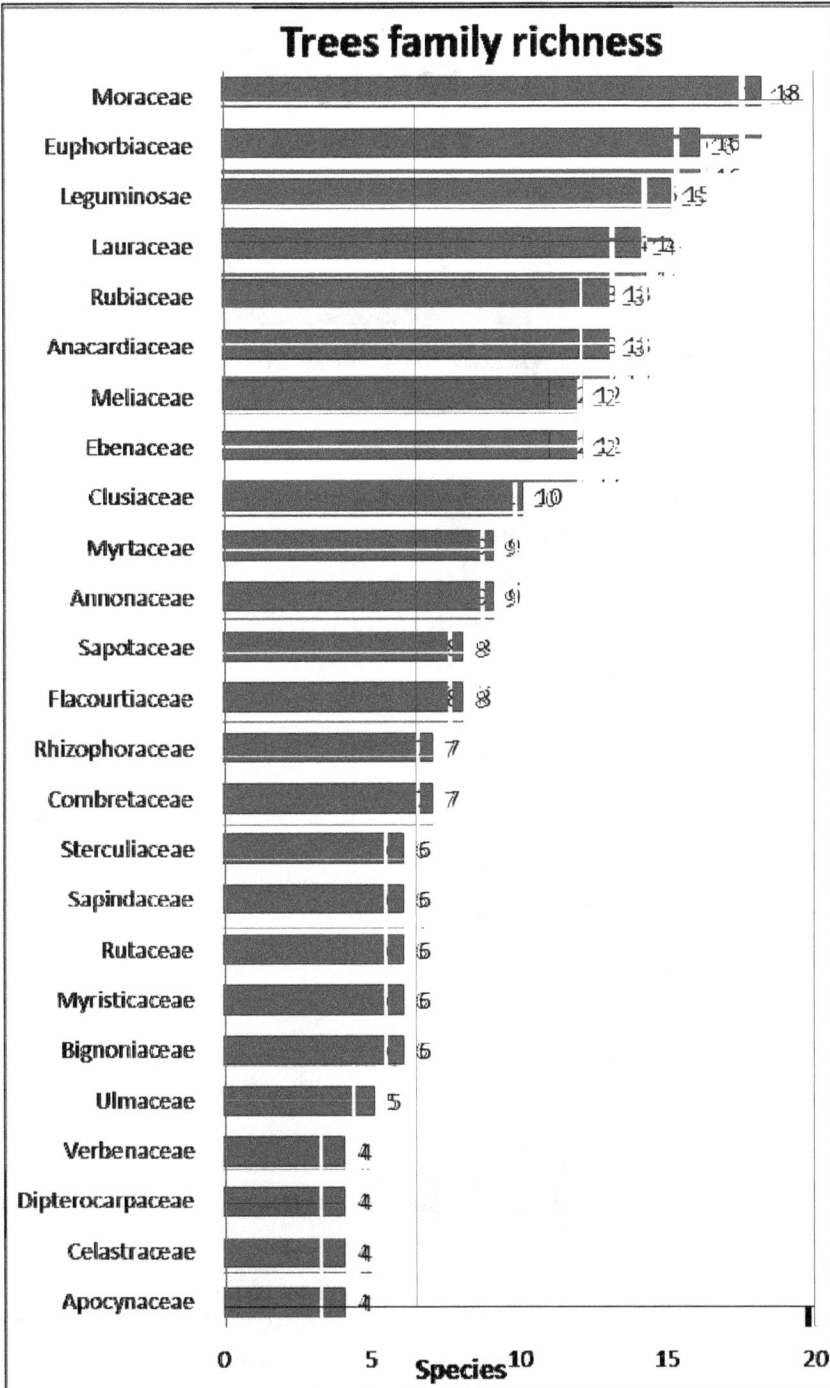

Figure 1.5: Richness of Families in Tree Species (Families with above 3 species shown).

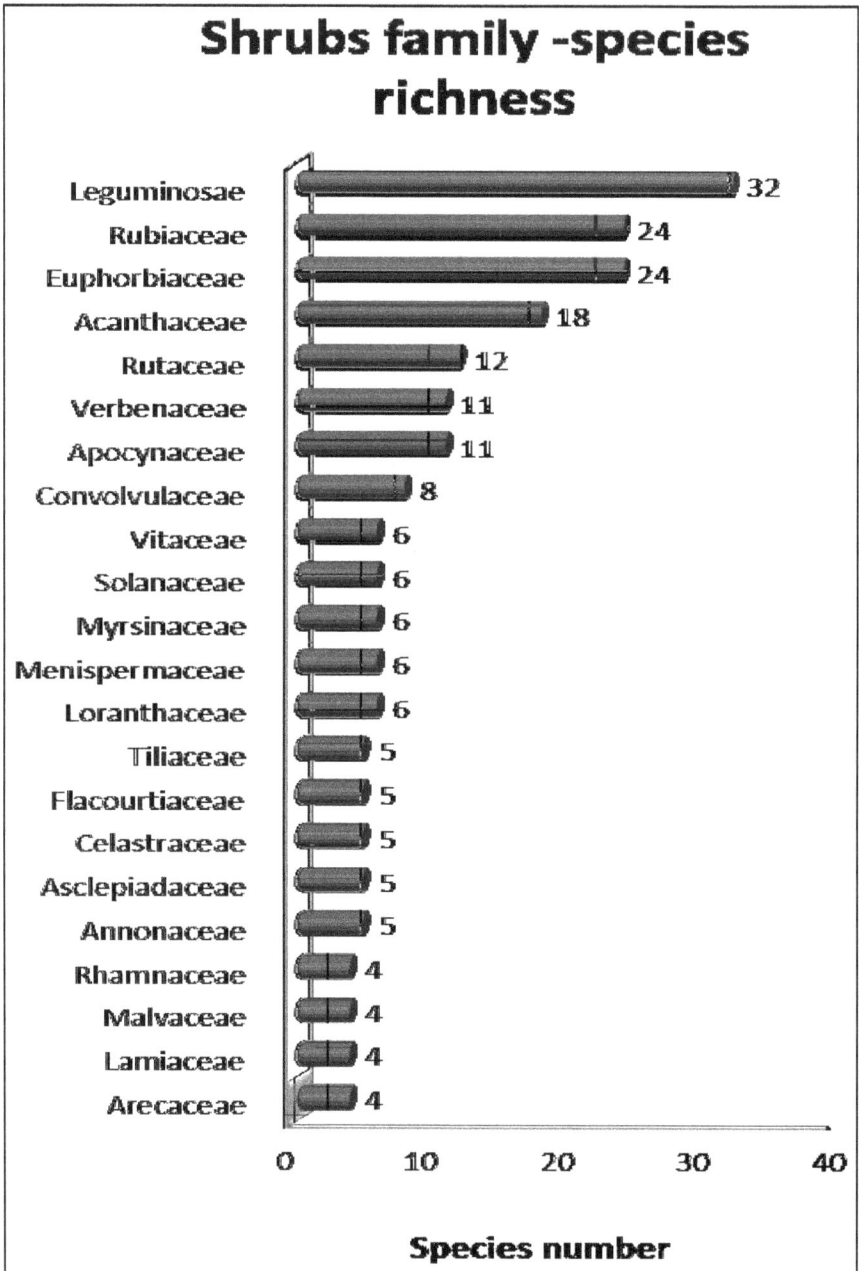

Shrubs family -species richness

Family	Species number
Leguminosae	32
Rubiaceae	24
Euphorbiaceae	24
Acanthaceae	18
Rutaceae	12
Verbenaceae	11
Apocynaceae	11
Convolvulaceae	8
Vitaceae	6
Solanaceae	6
Myrsinaceae	6
Menispermaceae	6
Loranthaceae	6
Tiliaceae	5
Flacourtiaceae	5
Celastraceae	5
Asclepiadaceae	5
Annonaceae	5
Rhamnaceae	4
Malvaceae	4
Lamiaceae	4
Arecaceae	4

Figure 1.6: Richness of Families in Shrub Species (Those with 4 or more species only shown).

Herbs family-species richness

Family	Species number
Poaceae	77
Cyperaceae	67
Orchidaceae	35
Scrophulariaceae	29
Rubiaceae	20
Asteraceae	20
Acanthaceae	20
Leguminosae	18
Commelinaceae	17
Eriocaulaceae	16
Lamiaceae	13
Araceae	13
Euphorbiaceae	11
Lythraceae	9
Balsaminaceae	9
Gentianaceae	7
Zingiberaceae	6
Urticaceae	6
Amaranthaceae	6
Onagraceae	5
Lentibulariaceae	5
Boraginaceae	5
Polygonaceae	4
Malvaceae	4
Hydrocharitaceae	4
Gesneriaceae	4
Apiaceae	4

Species number

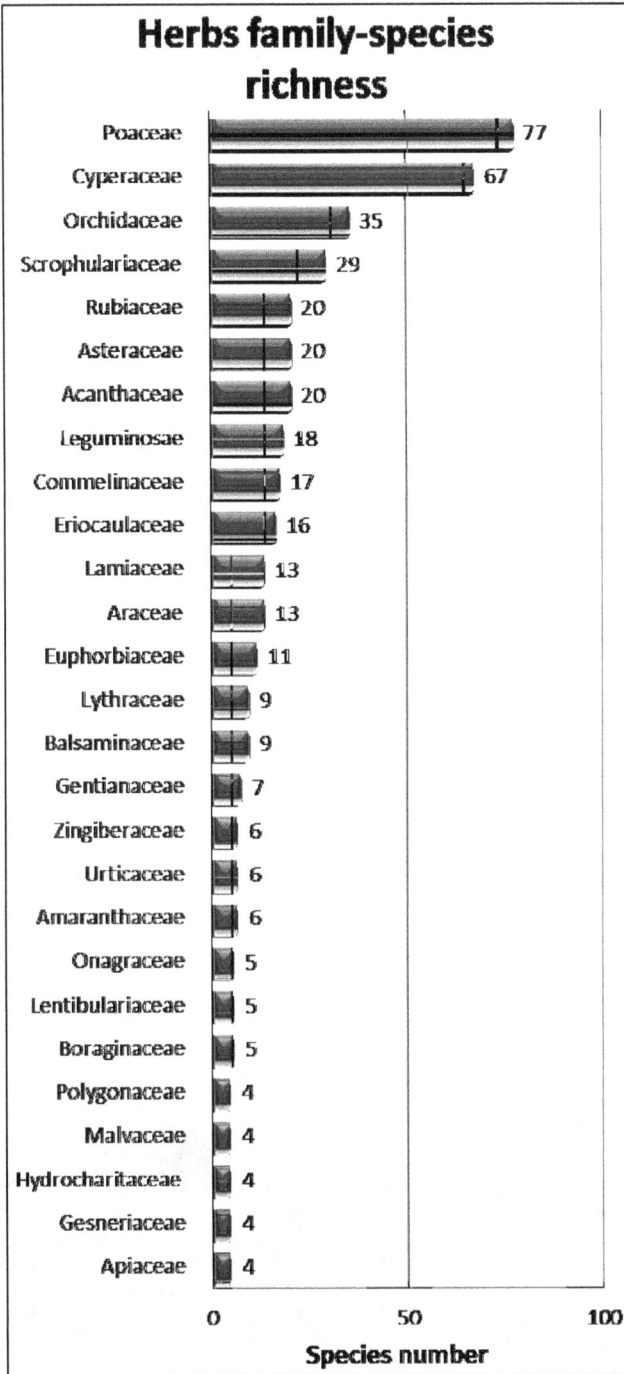

Figure 1.7: Richness of Families with Herb Species (Only families with 4 or more species given).

dark canopy of evergreen forest. Most of wetlands and very moist areas were under the dominance of Cyperaceae and to some extent under Scrophulariaceae.

Basal Area and Height

Details of transect-wise localities depicting tree species/transect, average height and estimated basal area/ha are given in the Table 1.3. Hadgeri-1 had the highest average tree height (21.82m) followed by Halsolli and Ambepal. These areas were characterised by lofty individuals of *Dipterocarpus indicus, Syzygium gardnerii* etc., the mature trees often attaining over 30 m. Most forests in Honavar and Siddapur taluks had greater heights owing to their predominantly evergreen and semi-evergreen forests.

The forests in general were mosaic of poor and mighty ones as far as tree heights and basal areas estimated/ha are concerned. Lowest average heights were seen in savanna and disturbed moist deciduous forests (*e.g.*, hill top savannas of Sirsi, Siddapur and stretches of forests in Joida which were under extensive shifting cultivation until end of the 19th century). Teak mixed forests and highly disturbed semi-evergreen forests (*e.g.*,Talekere) had lower height. Basal areas were also higher for *Dipterocarpus indicus* dominated areas of Karikan (85.41 sq.m/ha) and Kathlekan swamps dominated by swamp species like *Gymna cranthera* and in the nearby by other immense sized *Calophyllum tomentosum, Lophopetalum wightianum, Dipterocarpus indicus, Palaquium ellipticum,* (Figures 1.8 and 1.9) etc. Kushavali of Joida with large

Figure 1.8: *Lophopetalum wightianum* **in Kathalekan, Siddapur.**

Table 1.3: Transect-wise Numbers of Tree Species, Average Height and Estimated Basal Area per ha and Biomass-Carbon Sequestration

Sl.No.	Locality Name	Taluk	Tree Species	Average Height	Basal Area (m²/ha)	Above Ground Biomass (t/ha)	Below Ground Biomass (t/ha)	Total Biomass (t/ha)	Carbon Sequestration (t/ha)
1.	Asolli-1	Ankola	23	17.1	38.28	256.70	66.74	323.44	161.72
2.	Asolli-2	Ankola	33	17.4	38.73	259.75	67.53	327.28	163.64
3.	Hosakere	Ankola	30	16.0	37.62	252.28	65.59	317.87	158.94
4.	S1-Katangadde-Agasur	Ankola	40	9.6	9.08	58.75	15.28	74.03	37.01
5.	S2-Balikoppa-Badgon	Ankola	31	13.6	20.39	135.43	35.21	170.65	85.32
6.	S3-Hegdekoppa-Kasinmakki	Ankola	35	14.6	30.41	203.37	52.88	256.25	128.12
7.	S4-Vajralli-Ramanguli	Ankola	31	13.9	18.8	124.65	32.41	157.06	78.53
8.	Kachinabatti	Ankola	13	15.5	18.39	121.90	31.69	153.60	76.80
9.	Maabagi	Ankola	33	16.4	40.78	273.67	71.15	344.82	172.41
10.	Dakshinakoppa	Bhatkal	12	16.1	34.83	233.35	60.67	294.02	147.01
11.	Gujmavu (semi evergreen)	Bhatkal	33	15.8	32.97	220.75	57.40	278.15	139.07
12.	Hudil (evergreen)	Bhatkal	14	17.3	35.82	240.02	62.40	302.42	151.21
13.	Hudil (semi evergreen)	Bhatkal	27	15.8	46.11	309.81	80.55	390.36	195.18
14.	Golehalli	Haliyal	16	9.54	15.64	103.20	26.83	130.03	65.02
15.	Kudalgi-Tatigeri	Haliyal	12	10.98	17.79	117.84	30.64	148.48	74.24
16.	Magvad	Haliyal	19	14.9	23.90	159.21	41.39	200.60	100.30
17.	Sambrani	Haliyal	11	12.59	31.91	213.54	55.52	269.06	134.53
18.	Yadoga	Haliyal	13	14.3	26.30	175.53	45.64	221.17	110.58
19.	Ambepal-1	Honavar	25	19.2	31.49	210.66	54.77	265.43	132.72

Contd...

Table 1.3—*Contd...*

Sl.No.	Locality Name	Taluk	Tree Species	Average Height	Basal Area (m²/ha)	Above Ground Biomass (t/ha)	Below Ground Biomass (t/ha)	Total Biomass (t/ha)	Carbon Sequestration (t/ha)
20.	Ambepal-2	Honavar	32	19.6	48.80	328.07	85.30	413.36	206.68
21.	Chaturmukhabasti	Honavar	23	15.0	27.76	185.38	48.20	233.58	116.79
22.	Gersoppa	Honavar	31	18.2	30.13	201.47	52.38	253.85	126.93
23.	Gundabala	Honavar	32	15.4	29.05	194.16	50.48	244.65	122.32
24.	Hadageri-1	Honavar	23	21.8	53.69	361.20	93.91	455.11	227.56
25.	Hadageri-2	Honavar	19	19.3	45.53	305.85	79.52	385.38	192.69
26.	Halsolli	Honavar	9	20.5	30.64	204.93	53.28	258.21	129.11
27.	Hessige-1	Honavar	28	18.1	44.25	297.21	77.27	374.48	187.24
28.	Hessige-2	Honavar	27	16.6	48.87	328.52	85.42	413.94	206.97
29.	Hessige-3	Honavar	25	16.9	31.46	210.47	54.72	265.19	132.60
30.	Hessige-4	Honavar	30	17.4	51.56	346.77	90.16	436.93	218.46
31.	Kadnir	Honavar	24	16.0	38.17	255.96	66.55	322.51	161.25
32.	Karikan-lower slope	Honavar	28	13.6	41.87	281.10	73.09	354.19	177.09
33.	Karikan-semievergreen	Honavar	23	14.4	33.98	227.57	59.17	286.74	143.37
34.	Karikan-temple side-diptero patch	Honavar	21	17.9	85.41	576.29	149.83	726.12	363.06
35.	Mahime	Honavar	18	16.8	30.44	203.57	52.93	256.50	128.25
36.	Sharavathy-viewpoint	Honavar	28	17.7	34.70	232.47	60.44	292.92	146.46
37.	Tulsani-1	Honavar	27	17.3	36.44	244.26	63.51	307.77	153.89
38.	Tulsani-2	Honavar	23	17.1	30.86	206.44	53.68	260.12	130.06
39.	Castlerock IB	Joida	28	16.0	55.68	374.67	97.41	472.09	236.04

Contd...

Table 1.3–*Contd...*

Sl.No.	Locality Name	Taluk	Tree Species	Average Height	Basal Area (m²/ha)	Above Ground Biomass (t/ha)	Below Ground Biomass (t/ha)	Total Biomass (t/ha)	Carbon Sequestration (t/ha)
40.	Castlerock-moist-dec.	Joida	22	9.4	12.16	79.64	20.71	100.34	50.17
41.	Castlerock-semi everg	Joida	24	15.4	26.27	175.31	45.58	220.89	110.44
42.	Desaivada-Nandgadde	Joida	12	16.87	36.96	247.81	64.43	312.24	156.12
43.	Gavni-Kangihole-Joida	Joida	35	15.2	48.70	327.37	85.12	412.49	206.25
44.	Ivolli-Castlerock	Joida	19	13.7	33.99	227.62	59.18	286.80	143.40
45.	Joida-deciduous	Joida	21	16.9	39.44	264.62	68.80	333.42	166.71
46.	Kushavali	Joida	30	16.4	75.04	505.93	131.54	637.48	318.74
47.	Shivpura	Joida	12	15.90	33.79	226.26	58.83	285.09	142.55
48.	Gopishetta	Karwar	23	15.1	32.21	215.58	56.05	271.63	135.82
49.	Goyar-moist dec	Karwar	18	15.0	37.99	254.77	66.24	321.01	160.50
50.	Kalni-goyar	Karwar	32	17.3	45.05	302.66	78.69	381.35	190.67
51.	Karwar-moist dec	Karwar	17	10.8	13.48	88.61	23.04	111.65	55.82
52.	Devimane-Campsite	Kumta	36	16.7	42.99	288.63	75.04	363.67	181.84
53.	Devimane-Sirsi side	Kumta	30	14.4	40.28	270.31	70.28	340.59	170.30
54.	Devimane-temple	Kumta	29	14.8	39.54	265.30	68.98	334.27	167.14
55.	Devimane-with myristicas	Kumta	30	15.0	45.86	308.15	80.12	388.27	194.14
56.	Hulidevarakodlu	Kumta	34	18.2	43.53	292.35	76.01	368.36	184.18
57.	Kalve	Kumta	28	16.2	27.38	182.82	47.53	230.35	115.17
58.	Kalve-moist dec.	Kumta	22	14.3	28.76	192.18	49.97	242.14	121.07
59.	Kandalli-Devimane	Kumta	28	16.14	41.54	278.84	72.50	351.33	175.67

Contd...

Table 1.3–*Contd...*

Sl.No.	Locality Name	Taluk	Tree Species	Average Height	Basal Area (m²/ha)	Above Ground Biomass (t/ha)	Below Ground Biomass (t/ha)	Total Biomass (t/ha)	Carbon Sequestration (t/ha)
60.	Mastihalla-Devimane arch	Kumta	26	15.27	48.04	322.92	83.96	406.87	203.44
61.	Mathali-Kandalli-Devimane	Kumta	29	15.6	41.61	279.34	72.63	351.96	175.98
62.	Soppinahosalli	Kumta	15	14.5	25.43	169.60	44.09	213.69	106.85
63.	Surjaddi	Kumta	28	17.3	35.75	239.58	62.29	301.88	150.94
64.	Surjaddi-Morse	Kumta	30	17.2	29.40	196.53	51.10	247.63	123.82
65.	Attiveri-teakmixed-drydec	Mundgod	20	10.1	11.85	77.54	20.16	97.70	48.85
66.	Godnal	Mundgod	13	15.6	43.09	289.36	75.23	364.59	182.30
67.	Gunjavathi	Mundgod	9	14.3	20.44	135.78	35.30	171.08	85.54
68.	Karekoppa-Gunjavathi	Mundgod	11	17.0	36.72	246.15	64.00	310.15	155.08
69.	Katur	Mundgod	15	16.69	28.05	187.35	48.71	236.06	118.03
70.	Katur to Gunjavathi	Mundgod	17	9.07	29.08	194.37	50.54	244.91	122.46
71.	G1-Kathalekan-nonswamp	Siddapur	41	14.4	32.31	216.25	56.23	272.48	136.24
72.	G2-Kathalekan-nonswamp	Siddapur	39	16.2	39.14	262.58	68.27	330.86	165.43
73.	G3-Kathalekan-nonswamp	Siddapur	37	15.6	45.41	305.09	79.32	384.42	192.21
74.	G4-Kathalekan –nonswamp	Siddapur	38	16.7	35.87	240.42	62.51	302.93	151.46
75.	G5-Kathalekan-nonswamp	Siddapur	39	14.1	39.84	267.28	69.49	336.77	168.39
76.	Kathalekan-savanna	Siddapur	5	6.1	1.59	7.98	2.08	10.06	5.03
77.	G6-Kathalekan-nonswamp	Siddapur	23	17.8	50.86	342.04	88.93	430.97	215.48
78.	G7-Kathalekan-nonswamp	Siddapur	44	16.8	28.22	188.52	49.02	237.54	118.77
79.	G8-Kathalekan- nonswamp	Siddapur	34	16.1	41.24	276.80	71.97	348.77	174.38

Contd...

Table 1.3–*Contd...*

Sl.No.	Locality Name	Taluk	Tree Species	Average Height	Basal Area (m²/ha)	Above Ground Biomass (t/ha)	Below Ground Biomass (t/ha)	Total Biomass (t/ha)	Carbon Sequestration (t/ha)
80.	G9-Kathalekan-nonswamp	Siddapur	18	16.0	39.63	265.88	69.13	335.01	167.51
81.	Hartebailu-soppinabetta	Siddapur	23	11.5	17.80	117.87	30.65	148.51	74.26
82.	Hutgar	Siddapur	25	15.9	30.54	204.22	53.10	257.32	128.66
83.	Joginmath-1	Siddapur	35	17.1	31.48	210.65	54.77	265.42	132.71
84.	Joginmath_2-semievergreen	Siddapur	25	17.7	42.12	282.74	73.51	356.25	178.12
85.	Kathalekan-1	Siddapur	44	16.8	28.22	188.49	49.01	237.50	118.75
86.	Kathalekan-2	Siddapur	39	16.7	30.67	205.10	53.33	258.43	129.21
87.	Kathalekan –swamp-1	Siddapur	37	16.7	43.16	289.85	75.36	365.21	182.60
88.	Kathalekan –swamp-2	Siddapur	27	15.1	40.02	268.53	69.82	338.34	169.17
89.	Kathalekan –swamp-3	Siddapur	32	16.0	70.57	475.65	123.67	599.32	299.66
90.	Kathalekan –swamp-4	Siddapur	29	17.0	61.10	411.45	106.98	518.43	259.21
91.	Kathalekan –swamp-5	Siddapur	21	15.4	43.47	291.92	75.90	367.81	183.91
92.	Kathalekan –swamp-6	Siddapur	30	15.8	40.15	269.43	70.05	339.49	169.74
93.	Kathalekan –swamp-7	Siddapur	37	15.4	31.80	212.78	55.32	268.10	134.05
94.	Kathalekan –swamp-8	Siddapur	29	18.2	55.05	370.41	96.31	466.71	233.36
95.	Kathalekan –swamp-9	Siddapur	33	18.4	64.84	436.80	113.57	550.37	275.18
96.	Kathalekan-3	Siddapur	45	16.1	29.13	194.71	50.62	245.33	122.67
97.	Malemane-1	Siddapur	33	16.6	37.54	251.70	65.44	317.15	158.57
98.	Malemane-2	Siddapur	33	18.5	38.41	257.63	66.98	324.62	162.31
99.	Malemane-3	Siddapur	28	17.6	42.63	286.24	74.42	360.66	180.33

Contd...

Table 1.3–*Contd...*

Sl.No.	Locality Name	Taluk	Tree Species	Average Height	Basal Area (m²/ha)	Above Ground Biomass (t/ha)	Below Ground Biomass (t/ha)	Total Biomass (t/ha)	Carbon Sequestration (t/ha)
100.	Siddapur evergreen	Siddapur	26	18.0	30.20	201.95	52.51	254.46	127.23
101.	Talekere	Siddapur	14	10.6	10.12	65.82	17.11	82.94	41.47
102.	Bugadi-Bennehole	Sirsi	36	15.21	56.15	377.90	98.25	476.15	238.08
103.	Gondsor-sampekattu	Sirsi	10	8.7	3.74	22.52	5.86	28.37	14.19
104.	Hulekal-Sampegadde-Hebre	Sirsi	40	15.40	50.93	342.47	89.04	431.51	215.75
105.	Kanmaski-Vanalli	Sirsi	26	15.2	57.17	384.82	100.05	484.87	242.43
106.	Khurse	Sirsi	26	11.2	22.39	149.01	38.74	187.75	93.87
107.	Masrukuli	Sirsi	15	14.9	42.36	284.41	73.95	358.36	179.18
108.	Hiresara-bettaland	Yellapur	14	11.9	41.73	280.12	72.83	352.95	176.47
109.	S5-Gidgar-Yemmalli	Yellapur	39	19.1	43.54	292.39	76.02	368.41	184.21
110.	S6-Tarukunte-Birgadde	Yellapur	41	19.5	84.15	567.73	147.61	715.34	357.67
111.	S7-Arlihonda-Nandvalli	Yellapur	48	15.5	30.55	204.32	53.12	257.44	128.72
112.	S8-Yellapur-Mavalli	Yellapur	39	17.3	35.59	238.49	62.01	300.50	150.25
113.	S9-Kiruvatti	Yellapur	16	16.6	11.99	78.48	20.41	98.89	49.44
114.	Hasrapal-evergreen	Yellapur	24	19.1	34.79	233.05	60.59	293.65	146.82
115.	Hulimundgi-semievergreen	Yellapur	27	17.3	33.63	225.23	58.56	283.79	141.89
116.	Lalguli-moist-dec	Yellapur	15	16.9	42.32	284.10	73.87	357.96	178.98

Figure 1.9: *Dipterocarpus indicus* **in Karikan Sacred Grove.**

sized *Dysoxylum malabaricum, Holigarna grahamii* etc., had higher basal area (75.08 sq.m/ha). Hill slopes and sacred groves had higher basal areas. Places like Kanmaski-Vanalli and Bugadi of Sirsi, characterised by *Diospyros candolleana, Tricalysia spearocarpa, Pterygota alata* had higher basal areas of 57.17 sq.m/ha and 56.15 sq.m/ha respectively. Lowest basal areas were for savannised places (such as Gondsur-Sampekatte, in coastal stretches of Ankola (Ankola and Ramanguli ranges) and in samples of deciduous to dry deciduous forests between Kirwatti and Kalghatgi. The bettalands, allotted to areca gardeners for leaf manure extraction also have unsatisfactory biomass.

Biomass and Carbon Sequestration in Uttara Kannada Forests

Global climate change is one of the major causes of concerns of this century, the leading reason being the phenomenal burning of fossil fuels, releasing enormous quantities of carbon to the atmosphere. Land-use changes such as the conversion of forests to croplands may also contribute to increasing atmospheric carbon, forests being one of the major storehouses of carbon since plants absorb atmospheric carbondioxide during photosynthesis and fix the carbon in sugars, starches, cellulose, lignin and numerous other bio-molecules, thereby transferring carbon from atmosphere into the biological systems. Approximately 40 per cent of terrestrial carbon storage is in the tropical forest vegetation biomass, and 30-35 per cent of land surface photosynthesis happens here (Dixon *et al.*, 1994; Malhi and Grace, 2000). If such forests are disturbed or destroyed much more carbon is released than fixed (Palm *et al.*, 1986). While estimates of standing biomass in forests help to understand the carbon stocks, the knowledge of dynamics is useful to assess C-fixation potential of the stand and categorize them as C-source, C-sink or in C-steady state (Bhat *et al.*, 2002a, 2002b; Bhat and Ravindranath, 2011).

India ranks 10th amongst the most forested nations of the world (FAO, 2005) with 23.4 percent (76.87 million ha) of its geographical area under forest and tree cover (FSI, 2008). These forests provide various critical ecosystem goods and services for the population of the country. The role of forests in carbon storage and sequestration has increased appreciation of their importance manifold bringing them to the centre-stage of climate change mitigation strategies. Over the past few decades, national policies of India aimed at conservation and sustainable management of forests have transformed India's forests into a net sink of CO_2. The biomass carbon stock in India's forests was estimated at 7.94 MtC during 1880 and nearly half of that after a period of 100 years (Richards and Flint, 1994). The earliest available estimates for forest carbon stocks (biomass and soil) for 1986, were in the range of 8.58 to 9.57 GtC (Ravindranath *et al.*, 1997; Haripriya, 2003; Chhabra and Dadhwal, 2004). As per FAO (2005), the total forest carbon stocks in India have increased over a period of 20 years (1986-2005) to 10.01 GtC. The carbon stock for the period 2006–30 was projected to increase substantially with forest cover becoming more or less stable, and new forest carbon accretions coming from the current initiatives of afforestation and reforestation programme (Ravindranath *et al.*, 2008).

It is interesting to recall some earlier studies of biomass and carbon stocks in the Uttara Kannada forests. Prasad *et al.* (1987) estimated the average total standing biomass for reserve and minor forests at 248.68 and 142.54 t/ha respectively (minor forests set aside for meeting the biomass needs of village communities were exhaustively used and greatly degraded). The average annual productivity for these was estimated at 5.395 and 2.596 t/ha/yr respectively for reserved and minor forests. Bhat *et al.* (2000; 2002a; 2002b) monitored carbon stock dynamics in Uttara Kannada district for 10 years on 8 one ha sample forest plots of different management categories. Overall carbon stocks increased at an average rate of 1.008 t/ha/yr. The minor forests, under human pressure, had negative growth of 0.237 t/ha/yr whereas reserve forests had carbon assimilation rate of 1.31 tons/ha/yr. Monitoring for 25 years (from 1984 to 2009) in six 1-ha permanent forest plots in Uttara Kannada, under different levels of anthropogenic pressure, revealed that the above-ground showed that carbon accumulation was to the tune of 1.13 t C /ha /yr, of which, 0.58 ± 1.18 t C /ha/year was contributed by surviving trees and 0.55 ± 0.33 t C/ha/year was added by recruits (Bhat and Ravindranath, 2011). Study of relic evergreen forests with swampy areas in Kathalekan of Siddapur taluk showed higher above ground biomass (349.52 ± 110.79 tons/ha) and carbon storage (174.76 ± 55.39 tons/ha) for forests alongside streams and swamps and lesser above ground biomass (263.32 ± 42.04 tons/ha) and carbon storage of 131.66 ± 21.02 tons/ha for forests away from these water courses (Chandran *et al.*, 2010). Based on the basal area study sites are grouped in to High Disturbance (< 20 m²/ha), Moderate Disturbance (20 – 40 m²/ha) and Low Disturbance (> 40 m²/ha) localities. It was observed that out of the total studied localities, 10 were highly disturbed, 56 were moderately disturbed and 50 localities had low disturbances (Figure 1.10).

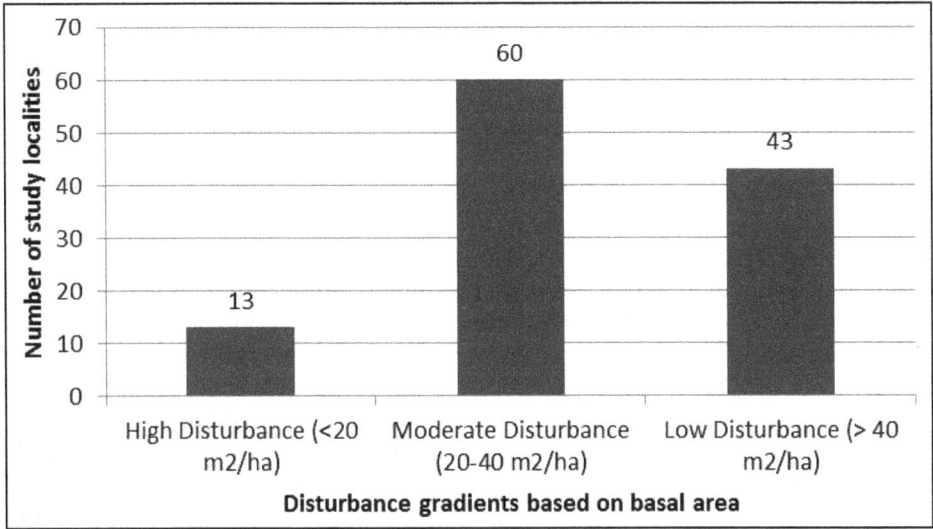

Figure 1.10: Basal Area/ha Based Sample Study Sites.

The total biomass for all the studied localities in the district as shown in Table 1.3 is the sum of estimates of above-ground and below-ground biomass for any particular site. The study sites in Kathalekan and Karikan sacred groves had the highest total biomass (749.08 and 726.12 t/ha respectively) and carbon storage (374.54 and 363.06 t/ha respectively). These forests have been protected for long because of the cultural and religious significance attached to them and hence, were relatively less disturbed than others. This allows the trees to grow to their fullest and accumulate significantly more biomass than in most other areas, prone to ongoing human pressures, or in combination with disturbances in the past (as in a savannized land). The congregation of RET species and rarer endemics such as *Dipterocarpus indicus* (Endangered), *Syzygium travancoricum* (Critically Endangered), *Myristica magnifica* (Endangered), *Myristica fatua* and *Hopea ponga* (Endangered), *Gymnacranthera canarica* (Vulnerable) along with other climax evergreen species like *Holigarna*.

Tree Species Richness, Diversity and Dominance

Kathalekan non-swamp forests were notable for their higher Shannon diversity values for trees (within 3-4), compared to swamp areas in the same may be due to special adaptations required for trees to survive in hypoxic soil conditions. Interestingly *Dipterocarpus* dominated portion of Karikan, a non-swamp sacred forest, despite high basal area (85.41 sq.m/ha), also had lower diversity of 2.24. Most other non-swamp evergreen-semi-evergreen forests had diversity values between 3 and 4. The moist deciduous forests in the rugged terrain of Ankola-Yellapur areas had higher diversity, compared to such forests in plainer areas. This is due to greater heterogeneity of the hilly landscapes. Lower Shanon diversity was found in dry deciduous and highly disturbed forests such Desaivada-Nandgadde (1.50) of Joida,

Gunjavathi of Mundgod (1.51), Sambrani (1.61) of Haliyal, Katur (1.70) of Mundgod, etc. These forests were not only disturbed but were extensively used for teak monoculturing. These forests had also prolific growth of weeds such as *Eupatorium sp* and several thorny shrubs. Some evergreen forests (*e.g.*, Talekere) dominated by *Hopea ponga* had lower Shannon diversity (1.47) and highest Simpson dominance (0.43). Hudil-evergreen and Tulsani-2 had higher Simpson dominance and lower diversities due to more of *Knema attenuata*, a Western Ghats endemic (Table 1.4).

Evergreenness and Endemism

More the evergreenness of a forest greater are the endemics contained in them (Table 1.5 and Figure 1.11). Seven transects had 100 per cent evergreeness (all the tree individuals being evergreen) and in 60 transects evergreenness was above 90 per cent. 16 transects where only 50-90 per cent of trees were evergreen are considered here as semi-evergreen forests. Remaining transects with evergreeness below 50 per cent are considered moist to dry deciduous, the latter practically without any evergreen species or poor in evergreens (Figure 1.12). The southern forests (of Bhatkal, Honavar, Siddapur and Kumta) tend to have more evergreenness than central (of Sirsi, Ankola and Yellapur) and of northern forests (Karwar, Supa taluks). Mundgod and Haliyal in the north-east are dominated by deciduous forests. Eastern parts of Sirsi and Yellapur tend to be of deciduous nature (Figure 1.12). The high endemism areas for

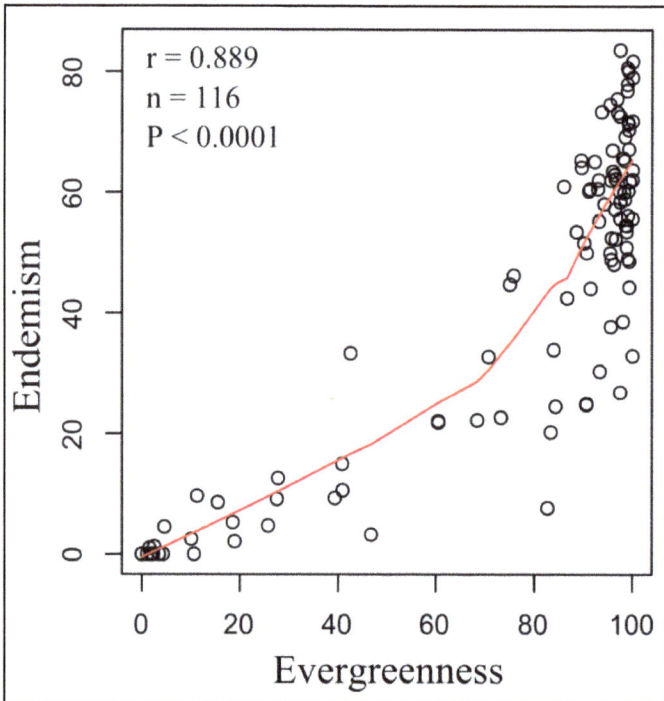

Figure 1.11: Correlation between Forest Stand Evergreenness and Western Ghat Endemism.

Table 1.4: Species Richness, Diversity, Dominance and Evenness

Sl.No.	Locality Name	Taluk	Sps. Richness	Shannon	Simpson Dominance	Simpson Diversity	Pielou Evenness
1.	Asolli-1	Ankola	5.03	2.78	0.08	0.92	0.89
2.	Asolli-2	Ankola	6.56	2.89	0.09	0.91	0.83
3.	Hosakere	Ankola	5.94	2.62	0.15	0.85	0.77
4.	S1-Katangadde-Agasur	Ankola	8.19	3.34	0.05	0.95	0.90
5.	S2-Balikoppa-Badgon	Ankola	6.28	2.73	0.11	0.89	0.80
6.	S3-Hegdekoppa-Kasinmakki	Ankola	7.10	3.08	0.07	0.93	0.87
7.	S4-Vajralli-Ramanguli	Ankola	6.28	2.90	0.09	0.91	0.85
8.	Kachinabatti	Ankola	2.96	2.14	0.15	0.85	0.83
9.	Maabagi	Ankola	6.99	3.19	0.05	0.95	0.91
10.	Dakshinakoppa	Bhatkal	2.78	1.74	0.31	0.69	0.70
11.	Gujmavu (semi evergreen)	Bhatkal	6.77	2.82	0.11	0.89	0.81
12.	Hudil (evergreen)	Bhatkal	2.87	1.61	0.37	0.63	0.61
13.	Hudil (semi evergreen)	Bhatkal	5.30	2.77	0.10	0.90	0.84
14.	Golehalli	Haliyal	3.46	2.15	0.20	0.80	0.78
15.	Kudalgi-Tatigeri	Haliyal	2.38	1.71	0.25	0.75	0.69
16.	Magvad	Haliyal	3.80	1.95	0.26	0.74	0.66
17.	Sambrani	Haliyal	2.32	1.61	0.29	0.71	0.67
18.	Yadoga	Haliyal	3.12	2.16	0.17	0.83	0.84
19.	Ambepal-1	Honavar	5.31	2.82	0.08	0.92	0.88
20.	Ambepal-2	Honavar	6.69	3.08	0.06	0.94	0.89
21.	Chaturmukhabasti	Honavar	5.42	2.71	0.10	0.90	0.86
22.	Gersoppa	Honavar	7.16	3.07	0.07	0.93	0.90

Table 1.4–*Contd...*

Sl.No.	Locality Name	Taluk	Sps. Richness	Shannon	Simpson Dominance	Simpson Diversity	Pielou Evenness
23.	Gundabala	Honavar	6.98	3.08	0.07	0.93	0.89
24.	Hadageri-1	Honavar	4.91	2.76	0.08	0.92	0.88
25.	Hadageri-2	Honavar	4.06	2.39	0.13	0.87	0.81
26.	Halsolli	Honavar	2.13	1.62	0.26	0.74	0.74
27.	Hessige-1	Honavar	5.84	2.79	0.08	0.92	0.84
28.	Hessige-2	Honavar	5.20	2.79	0.09	0.91	0.85
29.	Hessige-3	Honavar	5.04	2.49	0.14	0.86	0.77
30.	Hessige-4	Honavar	6.44	3.11	0.06	0.94	0.91
31.	Kadnir	Honavar	5.03	2.39	0.18	0.82	0.75
32.	Karikan-lower slope	Honavar	5.49	2.56	0.13	0.87	0.77
33.	Karikan-semievergreen	Honavar	4.48	2.57	0.13	0.87	0.82
34.	Karikan-temple side-diptero patch	Honavar	4.40	2.24	0.18	0.82	0.74
35.	Mahime	Honavar	4.09	2.41	0.13	0.87	0.83
36.	Sharavathy-viewpoint	Honavar	5.84	2.75	0.10	0.90	0.82
37.	Tulsani-1	Honavar	5.58	2.78	0.09	0.91	0.84
38.	Tulsani-2	Honavar	4.63	1.87	0.35	0.65	0.59
39.	Castlerock IB	Joida	5.50	3.00	0.06	0.94	0.90
40.	Castlerock-moist-dec.	Joida	4.44	2.32	0.16	0.84	0.75
41.	Castlerock-semi everg	Joida	4.97	2.55	0.14	0.86	0.80
42.	Desaivada-Nandgadde	Joida	2.21	1.50	0.30	0.70	0.60
43.	Gavni-Kangihole-Joida	Joida	6.79	2.94	0.08	0.92	0.83

Contd...

Table 1.4–*Contd...*

Sl.No.	Locality Name	Taluk	Sps. Richness	Shannon	Simpson Dominance	Simpson Diversity	Pielou Evenness
44.	Ivolli-Castlerock	Joida	3.70	1.94	0.26	0.74	0.66
45.	Joida-deciduous	Joida	4.77	2.66	0.09	0.91	0.88
46.	Kushavali	Joida	6.28	2.63	0.14	0.86	0.77
47.	Shivpura	Joida	2.52	1.89	0.19	0.81	0.76
48.	Gopishetta	Karwar	4.73	2.32	0.15	0.85	0.74
49.	Goyar-moist dec	Karwar	3.74	2.22	0.17	0.83	0.77
50.	Kalni-goyar	Karwar	6.73	2.97	0.08	0.92	0.86
51.	Karwar-moist dec	Karwar	4.23	2.58	0.10	0.90	0.91
52.	Devimane-Campsite	Kumta	7.02	3.01	0.08	0.92	0.84
53.	Devimane-Sirsi side	Kumta	6.04	2.73	0.10	0.90	0.80
54.	Devimane-temple	Kumta	5.52	2.69	0.11	0.89	0.80
55.	Devimane-with myristicas	Kumta	5.96	2.76	0.10	0.90	0.81
56.	Hulidevarakodlu	Kumta	6.97	2.87	0.10	0.90	0.81
57.	Kalve	Kumta	6.05	2.76	0.12	0.88	0.83
58.	Kalve-moist dec.	Kumta	4.64	2.53	0.12	0.88	0.82
59.	Kandalli-Devimane	Kumta	5.55	2.31	0.23	0.77	0.69
60.	Mastihalla-Devimane arch	Kumta	5.28	2.61	0.13	0.87	0.80
61.	Mathali-Kandalli-Devimane	Kumta	5.74	2.62	0.14	0.86	0.78
62.	Soppinahosalli	Kumta	3.34	2.34	0.13	0.87	0.87
63.	Surjaddi	Kumta	5.60	2.87	0.07	0.93	0.86
64.	Surjaddi-Morse	Kumta	6.05	2.79	0.09	0.91	0.82

Contd...

Table 1.4—*Contd...*

Contd...

Sl.No.	Locality Name	Taluk	Sps. Richness	Shannon	Simpson Dominance	Simpson Diversity	Pielou Evenness
65.	Attiveri-teakmixed-drydec	Mundgod	4.18	2.31	0.16	0.84	0.77
66.	Godnal	Mundgod	2.74	1.84	0.23	0.77	0.72
67.	Gunjavathi	Mundgod	2.08	1.51	0.32	0.68	0.69
68.	Karekoppa-Gunjavathi	Mundgod	2.32	1.94	0.19	0.81	0.81
69.	Katur	Mundgod	3.35	1.70	0.36	0.64	0.63
70.	Katur to Gunjavati	Mundgod	3.67	2.43	0.12	0.88	0.86
71.	G1-Kathalekan-nonswamp	Siddapur	7.91	3.39	0.04	0.96	0.91
72.	G2-Kathalekan-nonswamp	Siddapur	7.37	3.18	0.06	0.94	0.87
73.	G3-Kathalekan-nonswamp	Siddapur	7.75	3.42	0.04	0.96	0.95
74.	G4-Kathalekan -nonswamp	Siddapur	7.58	3.12	0.07	0.93	0.86
75.	G5-Kathalekan-nonswamp	Siddapur	7.51	3.16	0.06	0.94	0.86
76.	Kathalekan-savanna	Siddapur	1.82	1.52	0.23	0.77	0.95
77.	G6-Kathalekan-nonswamp	Siddapur	4.31	2.39	0.14	0.86	0.76
78.	G7-Kathalekan-nonswamp	Siddapur	8.69	3.36	0.05	0.95	0.89
79.	G8-Kathalekan- nonswamp	Siddapur	6.98	2.91	0.10	0.90	0.82
80.	G9-Kathalekan-nonswamp	Siddapur	3.25	1.77	0.32	0.68	0.61
81.	Hartebailu-soppinabetta	Siddapur	4.38	2.24	0.18	0.82	0.71
82.	Hutgar	Siddapur	5.22	2.53	0.13	0.87	0.79
83.	Joginmath-1	Siddapur	7.11	3.05	0.07	0.93	0.86
84.	Joginmath_2-semievergreen	Siddapur	5.42	2.84	0.08	0.92	0.88
85.	Kathalekan-1	Siddapur	8.69	3.36	0.05	0.95	0.89

Table 1.4—*Contd...*

Sl.No.	Locality Name	Taluk	Sps. Richness	Shannon	Simpson Dominance	Simpson Diversity	Pielou Evenness
86.	Kathalekan-2	Siddapur	7.56	3.31	0.05	0.95	0.90
87.	Kathalekan –swamp-1	Siddapur	7.53	2.92	0.09	0.91	0.81
88.	Kathalekan –swamp-2	Siddapur	5.58	2.67	0.11	0.89	0.81
89.	Kathalekan –swamp-3	Siddapur	6.51	2.74	0.11	0.89	0.79
90.	Kathalekan –swamp-4	Siddapur	5.89	2.52	0.15	0.85	0.75
91.	Kathalekan –swamp-5	Siddapur	4.43	2.55	0.11	0.89	0.84
92.	Kathalekan –swamp-6	Siddapur	6.01	2.76	0.13	0.87	0.81
93.	Kathalekan –swamp-7	Siddapur	7.74	3.08	0.08	0.92	0.85
94.	Kathalekan –swamp-8	Siddapur	5.79	2.69	0.10	0.90	0.80
95.	Kathalekan –swamp-9	Siddapur	6.67	3.04	0.07	0.93	0.87
96.	Kathalekan-3	Siddapur	8.59	3.38	0.05	0.95	0.89
97.	Malemane-1	Siddapur	6.42	3.16	0.05	0.95	0.90
98.	Malemane-2	Siddapur	6.40	3.08	0.07	0.93	0.88
99.	Malemane-3	Siddapur	5.69	2.82	0.08	0.92	0.85
100.	Siddapur evergreen	Siddapur	5.72	2.81	0.08	0.92	0.86
101.	Talekere	Siddapur	2.98	1.47	0.43	0.57	0.56
102.	Bugadi-Bennehole	Sirsi	7.17	3.00	0.08	0.92	0.84
103.	Gondsor-sampekattu	Sirsi	2.67	2.01	0.16	0.84	0.87
104.	Hulekal-Sampegadde-Hebre	Sirsi	7.92	3.14	0.07	0.93	0.85
105.	Kanmaski-Vanalli	Sirsi	4.76	2.80	0.08	0.92	0.86
106.	Khurse	Sirsi	4.99	2.44	0.14	0.86	0.75
107.	Masrukuli	Sirsi	3.39	2.04	0.21	0.79	0.75

Contd...

Table 1.4–*Contd...*

Sl.No.	Locality Name	Taluk	Sps. Richness	Shannon	Simpson Dominance	Simpson Diversity	Pielou Evenness
108.	Hiresara-bettaland	Yellapur	2.85	1.90	0.24	0.76	0.72
109.	S5-Gidgar-Yemmalli	Yellapur	7.94	3.09	0.08	0.92	0.84
110.	S6-Tarukunte-Birgadde	Yellapur	8.36	3.36	0.05	0.95	0.90
111.	S7-Arlihonda-Nandvalli	Yellapur	9.83	3.47	0.05	0.95	0.90
112.	S8-Yellapur-Mavalli	Yellapur	7.94	3.12	0.08	0.92	0.85
113.	S9-Kiruvatti	Yellapur	3.13	1.83	0.25	0.75	0.66
114.	Hasrapal-evergreen	Yellapur	5.16	2.91	0.07	0.93	0.91
115.	Hulimundgi-semievergreen	Yellapur	5.84	2.85	0.08	0.92	0.87
116.	Lalguli-moist-dec	Yellapur	3.53	2.27	0.14	0.86	0.84

trees are towards the Ghat areas of Bhatkal, Honavar, Siddapur and Kumta coinciding with higher occurrence of evergreen forests (Figure 1.13)

Table 1.5: Percentage of Western Ghat Endemism and Evergreenness in the Forest Samples

Sl.No.	Locality Name	Taluk	Per cent W Ghats (Endemism)	Per cent Evergreeness
1.	Asolli-1	Ankola	55.70	100.00
2.	Asolli-2	Ankola	73.28	96.95
3.	Hosakere	Ankola	55.30	93.18
4.	S1-Katangadde-Agasur	Ankola	22.22	68.38
5.	S2-Balikoppa-Badgon	Ankola	2.52	10.08
6.	S3-Hegdekoppa-Kasinmakki	Ankola	15.00	40.83
7.	S4-Vajralli-Ramanguli	Ankola	12.61	27.73
8.	Kachinabatti	Ankola	8.62	15.52
9.	Maabagi	Ankola	22.68	73.20
10.	Dakshinakoppa	Bhatkal	7.69	82.69
11.	Gujmavu (semi evergreen)	Bhatkal	63.72	100.00
12.	Hudil (evergreen)	Bhatkal	80.65	98.92
13.	Hudil (semi evergreen)	Bhatkal	30.37	93.33
14.	Golehalli	Haliyal	0	0
15.	Kudalgi-Tatigeri	Haliyal	0	0
16.	Magvad	Haliyal	0.00	1.75
17.	Sambrani	Haliyal	0	0.0
18.	Yadoga	Haliyal	0.00	10.64
19.	Ambepal-1	Honavar	48.91	95.65
20.	Ambepal-2	Honavar	62.14	100.00
21.	Chaturmukhabasti	Honavar	32.76	70.69
22.	Gersoppa	Honavar	50.00	95.45
23.	Gundabala	Honavar	32.94	100.00
24.	Hadageri-1	Honavar	54.55	98.86
25.	Hadageri-2	Honavar	53.57	98.81
26.	Halsolli	Honavar	79.07	100.00
27.	Hessige-1	Honavar	49.02	99.02
28.	Hessige-2	Honavar	44.30	99.33
29.	Hessige-3	Honavar	24.79	90.60
30.	Hessige-4	Honavar	37.78	95.56
31.	Kadnir	Honavar	67.01	95.88
32.	Karikan-lower slope	Honavar	62.04	95.62

Contd...

Table 1.5–*Contd...*

Sl.No.	Locality Name	Taluk	Per cent W Ghats (Endemism)	Per cent Evergreeness
33.	Karikan-semievergreen	Honavar	62.96	96.30
34.	Karikan-temple side-diptero patch	Honavar	75.53	96.81
35.	Mahime	Honavar	25.00	90.63
36.	Sharavathy-viewpoint	Honavar	65.69	98.04
37.	Tulsani-1	Honavar	64.15	89.62
38.	Tulsani-2	Honavar	83.62	97.41
39.	Castlerock IB	Joida	71.85	100.00
40.	Castlerock-moist-dec.	Joida	5.31	18.58
41.	Castlerock-semi everg	Joida	24.51	84.31
42.	Desaivada-Nandgadde	Joida	0	0
43.	Gavni-Kangihole-Joida	Joida	50.00	90.67
44.	Ivolli-Castlerock	Joida	65.12	92.25
45.	Joida-deciduous	Joida	10.61	40.91
46.	Kushavali	Joida	38.61	98.02
47.	Shivpura	Joida	0	0
48.	Gopishetta	Karwar	4.76	25.71
49.	Goyar-moist dec	Karwar	0.00	1.06
50.	Kalni-goyar	Karwar	61.00	86.00
51.	Karwar-moist dec	Karwar	0.00	2.27
52.	Devimane-Campsite	Kumta	60.27	91.10
53.	Devimane-Sirsi side	Kumta	51.64	90.16
54.	Devimane-temple	Kumta	58.49	97.48
55.	Devimane-with myristicas	Kumta	74.62	95.38
56.	Hulidevarakodlu	Kumta	53.51	88.60
57.	Kalve	Kumta	62.07	93.10
58.	Kalve-moist dec.	Kumta	3.26	46.74
59.	Kandalli-Devimane	Kumta	69.23	98.46
60.	Mastihalla-Devimane arch	Kumta	62.28	96.49
61.	Mathali-Kandalli-Devimane	Kumta	67.18	99.24
62.	Soppinahosalli	Kumta	4.55	4.55
63.	Surjaddi	Kumta	65.32	89.52
64.	Surjaddi-Morse	Kumta	65.29	98.35
65.	Attiveri-teakmixed-drydec	Mundgod	0.00	4.26
66.	Godnal	Mundgod	1.25	2.50
67.	Gunjavathi	Mundgod	0.00	2.13
68.	Karekoppa-Gunjavathi	Mundgod	0.00	0.00

Contd...

Table 1.5–*Contd...*

Sl.No.	Locality Name	Taluk	Per cent W Ghats (Endemism)	Per cent Evergreeness
69.	Katur	Mundgod	1	1.54
70.	Katur to Gunjavati	Mundgod	0	0
71.	G1-Kathalekan-nonswamp	Siddapur	54.61	98.58
72.	G2-Kathalekan-nonswamp	Siddapur	61.85	99.42
73.	G3-Kathalekan-nonswamp	Siddapur	60.58	91.35
74.	G4-Kathalekan -nonswamp	Siddapur	58.91	98.45
75.	G5-Kathalekan-nonswamp	Siddapur	55.70	97.47
76.	Kathalekan-savanna	Siddapur	0.00	0.00
77.	G6-Kathalekan-nonswamp	Siddapur	50.91	98.79
78.	G7-Kathalekan-nonswamp	Siddapur	52.48	95.74
79.	G8-Kathalekan- nonswamp	Siddapur	60.18	97.35
80.	G9-Kathalekan-nonswamp	Siddapur	77.96	98.92
81.	Hartebailu-soppinabetta	Siddapur	44.74	75.00
82.	Hutgar	Siddapur	60.61	92.93
83.	Joginmath-1	Siddapur	26.89	97.48
84.	Joginmath_2-semievergreen	Siddapur	20.24	83.33
85.	Kathalekan-1	Siddapur	52.48	95.74
86.	Kathalekan-2	Siddapur	44.08	91.45
87.	Kathalekan –swamp-1	Siddapur	71.43	99.16
88.	Kathalekan –swamp-2	Siddapur	71.70	99.06
89.	Kathalekan –swamp-3	Siddapur	72.65	97.44
90.	Kathalekan –swamp-4	Siddapur	80.17	99.14
91.	Kathalekan –swamp-5	Siddapur	76.92	98.90
92.	Kathalekan –swamp-6	Siddapur	70.40	99.20
93.	Kathalekan –swamp-7	Siddapur	56.19	99.05
94.	Kathalekan –swamp-8	Siddapur	81.75	100.00
95.	Kathalekan –swamp-9	Siddapur	60.33	99.17
96.	Kathalekan-3	Siddapur	33.93	83.93
97.	Malemane-1	Siddapur	48.63	99.32
98.	Malemane-2	Siddapur	63.51	95.95
99.	Malemane-3	Siddapur	60.00	98.26
100.	Siddapur evergreen	Siddapur	48.10	96.20
101.	Talekere	Siddapur	73.42	93.67
102.	Bugadi-Bennehole	Sirsi	46.21	75.76
103.	Gondsor-sampekattu	Sirsi	0.00	3.45
104.	Hulekal-Sampegadde-Hebre	Sirsi	57.25	96.38

Contd...

Table 1.5–*Contd...*

Sl.No.	Locality Name	Taluk	Per cent W Ghats (Endemism)	Per cent Evergreeness
105.	Kanmaski-Vanalli	Sirsi	58.12	94.24
106.	Khurse	Sirsi	9.33	39.33
107.	Masrukuli	Sirsi	9.68	11.29
108.	Hiresara-bettaland	Yellapur	2.11	18.95
109.	S5-Gidgar-Yemmalli	Yellapur	9.17	27.50
110.	S6-Tarukunte-Birgadde	Yellapur	42.50	86.67
111.	S7-Arlihonda-Nandvalli	Yellapur	21.85	60.50
112.	S8-Yellapur-Mavalli	Yellapur	33.33	42.50
113.	S9-Kiruvatti	Yellapur	0.00	0.00
114.	Hasrapal-evergreen	Yellapur	52.33	96.51
115.	Hulimundgi-semievergreen	Yellapur	22.09	60.47
116.	Lalguli-moist-dec	Yellapur	0.00	1.89

A total of 76 Western Ghat endemic tree species were found in the study areas. Altogether 127 endemic trees were endemic to Western Ghat-Sri Lanka biodiversity hot spot (45.6 per cent endemism). Western Ghat endemic shrub species numbered to 39 and together with Sri Lanka the shrub species were 82. Herb layer had 76 Western Ghat endemics and together with Sri Lanka endemics rise to 137 species. Highest evergreen forests (100 per cent) were found in Dipterocarpus forests of Asolli 1 (Ankola), Ambepal-2 (Honnavar), Kathalekan Swamp grid 8-T3 (Siddapur)and non-Dipterocarpus forests of Gujmaav of Bhatkal, Gundabala (Honnavar), Halsolli (Honnavar) and Castlerock IB (Joida). These were either Kans or less disturbed areas in areas with difficult access. Most of the Kathlekan, Karikan, Malemane, Gersoppa and Devimane area forests were higher evergreen forests as fire was absent and protected due to reserved status. All Mundgod taluk transects along with Goyar (Karwar), Magvad (Haliyal), etc., which were moist to dry deciduous had very negligible to zero endemism. Endemism is seen as a factor closely correlated to forest evergreenness. Nearly 50 per cent of total transects were having 50 per cent and above tree endemism; such forests had evergreenness of 90 per cent and above. The deciduous forests had hardly any Western Ghat endemics. Endemism is the first casualty even in high rainfall areas under heavy human disturbances. The endemics tend to decline in the wake of fire, logging, grazing and such disturbances leading to endemic poor secondary evergreen forests and finally into deciduous ones devoid of any endemics. Even in endemic rich forests the more sensitive ones such as *Syzygium travancoricum, Dipterocarpus indicus, Palaquium ellipticum, Madhuca bourdilloni, Myristica* spp., vanish early with disturbances. The habitats of these species are rich watershed areas giving rise to perennial streams.

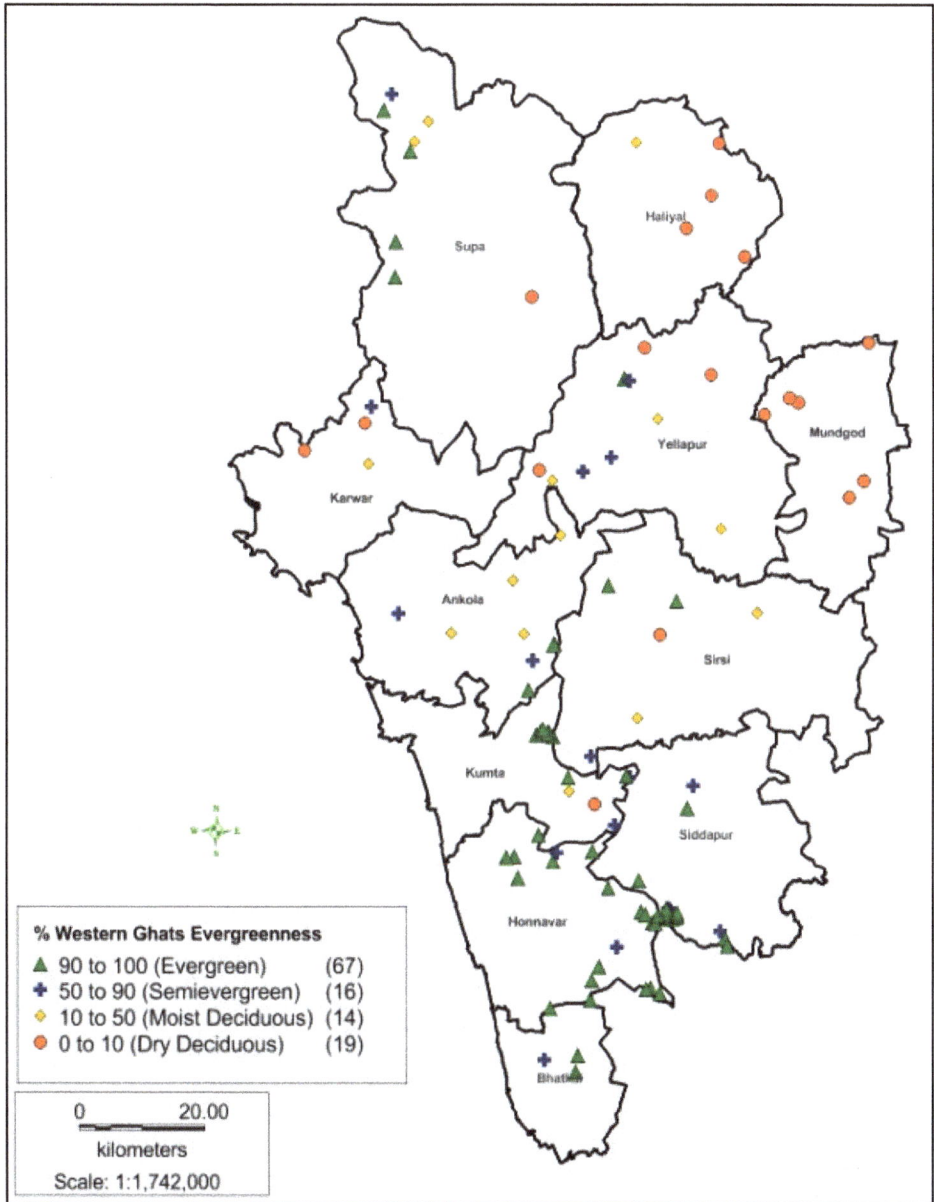

Figure 1.12: Uttara Kannada Map Showing Percentage of Tree Community Evergreeness in the Samples Studied.

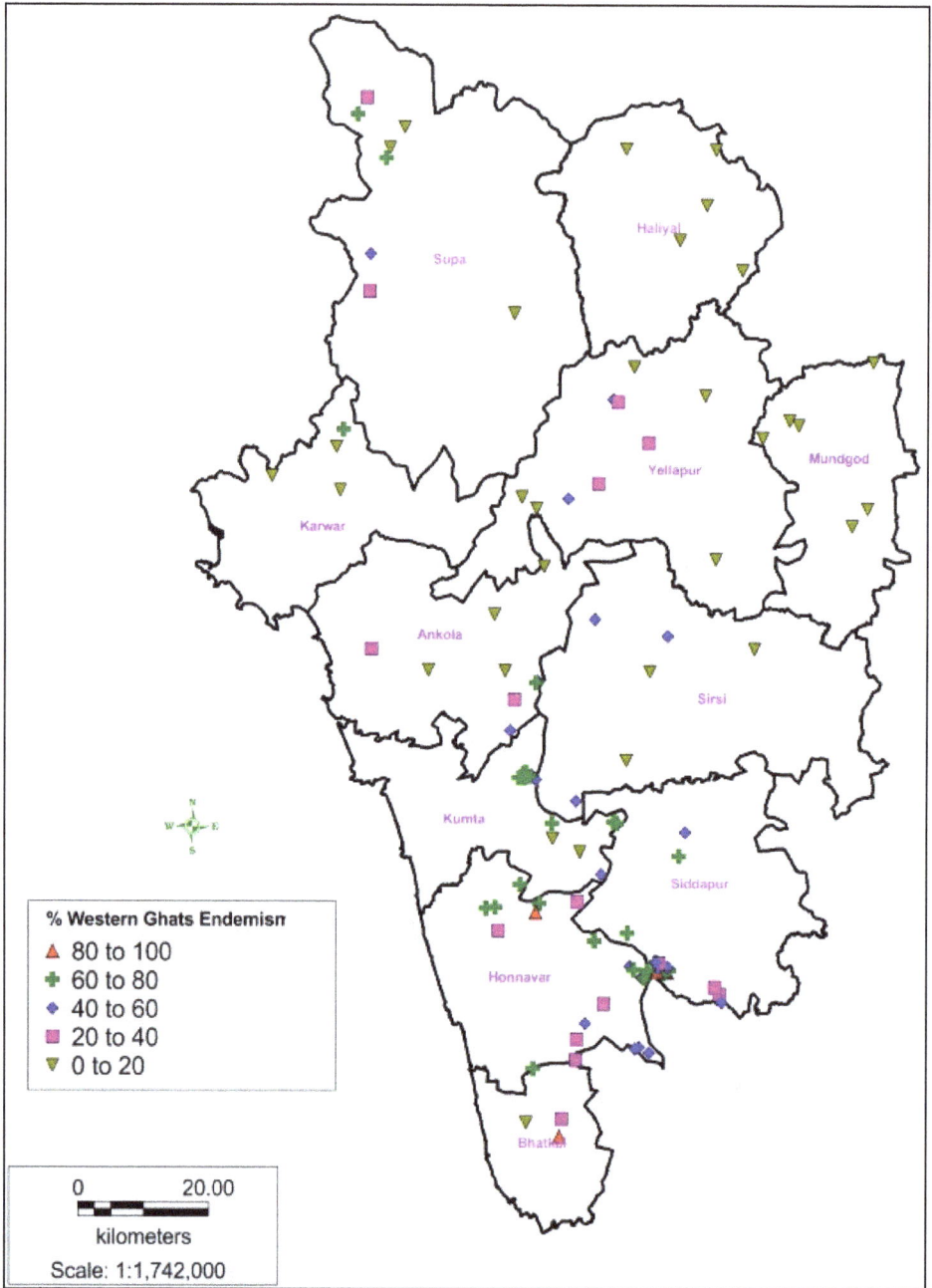

**Figure 1.13: Uttara Kannada Map Showing Forest Samples with
Tree Endemism (of Western Ghats).**

Important Value Index (IVI)

Important value index is an important parameter to estimate the dominant species in an area taking into account its basal area, density and frequency. Higher the IVI greater is the dominance of that species. Asolli-1 of Ankola has higher IVI for *Dipterocarpus indicus* (50.15) followed by *Knema attenuata* (30.77) and *Holigarna grahamii* (26.25); all of these are Western Ghat endemics. *Dipterocarpus indicus* is red-listed by IUCN as Endangered. The dominance of evergreen tree species like *Olea dioca, Aporosa lindleyana Holigarna arnottiana* in a forest indicates secondary nature of the forest. Greater human pressures in such forests, especially in the form of forest burning, such as at Maabgi of Ankola, might have increased deciduous species like *Terminalia alata, Vitex altissima* and *Dillenia pentagyna* etc. Higher values of biomass/C-stocks are associated with less human or natural disturbances or better site qualities (Lugo and Brown,1992; Brown, 1997). Whereas the undisturbed parts of Kathalekan had high biomass, within the forest interior some hills were savannized in the past, a sample there having least biomass and carbon storage of 10.06 t/ha and 5.03 t/ha respectively). In Gondsor-Sampekattu savanna of Joida the values were higher; with low biomass of 28.37 t/ha and carbon sequestration of 14.19 t/ha this site was only next higher in hierarchy among the 116 transects (Figure 1.14).

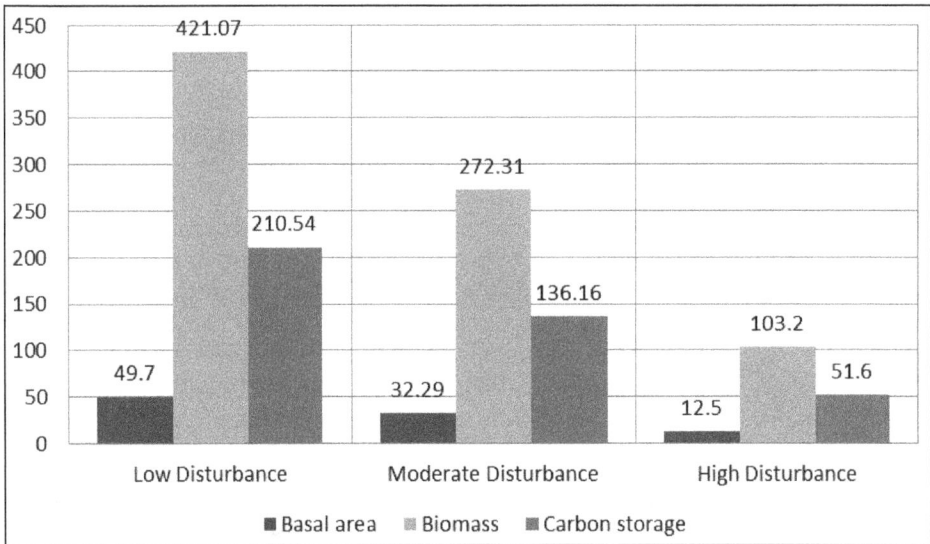

Figure 1.14: Average Basal Area, Biomass and Carbon Storage in Samples of different Disturbance Gradients.

Prioritisation of Forests for Biodiversity Conservation

Role of Endemism in Conservation Priorities

Species richness and endemism are two key attributes of biodiversity that reflect the complexity and uniqueness of natural ecosystems (Caldecott *et al.*, 1996). Myers *et al.* (2000) strongly favour identification and prioritisation of 'hotspots', or areas featuring exceptional concentrations of endemic species and experiencing exceptional loss of habitat. Their focus is more on species, rather than populations or other taxa, as the most prominent and readily recognizable form of biodiversity. Concentrating a large proportion of conservation support on these areas would go far to stem the mass extinction of species that is now underway. Nelson *et al.* (1990), based on forest studies in Brazilian Amazonia, realized the importance of locating true concentrations of plant endemism for selecting priority conservation areas to guarantee preservation of unique species. A study on 19 species of endemic mammals and birds in Mexico made Peterson *et al.* (2000) favour setting of regional conservation priorities based on combinations of modeling individual endemic species' distributions, evaluating regional concentrations of species richness, and using complementarity of areas by maximizing inclusion of species in the overall system. The optimized reserve system identified by this approach is stated to have performed 33–58 per cent better than existing protected areas in inclusion of the endemic species. Therefore they favoured making necessary adjustments in the existing systems through incorporation of endemic areas. Strengthening such observations Stattersfield *et al.* (1998) conclude that the 218 endemic bird areas identified by Birdlife International provide a reasonable overlap with the biodiversity hotspots identified by other conservation organisations, and are a focus for conservation action. Burlakova *et al.* (2011) advocated the need for adopting species rarity and endemism in aquatic domains also for the conservation of regionally rare and endemic fresh water molluscs in Texas.

Reviewing the role of endemism, Meadows (2008) stated that ecoregions rich in endemics are also rich in overall species. For example, the 10 percent of the world's land area with the most endemics also has more than 60 percent of all terrestrial vertebrate species. Likewise10 percent of land with the greatest number of endemic amphibians and reptiles also contains more than 70 percent of all terrestrial vertebrate species. In addition, ecoregions rich in endemics of any one vertebrate class are also rich in endemics of the other three classes. At the same time many researchers on vertebrate conservation also content that their findings may not apply to nonvertebrates and that endemism is only one criterion for planning. However, using endemism along with other factors to identify global priorities focuses conservation in critical regions, where on-the-ground efforts will yield the greatest payoffs for biodiversity.

Humid tropical forests, like the rain forests, are richest systems in biodiversity. Regions of high rainfall also have large volumes of water in the river flow (World Water Assessment Programme, 2012). The confluence of rainforests and hydropower potential have prompted many nations with large areas of tropical rainforest - including Brazil, Peru, Colombia, the Democratic Republic of the Congo, Vietnam, and Malaysia - plan to expand their hydropower energy capacity. It is generally assumed that deforestation will have a positive effect on river discharge and energy generation resulting from declines in evapotranspiration (ET) associated with forest conversion. Study in the Xingu River basin of Amazonian Brazil using hydrological and climate models showed that simulated deforestation of 20 per cent and 40 per cent within this basin increased discharge by 4–8 per cent and 10–12 per cent, which could make similar increases in energy generation from a very large hydropower station planned in the river. When indirect effects were considered, simulated deforestation inhibited rainfall in the Xingu basin by 6 to 36 per cent, thus offsetting the likely gains (Stickler *et al.*, 2013). Moreover the loss of top soil and landslides and sedimentation in the downstream areas following deforestation are also to be considered. Forest decline can as well upset microclimate conditions and cause disappearance of scores of sensitive species.

The Western Ghats together with Sri Lanka constitute one of the 34 Biodiversity Hotspots of the world in view of exceptionally rich biodiversity, high degree of endemism and at the same time undergoing tremendous threat from human activities. The original extent of this combined Hotspot was 189,611 km². Of the hotspot vegetation what remains today is merely 43,611 km² area. Tremendous population pressure and biomass needs have created heavy fragmentation of Western Ghat forests. Both these regions in this hotspot together continue to shelter still 3,049 endemic plant species, 10 endemic threatened birds, 14 endemic threatened mammals, 87 threatened amphibians and so on (Conservation International, 2013).

Protected Areas in Uttara Kannada

Karnataka has five National Parks and 21 Wildlife Sanctuaries. Uttara Kannada district has mainly two important protected areas namely **Anshi National Park** and **Dandeli Wildlife Sanctuary**. These two PAs are brought together under **Dandeli-Anshi Tiger Reserve** with focus on tiger conservation. The DATR presently covers an area of 1365 sq.km. in the taluks of Joida, Haliyal and Karwar. Admittedly, we were not given permission to carry out forest ecological studies within this Tiger Reserve. Hence we have relatively lesser sampling areas within these taluks. Recently (in 2011) **Attivery Bird Sanctuary** was declared in Mundgod taluk covering 2.23 sq.km area, mainly composed of a reservoir and its peripheral areas.

Conservation Reserves are a new concept under the framework of Protected Areas under the Wildlife (Protection) Amendment Act of 2002. These reserves they seek to protect habitats that are under private ownership also, through active stakeholder participation. They are typically buffer zones or connectors and migration corridors between National Parks, Wildlife Sanctuaries and reserved protected forests in India. They are designated as conservation reserves if they are uninhabited and completely owned by the government but used for subsistence by communities, and community reserves if part of the lands are privately owned. Administration of such reserves would be through joint participation of forest officials and local bodies like gram sabhas and gram panchayats. They do not involve any displacement and protect user rights of communities. In Uttara Kannada, some such Conservation Reserves were set up by the Government of Karnataka, under the initiative of the Mr Anant Hedge Ashishar of Western Ghat Task Force, the Karnataka Forest Department, ATREE and SACON with technical inputs from Mr. Balachandra Hegde. Presence of endangered and endemic species, critical corridors connecting larger Western Ghats landscape and potential threats for the region etc., were considered for identifying conservation priority areas (Dandekar- http://sandrp.in/rivers/Novel_Conservation_reserves). Four such reserves were set out to protect Lion tailed macaque habitats, rare and endangered Myristica Swamps, Hornbill habitats and a riverain ecosystem (details are given in Table 1.6 and Figure 1.15).

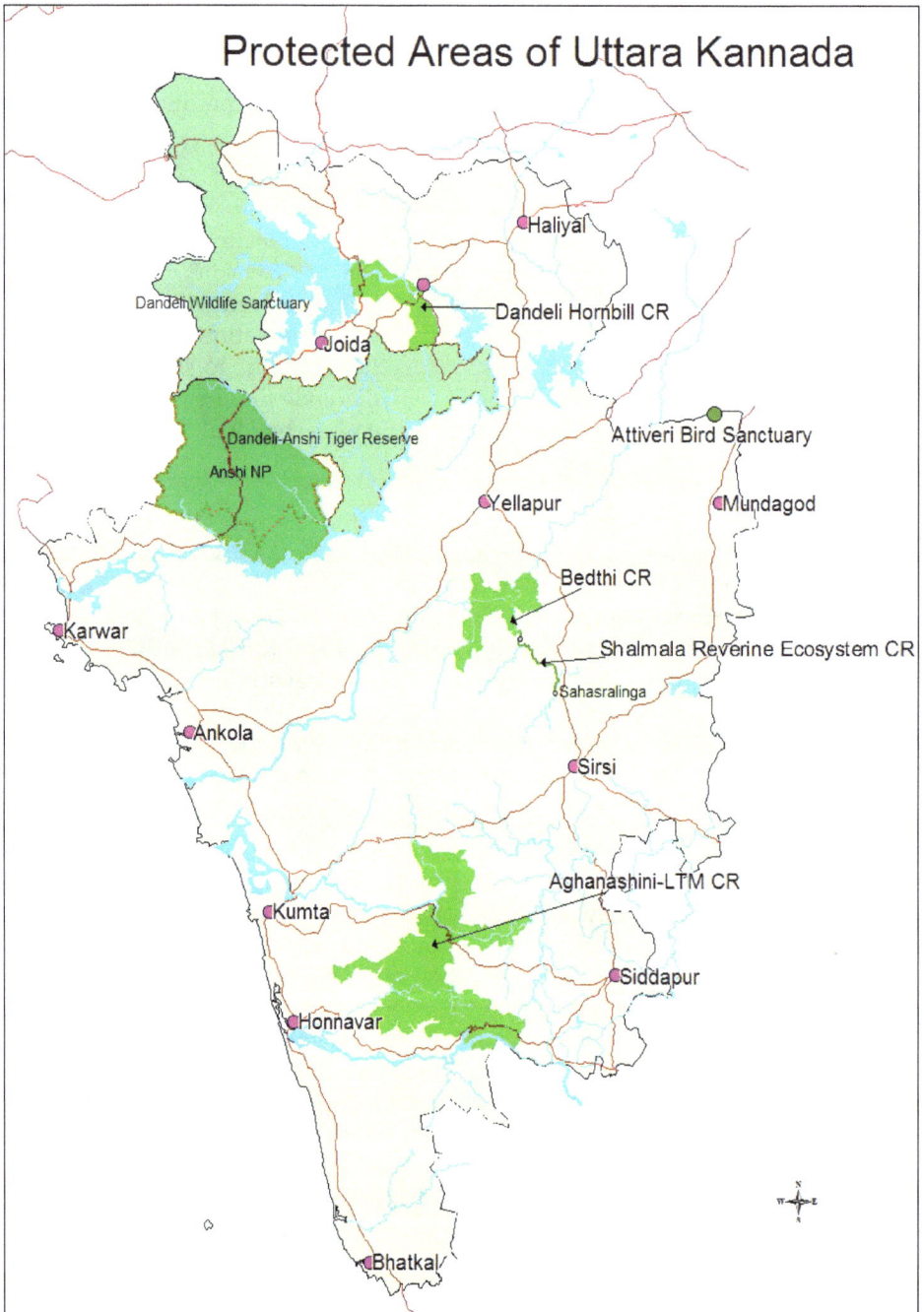

Figure 1.15: Protected Areas of Uttara Kannada.

Table 1.6: Details of Conservation Reserves in Uttara Kannada

Name	Area (sq.km)	Coservation Priority Species	Priority Locations
Aghanashini LTM Conservation Reserve	299.52	Lion tailed macaque, Myristica swamps	Unchalli Falls, Kathalekan, Muktihole
Bedthi Conservation Reserve	57.07	Hornbills *Coscinium fenestratum* (medicinal plant)	Magod Falls, Jenukallugudda, Bilihalla Valley, Konkikote
Shalmala Riparian Eco-system Conservation Reserve	4.89	Flora and fauna and as an important corridor in Western Ghats of Karnataka	
Hornbill Conservation Reserve	52.50	Hornbills	Kali River

Assigning Conservation Values through Correlation between Five Notable Parameters of Tree Communities in Uttara Kannada

To be helpful in preparing a composite conservation index for forest patches studied through 116 transects and covering the entire district we considered five important variables (per cent evergreenness, per cent endemism, basal area, tree height and Shannon diversity index) that were studied about these samples. The relative importance of these variables in assigning conservation values, based on the 116 sample studies is depicted in Figure 1.16. Some notable points on these variables are given below:

Evergreenness

Evergreen forests of the Western Ghats, due to various reasons, such as seats of high endemic diversity and high hydrological value is an important factor for assigning convservation values. We have considered here percentage of evergreen trees in the total tree population of the sample as evergreenness.

Endemism

Tree endemism is of overall importance in for assigining conservation values. Through an earlier study in the Western Ghats it was established that forest evergreenness in a stand, is a strong positive determinant of tree endemism in the same stand (Chandran, 1997). The find was carried forth beyond into the domain of endemism among fresh water fishes in the streams of Sharavathi River catchment (Sreekantha *et al.*, 2007) establishing that the number and percentage of endemic fishes among total fish fauna in a stream was directly correlated to the percentage of evergreenness and tree endemism in the catchment area forests of that particular stream. In Uttara Kannada amphibian studies highest species diversity (35 species), with high percentage of Western Ghat endemism (26 species, 74 per cent endemism) occurred in a mere 2.25 sq.km area of high evergreen forests (almost 100 per cent) characterised by Myristic swamps in Kathalekan of Siddapur taluk (Chandran *et al.*, 2010).

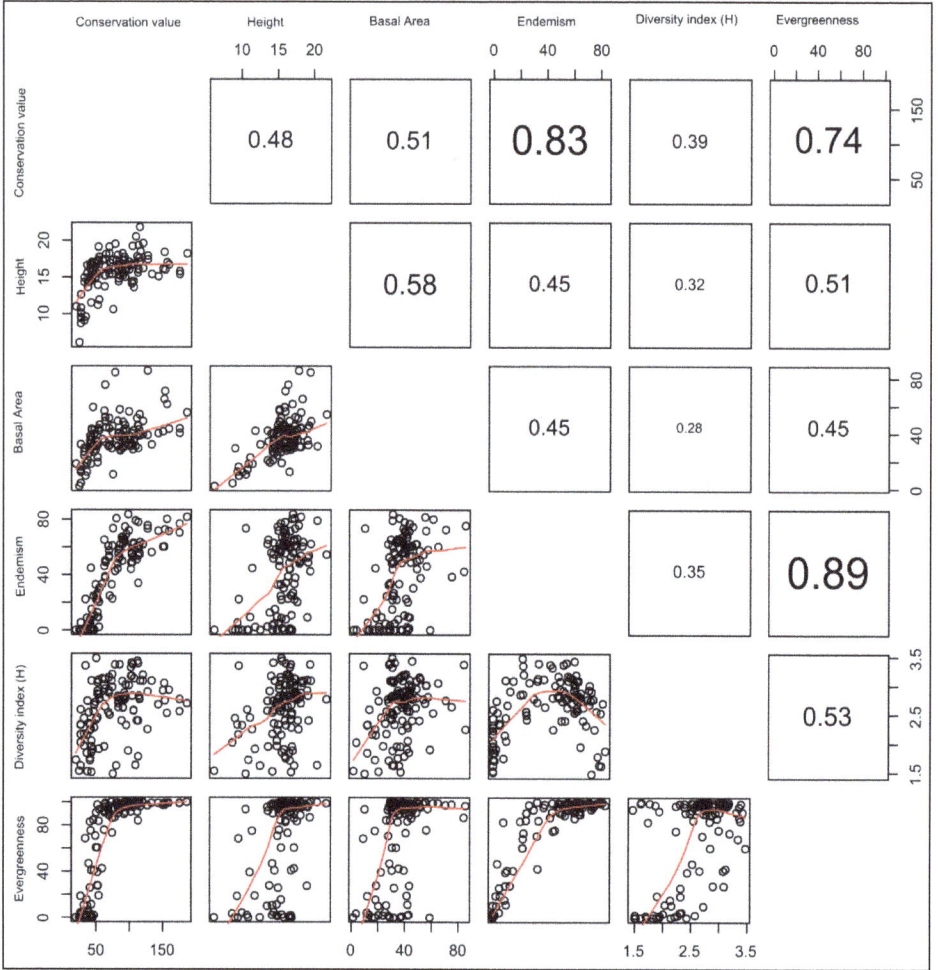

Figure 1.16: Relative Importance of Tree Community Parameters for Assigning Conservation Values (n=116, P< 0.0001).

Tree Heights

World over tallness of forest stands is considered indicative of the old growth nature and merits high conservation value. Human impacts on tall forests through logging, either clear felling or selection felling would result in regenerated trees of lesser heights, as the competition for light is minimised.

Basal Area

Basal area of trees per hectare expressed sq.m is a standard expression in forest ecological studies worldwide about the relative growth of forests. It is a major factor for estimating forest stand biomass and therefore also for estimation of carbon sequestration.

Diversity Value

It is generally accepted that higher diversity in general goes hand in hand with conservation importance.

It is evident from the analysis that the highest correlation (0.89) was between ' per cent endemism' and ' per cent evergreenness'. This can be also justified from the fact that those transects which have 50 per cent or more endemism have evergreenness values as high as 90 per cent and more (Table 1.5). This was followed by the parameters basal area, tree height and Shannon diversity with which it had correlation in descending order of 0.51, 0.48 and 0.39 respectively. The percentage endemism is the most decisive factor in conferring higher conservation value to any ecosystem as endemics lost in their respective regions are irreplaceable. Higher endemism in a particular area indicates the presence of high sensitive species in that area, implying that there should be greater prioritisation in the conservation of endemic areas in any conservation programmes, not sidelining in any way the importance in conservation of any widely distributed but threatened species like elephant or tiger. The tree height and basal area can also be considered as factors contributing to the overall conservation value of the forest areas.

Principal Component Analysis

A PCA of the sample sites was carried out considering the observed and quantified characters of tree communities *viz.*, evergreenness, endemism (of Western Ghats), height and basal area. The PCA (Figure 1.17) shows that the first two axes accounted for 86.57 per cent of the cumulative variance explained by the four gradients extracted in the PCA analysis and the direction and length of each arrow indicated the direction and rate of maximum changes in each variable. The Eigen value for the first axis was 2.53, whereas for the second axis it was 0.92. The loading scores obtained in PCA also indicated that the prime factors influencing the forest vegetation are evergreenness and endemism than the height and basal area in first axis (Table 1.3). The high evergreen/endemism rich areas also host many rare and threatened species and essentially need to be prioritised for biodiversity conservation.

In a scheme of ranking sites representative of forest patches, areas of high tree endemism- which are essentially high evergreen areas, degree of endemism needs to be conferred higher value than height or basal area/ha. The latter two are, in the conditions of Uttara Kannada, much dependent on the degree of protection that a forest patch enjoys. Even moist deciduous forests, can attain much height and high basal area, but tree endemism (of Western Ghats) is scanty here even in the absence of human disturbances. At the same time height factor cannot be ignored in evergreen forests as we hardly get threatened tree species like *Dipterocarpus indicus, Myristica magnifica, Syzygium travancoricum* etc. in any dwarfish evergreen forest. Being on the positive side of conservation both height and basal area are to be given a proportionate score or rank points while preparing a composite index for conservation. Most of the areas which are negatively correlated with evergreen and endemism axis are mostly degraded areas or deciduous forests with hardly any Western Ghat tree endemism. The presence in the Western Ghats of a variety of life forms such as very sensitive fauna like endemic amphibians, fishes, birds or butterflies etc. and endemic primate

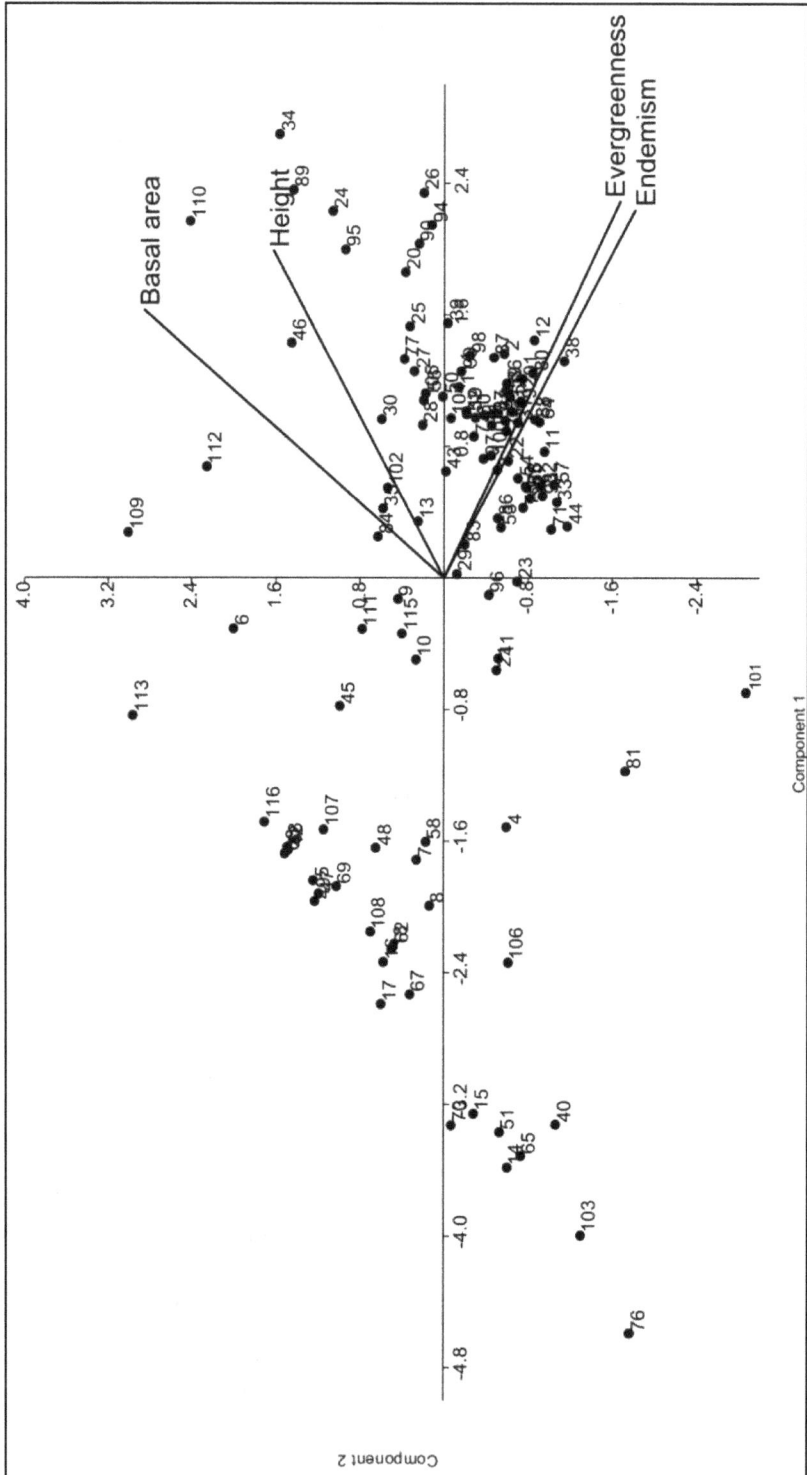

Figure 1.17: PCA Scatter Diagram Represents Four Variables with 116 Sampling Sites.

like Lion-tailed macaque is dependent on forest evergreenness and tree endemism. Recognizing the relative importance of high tree endemism (reflecting degree of evergreenness of the forests), forest canopy heights and basal areas/ha, on the basis of PCA carried out, we have formulated a conservation prioritisation scheme. The rationale for ranking is based on a cumulative score based on the relatively higher importance of endemism, followed by height and basal areas. In assigning conservation scores for each transect sample we have added the loading scores from Table 1.7 for each of the three parameters considered *viz.*, endemism, height and basal area. These scores are in addition to a value assigned to threatened tree species (highest for Critically Endangered, followed in importance by Endangered and Vulnerable species). Also taken into consideration is the value for diversity index (Shannon-Weiner) of each transect?

Table 1.7: Summary of PCA Analysis:
Eigen Value, Per cent of Variance and Loading Score

	Axis 1	Axis 2	Axis 3	Axis 4
Eigen value	2.53963	0.9233	0.4315	0.1054
Per cent variance	63.491	23.084	10.79	2.6355
		PCA view – loading score		
Endemism	0.8678	−0.4236	0.1347	0.2219
Evergreenness	0.8894	−0.3915	0.0305	−0.2342
Basal area	0.6321	0.6678	0.3928	−0.0153
Height	0.772	0.3805	−0.5081	0.0329

Evolving Criteria for a Composite Index for Forest Biodiversity Conservation Assessment in Uttara Kannada

For major mammals like tigers, elephants and such flagship species the Dandeli-Anshi Tiger Reserve, composed of Anshi National Park and Dandeli Wildlife Sanctuary, together cover 1365 sq.km or 13.3 per cent of the district itself. The DATR already covers good parts of Joida and western parts of Haliyal taluks and interior parts of Karwar taluk towards the Anshi Ghat. The forests covered under its domain include evergreen to semi-evergreen, moist and dry deciduous and savannah and scrub as well. We were not successful in getting necessary permission to make studies in this wildlife rich protected area. Such restrictions were not faced in Conservation Reserves and other administrative categories of forests. Now that we have already made 116 sample transects in the district and gathered data on the plant species diversity for each sample area, basal area, biomass, estimates of carbon sequestration, percentage of evergreenness and Western Ghat endemism and about the distribution of threatened species etc. we have attempted here to formulate a scientific basis for ranking of each representative sample site, through value assignments given to the key parameters *viz.*, 1. Endemism (reflects evergreenness of the forest stand); 2. Basal area of trees (indicator of biomass and carbon sequestered), 3. Canopy height; 4. Diversity index and 5. Presence of threatened tree species (based on IUCN Red List)

Criteria for Selection and Assignment of Scores for Evolving Composite Conservation Ranking of Forest Patches (Table 1.8 for details)

Tree Endemism

One of the prime reasons for Western Ghats constituting a Global Biodiversity Hotspot along with Sri Lanka is the high degree of endemism in the flora and fauna. As per cent of tree endemism is strongly linked to per cent of evergreenness, considering both the parameters for assigining scores would amount to bias against other parameters for evolving composite conservation index. We have assigned a score for tree endemism starting with a minimum of 5 for 20-30 per cent endemism for a sample adding additional 5 points for every 10 per cent interval.

Table 1.8: Criteria for Composite Index for Biodiversity Conservation Importance Ranking in Uttara Kannada

Parameter		Score	Parameter		Score
Average height (m)	14-15	5	**Basal area (mi)**	20-30	5
	15-16	7		30-40	7
	16-17	9		40-50	9
	17-18	11		50-60	11
	18-19	13		60-70	13
	19-20	15		70-80	15
	20-21	17		80-90	17
Endemism per cent	20-30	5	**Threatened species**	Vulnerable	10
	30-40	10		Endangered	20
	40-50	15		Critically Endangered	30
	50-60	20		New species	30
	60-70	25			
	70-80	30			
	80-90	35			
Diversity index(Shannon)	1-2	5			
	2-3	7			
	3-4	9			
Add value for all transects		20			

Basal Area

From a starting minimum score of 5 points for 20-30 sq.m basal area/ha 2 additional points are given for every 10 sq.m addition in basal area (30-40 sq.m, 40-50 sq.m and so on). The assignment is made bearing in mind the fact that selective logging of trees and other forms of extraction of biomass can reduce the basal area even for a high diversity forest rich in endemism.

Average Height of Trees

Most of Uttara Kannada falling in the high rainfall zone, except Mundgod and eastern parts of Haliyal and Yellapur, would support high statured trees, the tallest emergent exceeding 30 m and the lesser ones attaining anything from few meters to over 20 m. We have estimated the height of each tree within transect cum quadrat samples and arrived at the average height per sample. Undisturbed forests tend to have more heights than disturbed and secondary forests or savannas. As tall statured forests have greater conservation importance than dwarfer ones we have assigned a score ranging from a minmum of 5 for average tree height of 14-15 m with addition of 2 points for every 1 meter increment.

Diversity

A minimum score of 5 points has been given for Shannon-Weiner diversity index of 1-2, 7 points for 2-3 and 9 points for 3-4.

Threatened Species

Presence of any IUCN Red Listed tree species in any forest sample, notwithstanding any other parameter considered here automatically raises the conservation importance of that forest. This is despite the fact that many tree species are yet to be evaluated for their rarity by the IUCN. We have assigned 30 points for each Critically Endangered tree species, 20 points for an Endangered species and 10 points for a Vulnerable species. Any new tree species described from the forest in particular will gain for the hosting site yet another 30 points.

Additional Value

As it is difficult to rank an ecosystem sample holistically as many features go undervalued or unconsidered for ranking all sites have been uniformly given an add value of 20 points (Table 1.8 for ranking criteria). Results of application of the composite ranking system for forest conservation adopted here is presented in Table 1.9.

Significant Results

IUCN Red Listed Plants

The forest studies revealed the presence of six threatened tree species in the transect areas. Kathalekan Myristica swamps and other swamps in Siddapur and Honavar are extremely threatened habitats with threatened species like *Myristica magnifica, Gymnacranthera canarica, Syzygium travancoricum* and the new tree species *Semecarpus kathalekanensis* exclusive to the swamps. *Dipterocarpus indicus* is present in the relic kan forests of Siddapur, Honavar and very rarely in Ankola (Table 1.10).

Endemism

Details of tree endemism were given in Tables 1.5 and Figure 1.13. Endemics are exclusive trees to the Western Ghats. Concentration of endemic trees, expressed as percentage of endemism in the sample stands, is more in the southern evergreen forests of Siddapur and Honavar, followed by Kumta and Sirsi. Endemism tends to decline in the northern forests, as most of them are secondary in nature. The Myristica swamps, themselves highly threatened areas, are remarkable for the congregation of Western Ghat endemic trees: *e.g.,* Sample areas from Kathalekan swamps T1 (76.92 per cent), T2 (80.17), T3 (81.75 per cent), T4 (71.7 per cent), T6 (71.43 per cent), T8 (70.4

Table 1.9: Composite Conservation Index, Based on Total Site Ranking Score, for 116 Forest Samples

Sl.No.	Asolli-1 Asolli-2	Taluk	Score for Parameters					Add Value	Total
			Height	Basal Area	Endemism	Threatened sp.	Diversity		
1.	Hosakere	Ankola	11	7	20	50	7	20	115
2.	S1-Katangadde-Agasur	Ankola	13	7	30	50	7	20	127
3.	S2-Balikoppa-Badgon	Ankola	9	7	20	20	7	20	83
4.	S3-Hegdekoppa-Kasinmakki	Ankola			5		9	20	34
5.	S4-Vajralli-Ramanguli	Ankola	5	5			7	20	37
6.	Kachinabatti	Ankola	7	7			9	20	43
7.	Maabagi	Ankola	5			20	7	20	52
8.	Dakshinakoppa	Ankola	9				7	20	36
9.	Gujmavu (semi evergreen)	Ankola	11	9	5	20	9	20	74
10.	Hudil (evergreen)	Bhatkal	9	7			5	20	41
11.	Hudil (semi evergreen)	Bhatkal	9	7	25	25	7	20	93
12.	Golehalli	Bhatkal	13	7	35		5	20	80
13.	Kudalgi-Tatigeri	Bhatkal	9	9	10		7	20	55
14.	Magvad	Haliyal				7		20	27
15.	Sambrani	Haliyal						20	20
16.	Yadoga	Haliyal	7	5				20	32
17.	Ambepal-1	Haliyal		7				20	27
18.	Ambepal-2	Haliyal	7	5				20	32
19.	Chaturmukhabasti	Honavar	15	7	15	40	7	20	104
20.	Gersoppa	Honavar	17	9	25	40	9	20	120
21.	Gundabala	Honavar	7	5	10		7	20	49

Contd...

Table 1.9—*Contd...*

Sl.No.	Asolli-1 Asolli-2	Taluk	Score for Parameters					Add Value	Total
			Height	Basal Area	Endemism	Threatened sp.	Diversity		
22.	Hadageri-1	Honavar	13	7	20	20	9	20	89
23.	Hadageri-2	Honavar	9	5	10	20	9	20	73
24.	Halsolli	Honavar	17	11	20	40	7	20	115
25.	Hessige-1	Honavar	17	9	20	40	7	20	113
26.	Hessige-2	Honavar	17	7	30	30	5	20	109
27.	Hessige-3	Honavar	13	9	15		7	20	64
28.	Hessige-4	Honavar	11	9	15	20	7	20	82
29.	Kadnir	Honavar	11	7	5	20	7	20	70
30.	Karikan-lower slope	Honavar	13	11	10		9	20	63
31.	Karikan-semievergreen	Honavar	9	7	25	20	7	20	88
32.	Karikan-temple side-diptero patch	Honavar	5	9	25	40	7	20	106
33.	Mahime	Honavar	7	7	25	40	7	20	106
34.	Sharavathy-viewpoint	Honavar	13	17	30	40	7	20	127
35.	Tulsani-1	Honavar	11	7	5		7	20	50
36.	Tulsani-2	Honavar	13	7	25	20	7	20	92
37.	Castlerock IB	Honavar	13	7	25	20	7	20	92
38.	Castlerock-moist-dec.	Honavar	11	7	35	20	5	20	98
39.	Castlerock-semi everg	Joida	9	11	30	20	9	20	99
40.	Desaivada-Nandgadde	Joida					7	20	27
41.	Gavni-Kangihole-Joida	Joida	9	5	5		7	20	46

Contd...

Table 1.9–*Contd...*

Sl.No.	Asolli-1 / Asolli-2	Taluk	Height	Basal Area	Endemism	Threatened sp.	Diversity	Add Value	Total
			Score for Parameters						
42.	Ivolli-Castlerock	Joida	11	7			5	20	43
43.	Joida-deciduous	Joida	7	9	20		7	20	63
44.	Kushavali	Joida	5	7	25		5	20	62
45.	Shivpura	Joida	11	9			7	20	47
46.	Gopishetta	Joida	11	15	10		7	20	63
47.	Goyar-moist dec	Joida	9	11			5	20	45
48.	Kalni-goyar	Karwar	7	7			7	20	41
49.	Karwar-moist dec	Karwar	7	7			7	20	41
50.	Devimane-Campsite	Karwar	13	9	25		7	20	74
51.	Devimane-Sirsi side	Karwar					7	20	27
52.	Devimane-temple	Kumta	11	9	25	20	9	20	94
53.	Devimane-with myristicas	Kumta	7	9	20	20	7	20	83
54.	Hulidevarakodlu	Kumta	7	9	20	20	7	20	83
55.	Kalve	Kumta	7	9	30	20	7	20	93
56.	Kalve-moist dec.	Kumta	13	9	20		7	20	69
57.	Kandalli-Devimane	Kumta	9	5	25	20	7	20	86
58.	Mastihalla-Devimane arch	Kumta	7	5			7	20	39
59.	Mathali-Kandalli-Devimane	Kumta	9	9	30	20	7	20	95
60.	Soppinahosalli	Kumta	9	9	25	20	7	20	90
61.	Surjaddi	Kumta	9	9	25	20	7	20	90
62.	Surjaddi-Morse	Kumta	7	5			7	20	39

Contd...

Table 1.9–*Contd...*

Sl.No.	Asolli-1 / Asolli-2	Taluk	Score for Parameters					Add Value	Total
			Height	Basal Area	Endemism	Threatened sp.	Diversity		
63.	Attiveri-teakmixed-dry dec	Kumta	13	7	25	20	7	20	92
64.	Godnal	Kumta	11	7	25	20	7	20	90
65.	Gunjavathi	Mundgod					7	20	27
66.	Karekoppa-Gunjavathi	Mundgod	9	9			5	20	43
67.	Katur	Mundgod	7	5			5	20	37
68.	Katur to Gunjavati	Mundgod	11	7			5	20	43
69.	G1-Kathalekan-nonswamp	Mundgod	11	5			5	20	41
70.	G2-Kathalekan-nonswamp	Mundgod		5			7	20	32
71.	G3-Kathalekan-nonswamp	Siddapur	7	7	20	40	9	20	103
72.	G4-Kathalekan -nonswamp	Siddapur	9	7	25	50	9	20	120
73.	G5-Kathalekan-nonswamp	Siddapur	9	9	25	40	9	20	112
74.	Kathalekan-savanna	Siddapur	11	7	20	40	9	20	107
75.	G6-Kathalekan-nonswamp	Siddapur	5	9	20	40	9	20	103
76.	G7-Kathalekan-nonswamp	Siddapur					5	20	25
77.	G8-Kathalekan- nonswamp	Siddapur	13	11	20	40	7	20	111
78.	G9-Kathalekan-nonswamp	Siddapur	11	5	20	40	9	20	105
79.	Hartebailu-soppinabetta	Siddapur	9	9	25	80	7	20	150
80.	Hutgar	Siddapur	9	9	30	40	5	20	113
81.	Joginmath-1	Siddapur			15		7	20	42
82.	Joginmath_2-semievergreen	Siddapur	9	7	25		7	20	68
83.	Kathalekan-1	Siddapur	11	7	5		9	20	52

Contd...

Table 1.9—Contd...

Sl.No.	Asolli-1 Asolli-2	Taluk	Score for Parameters					Add Value	Total
			Height	Basal Area	Endemism	Threatened sp.	Diversity		
84.	Kathalekan-2	Siddapur	13	9	5		7	20	54
85.	Kathalekan –swamp-1	Siddapur	11	5	20	20	9	20	85
86.	Kathalekan –swamp-2	Siddapur	11	7	15	40	9	20	102
87.	Kathalekan –swamp-3	Siddapur	11	9	30	80	9	20	159
88.	Kathalekan –swamp-4	Siddapur	7	9	30	70	7	20	143
89.	Kathalekan –swamp-5	Siddapur	9	15	30	70	9	20	153
90.	Kathalekan –swamp-6	Siddapur	11	13	35	70	7	20	156
91.	Kathalekan –swamp-7	Siddapur	9	9	30	100	7	20	175
92.	Kathalekan –swamp-8	Siddapur	9	9	30	100	7	20	175
93.	Kathalekan –swamp-9	Siddapur	9	7	20	70	7	20	133
94.	Kathalekan-3	Siddapur	13	11	35	100	7	20	186
95.	Malemane-1	Siddapur	15	13	25	70	9	20	152
96.	Malemane-2	Siddapur	9	5	10	60	9	20	113
97.	Malemane-3	Siddapur	11	7	15	40	9	20	102
98.	Siddapur evergreen	Siddapur	15	7	25	40	9	20	116
99.	Talekere	Siddapur	13	9	25	40	7	20	114
100.	Bugadi-Bennehole	Siddapur	13	7	15	20	7	20	82
101.	Gondsor-sampekattu	Siddapur			30	20	5	20	75
102.	Hulekal-Sampegadde-Hebre	Sirsi	7	11	15		9	20	62
103.	Kanmaski-Vanalli	Sirsi					7	20	27
104.	Khurse	Sirsi	9	11	20		9	20	69

Contd...

Table 1.9–*Contd...*

Sl.No.	Asolli-1 / Asolli-2	Taluk	Score for Parameters					Add Value	Total
			Height	Basal Area	Endemism	Threatened sp.	Diversity		
105.	Masrukuli	Sirsi	7	11	20	20	7	20	85
106.	Hiresara-bettaland	Sirsi		5	20		7	20	52
107.	S5-Gidgar-Yemmalli	Sirsi	7	9			7	20	43
108.	S6-Tarukunte-Birgadde	Yellapur		9	20		5	20	54
109.	S7-Arlihonda-Nandvalli	Yellapur	15	9			9	20	53
110.	S8-Yellapur-Mavalli	Yellapur	17	17		15	9	20	78
111.	S9-Kiruvatti	Yellapur	9	7		5	9	20	50
112.	Hasrapal-evergreen	Yellapur	13	7		10	9	20	59
113.	Hulimundgi-semievergreen	Yellapur	11				5	20	36
114.	Lalguli-moist-dec	Yellapur	15	7		20	7	20	69
115.	Asolli-1	Yellapur	13	7		5	7	20	52
116.	Asolli-2	Yellapur	11	9			7	20	47

Table 1.10: The IUCN Red Listed Tree Species found in Various Transects

Red Listed Species	Family	Category	Locations	Taluk	Remarks
Gymnacranthera canarica	Myristicaceae	Vulnerable	Alsolli 1, Alsolli 2 Halsolli, Kathalekan G1, G2 Kathalekan swamp T1, T2, T3, T4, T5, T6	Ankola Honavar Siddapur Siddapur	Confined to Myristica swamps only
Myristica fatua	Myristicaceae	Endangered	Halsolli Kathalekan swamps T1,T2, T5, T9	Honavar Siddapur	Confined to Myristica swamps only. In relics of primary forests
Dipterocarpus indicus	Dipterocarpaceae	Endangered	Alsolli 1, Alsolli 2 Ambepal 1, Ambepal 2 Hadageri 1, Hadageri 2 Karikan lower slope Karikan s.evergreen Karikan templeside Kathalekan non-swamp grids G1, G2, G3, G4, G5, G6, G7, G8 Kathalekan swamp grids T1, T2, T3, T4, T5, T6, T7, T8, T9	Ankola Honavar Honavar Siddapur	New reports for Ankola in relics of primary forests. Northward range extention in Western Ghats
Hopea ponga	Dipterocarpaceae	Endangered	Widespread in evergreen forests Honavar, Kumta, Siddapur, Sirsi and Ankola and sparingly in Karwar and Yellapur	Honavar, Kumta, Siddapur, Sirsi, Ankola, Yellapur Karwar	
Vateria indica	Dipterocarpaceae	Endangered	Kathalekan 3	Siddapur	Planted widespread in the district; natural in Mattigar kan, Siddapur
Syzygium travancoricum	Myrtaceae	Critically Endangered	Kathalekan G8, Kathalekan swamp T3, T6, T8, T5	Siddapur	Also found very sparingly in Ankola Ghats. Range extention in Uttara Kannada reported for first time
Semecarpus kathalekanensis	Anacardiaceae	New tree species	Kathalekan swamps T1, T2	Siddapur	New tree species reported

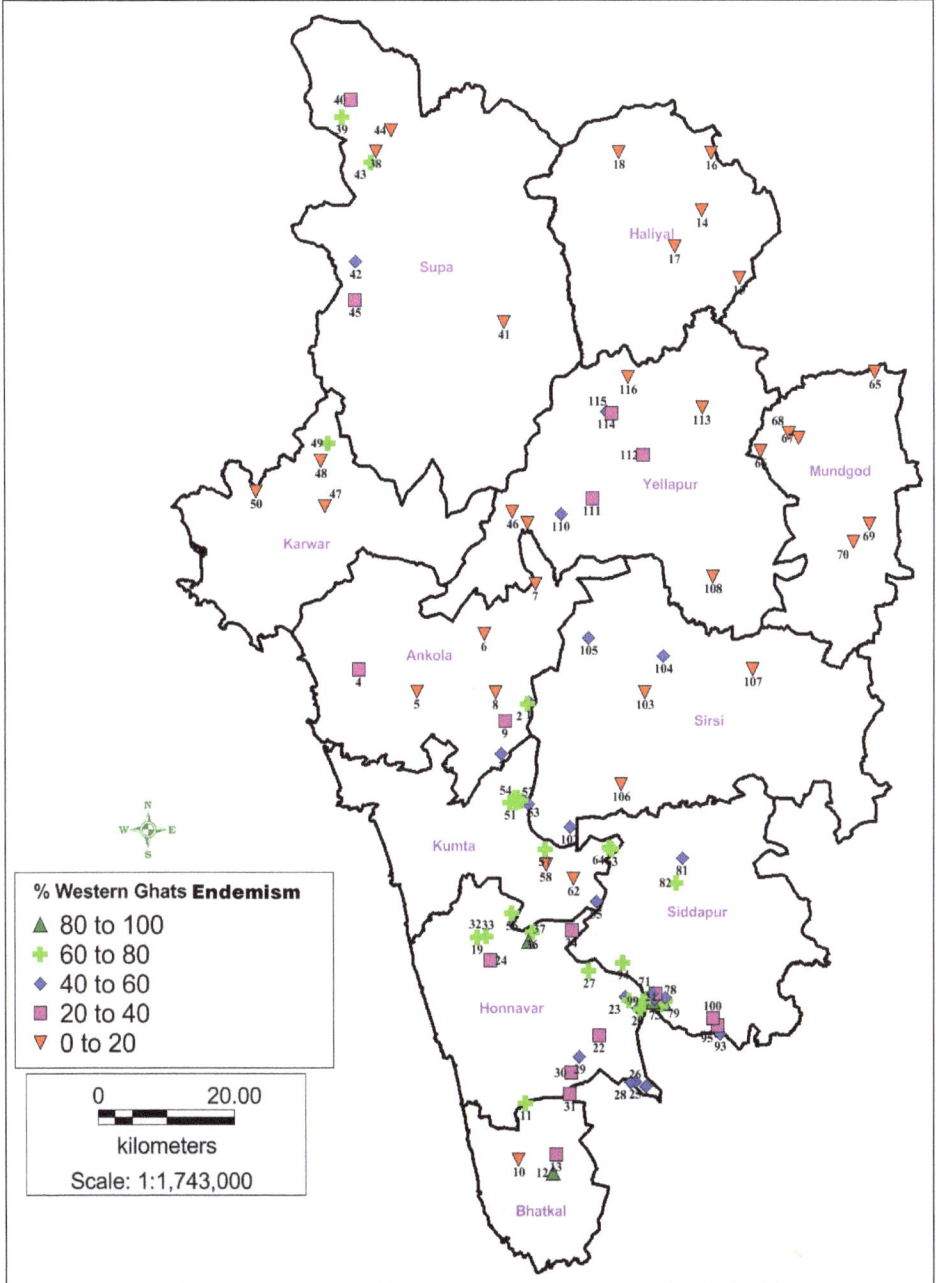

Figure 1.18: Tree Endemism Levels in different Transects of Uttara Kannada.

per cent) and T9 (72.65 per cent). Non-swamp samples from Kathalekan are also rich in tree endemism: *e.g.*, T8 (77.96 per cent). Some other high endemic areas are in Tulsani-2 of Honavar (83.2 per cent), Hudil in Bhatkal (80.65), Karikan temple side sample in Honavar (75 per cent), Halsolli in Honavar (79.07 per cent) and from Devimane Ghat in Kumta (74.62 per cent). The northern forests show drastic decline in endemism (Figure 1.18 for distribution of tree endemism).

High endemism concentration areas in the Siddapur and Honavar ghats covering the drainage areas of Sharavathi and Aghanashini rivers constituting the backbone of the Aghanashini Conservation Reserve, the main habitat for the Endangered primate Lion tailed macaque and for the presence of maximum Myristica swamps in the district (Figure 1.18). In a recent study 35 amphibian species were reported from Kathalekan, most of them in and around the Myristica swamps. Of these 74 per cent are endmics to the Western Ghats. *Philautus ponmudi* is Critically Endagered and its northernmost distribution range in Western Ghats ends in Kathalekan. Five of these species are Endangered and yet another 5 are Vulnerable. Several of them being data deficient also might figure in the threat categories of the Red List (Chandran *et al.*, 2010).

Tree Height Criteria

In Uttara Kannada forests make a mosaic of secondary ones, due to anthropogenic effects through centuries, in different stages of succession here and their enmeshing relics of primary forests like the kans. Several tall growing emergent, evergreen, endemic and non-endemic trees like *Artocarpus hirsuta, Dipterocarpus indicus, Syzygium travancroicum, S. gardneri, Dysoxylum malabaricum, Calophyllum tomentosum, Ficus nervosa, Lophopetalum wightianum* etc. are instrumental in maintaining tiered structure of the forests and in providing habitats for several birds (*e.g.*, Imperial pigeon, hornbills), bats and Lion-tailed macaque. Vertical compression of forests can adversely affect primary forest arboreal fauna.

Conservation Value and Basal Area

Old growth and primary forests tend to accumulate more biomass, as reflected in the girth of tree trunks. Naturally, long periods of undisturbed growth will increase tree dimensions as could be seen from the basal areas/ha. Several endemic tree species are associated with high basal area forests and increase their conservation importance. High basal area is index of high biomass and high levels of carbon sequestration. Human impacts in the form of logging can reduce stand basal areas. Selective cutting may not eliminate tree species as such in a mixed stand. Therefore the conservation value of the stand may not suffer seriously through some degree of forest exploitation. But forest exploitation through tree cutting and mutilation can severely alter the faunal composition by upsetting their habitat qualities. Forest stands of exceptional conservation values should have at least over 40 m²/ha of tree basal area in Uttara Kannada conditions.

Significance of Diversity

Biodiveristy, in terms of genetic, species and ecosystem diversity, is paramount in considerations of conservation. Biodiversity on the earth is getting seriously impacted due to various human activities, directly (through exploitation and alterations in habitats for human wants) and indirectly (through pollution of land and water and air and climatic changes). Protecting of biological diversity has economic and ethical grounds. The species have their own intrinsic values unrelated to human needs. In the complexity of biological communities loss of one species may have far reaching consequences affecting even the humans. As far as forest tree diversity in Uttara Kannada forests (expressed in Shannon diversity index of sample areas) is considered it is found that conservation values of tree community expressed as a composite value are not necessarily dependent on high diversity index, though there is tendency towards increase in conservation value with rising diversity index. Stand should not be however too poor (diversity value of 1-2 in our study areas). Several stands of climax vegetation are too specialised to their habitats (*e.g.*, Myristica swamps), so that they are not that rich in tree species but are seats of high Western Ghats endemism. The Myristica swamps and their immediate surroundings abound in endemic individuals of similar nature so much so the stand diversity as expressed in Shannon index ranges between 2-3 only, unlike many secondary disturbed forests which have diversities of 3-4.

Prioritsation of Forest Areas for Conservation

On the basis of the composite conservation index prepared for assessing the plant diversity (mainly trees) conservation value of stands in forests, based on 116 transects, each covering 2000 m² of forest, a map has been prepared and presented here (Figure 1.19). High and very high conservation ranking areas (conservation values of $\geq$ p 80) are along the Western Ghat regions in the south of the district, mainly in the taluks of Honavar, southern Siddapur (Kathalekan-Malemane area), a zone of high percentage evergreen forests with high degree of forest endemism. If faunal endemism is added to these areas the conservation values will continue to increase as compared to the northern taluks. Conservation values of moderate importance are found in sites throughout the district, scantily so in Mundgod, Joida and Haliyal taluks- despite the fact that these are of high conservation value for non-endemic mammals and flagship species like elephant and tiger. As bulk of the Dandeli-Anshi Tiger Reserve is in Joida, Karwar and Haliyal taluks (as we had no permission to work in such areas, our reliance is more on few samples studied earlier to the new regulations imposed on scientific studies in the Tiger Reserves within the State). As savannah, deciduous forests and more of open grassy areas exist especially in Joida taluk, as telltale marks of its shifting cultivation history, the grazing ecosystem fares better here ideally promoting tiger and panther and other carnivores through promotion of herbivorus prey animals. Not belittling in any way the importance of major mammal conservation in National Parks and Sanctuaries of Western Ghats, as far as Uttara Kannada is concerned scanty efforts were ever made to conserve pockets of high endemism, which is the major consideration for conferring Biodiversity Hot Spot status for the Ghats as a whole. Our studies on forest vegetation clearly reveal that high evergreen forest areas, with negligible timber values, unlike the deciduous forest zone of the district of north-eastern taluks, have greater endemic biodiverisity as reflected in the forest stands.

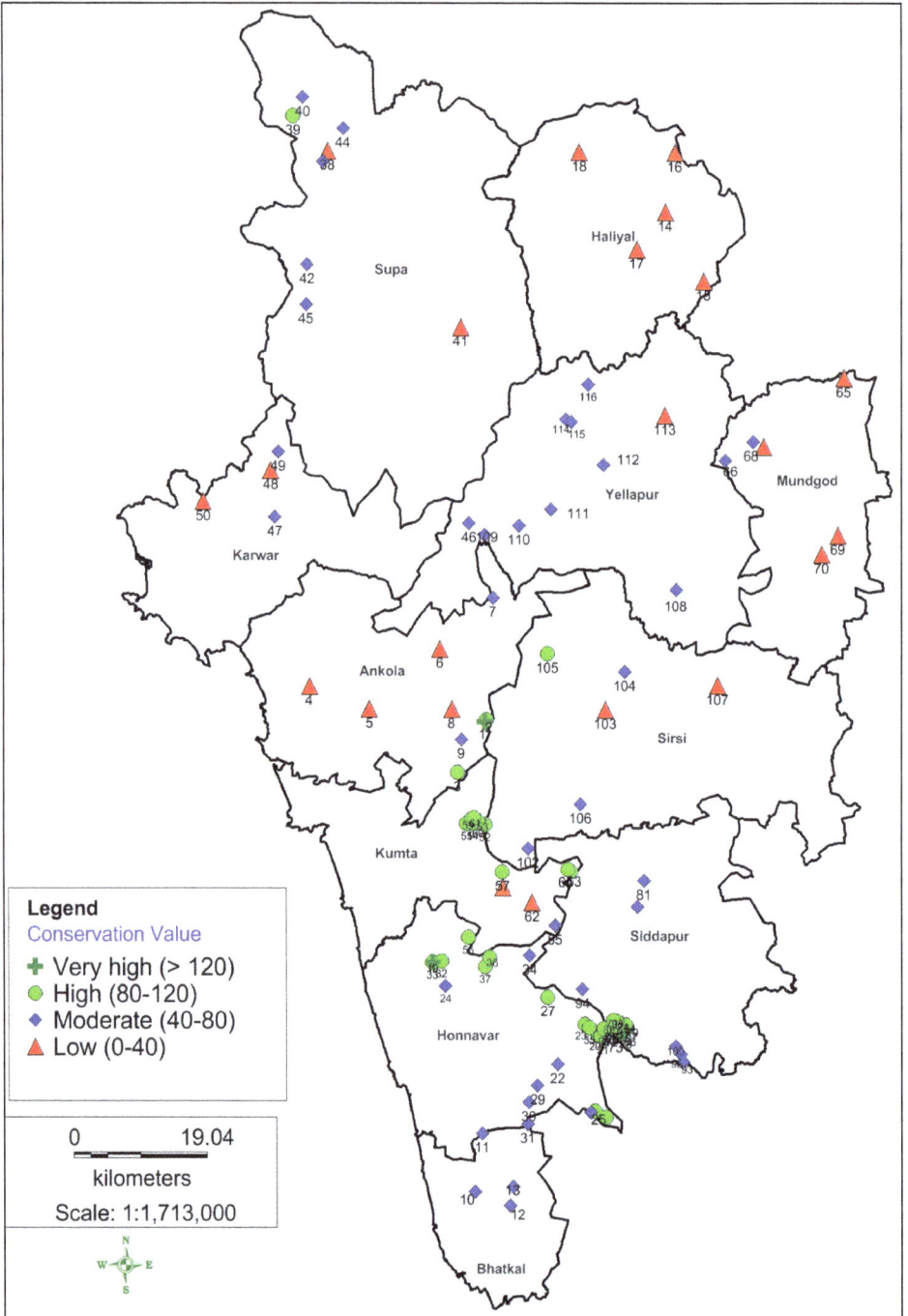

Figure 1.19: Tree Diversity Conservation Values of Forest Stands in Uttara Kannada (for non-economic parameters).

Moreover, in studies related to fresh water fishes of Sharavathi River tributaries in Shimoga taluk it was found earlier that the landscape elements in the catchment of the tributary plays a decisive role in especially fish endemism. A tributary, for instance like Yennehole, had 18 species of fishes of which 8 were Western Ghat endemics. The catchment area forests in Yennehole basin had 86 to 100 per cent evergreen trees where endemism varied between 46-58 per cent. Yet another tributary Nagodi with 19 fish species had also 8 endemic fishes, including a new species *Schistura nagodiensis*, and 68-99 per cent evergreen forests with 36-71 per cent tree endemism in the drainage basin. In contrast Nandihole tributary flowing through a degraded landscape of highly human affected forests, agricultural areas and monoculture tree plantations, with merely 0-16 per cent evergreenness and low endemism of 0-11 per cent in the catchment forests. Although there were 14 fish species in the river only 2 were endemics. The message is that the evergreenness of forests, which goes in harmony with endemism, plays a very crucial role for the entire forest and linked ecosystems in the Western Ghats, as landscape elements play decisive role in distribution of aquatic organisms like fishes making untrammelled nature a holistic system (Sreekantha *et al.*, 2007). High tree endemism areas of Kathalekan, especially the Myristica swamps and adjoining damp areas, had at least 35 species of amphibians. Already we have referred to the amphibian richness of the swampy Kathalekan evergreen forests with high percentage of tree endemism. Conservation values assigned for forest areas for prioritisation of conservation, using the composite index is given in Figure 1.19.

Nelson *et al.* (1990), Peterson *et al.* (2000) and Myers *et al.* (2000) strongly favour identification and prioritisation of 'hotspots', or areas featuring exceptional concentrations of endemic species and experiencing exceptional loss of habitat. Their focus is more on endemic species, rather than populations or other taxa, as the most prominent and readily recognizable form of biodiversity.

Using a grid system (preferably 1x1 km) of forest surveying we need to have a proper stock of the distribution of endemic tree species, and demarcate areas of high tree endemism for prioritization of conservation as such areas are also good for endemic faunal elements and for their hydrological importance. Centres of high floristic endemism (of especially trees) are also the centres of endemic fishes in the streams draining them, in addition to amphibians and birds.

The role of man-made plantations needs a re-evaluation, in the light of high soil erosion, weed infestation, poor hydrology and poor associated faunal diversity as compared to natural forests (Murthy *et al.*, 2002, 2005). The teak plantation areas in general, despite the high value of teak timber, were found to have lower biomass and needs enrichment planting by NTFP species, nectar species for honey bee promotion, Soil erosion from forests and forest plantations is a matter of grave concern. As rains are often very high (upwards of 3000 mm/per annum) in most places, and so much of rainfall within a short period of mostly four months a dense forest cover is required to check soil erosion and increase infiltration into the ground water. Here we recommend eventual conversion of deciduous forests and their degradation stages (except grasslands or grassy blanks, critical resources for grazing ecosystems) in heavy rainfall zone into evergreen forests. Poor grade tree plantations with eroded soils need to be restored with natural forest species through planting of saplings and dibbling of seeds.

Forest restoration in the catchment areas of rivers will improve perennial nature of streams ensuring perpetual inflow of clear water into the storage dams of hydroelectric projects in Sharavathi and Kali rivers than bringing into them an onrush of water turbid with soils down the poorly vegetated terrain. The active monsoon period being of four months it is necessary to increase residency of water within the watershed soils than releasing it en mass into the reservoirs or other downstream areas as surface water, which eventually get lost through faster evaporation in the prevailing climatic conditions.

The species chosen for forest enrichment/afforestation should have strong bearing on a. increase in endemism; b. more of ecologically site specific NTFP species; c. benefit to birds and bats and other frugivorous animals and d. favour populations of wild bees and create employment opportunities through bee-keeping and enhance pollination services of both cultivated crops and forest plants.

A system for assigning conservation values to the forest patches based on characteristics of tree communities has been adopted here. **Assignment of conservation priorities is based mainly on five variables of forests namely: a). per cent evergreenness, b). per cent endemism, c). basal area, d). tree height and e). Shannon diversity index.** Principal component analysis based on the first four variables revealed that evergreenness of the forests is strongly linked to the presence of endemic trees. Higher the evergreen components more endemics congregate in such areas. Basal area and tree heights are linked to other two factors – but not so strongly as these two are subjected to rapid fluctuations depending on human impacts. Relative correlation between these five factors was obtained through application of Pearson correlation matrix. A composite conservation index is prepared for the 116 forest samples using scores allotted to the factors per cent endemism, mean canopy height, basal area and diversity index. Additionally the presence of IUCN Red Listed trees, if any, were given high conservation score- the actual score depending on the category of threat.

Highest conservation values are more for forests towards the south from Sharavathi Valley (Kathalekan-Malemane-Gersopppa stretch to the Aghanashini valley in Siddapur and to a small extent in Sirsi). Incidentally this stretch of forests, having the northernmost populations in the Western Ghats of the Endangered primate Lion-tailed macaque, of Myristica swamps and *Dipterocarpus* trees, has been already declared by the Government of Karnataka as Aghanashini LTM Conservation Reserve.

The study reveals that there is only a thin line difference between rain forests and deserts. Whereas the heavy rainfall of coast and malnadu taluks can potentially promote loftiest evergreen forests of Western Ghats many locations are characterized by poorer vegetation- poorer in biomass and in conservation ranking. The poorest savanna site exists on a hill top ironically in the Kathalekan forests of highest conservation value, dotted with Myristica swamps, by presence of lofty *Dipterocarpus* threatened and endemic plant and animal species (especially amphibians and LTM). Whereas the swamp forest samples of Kathalekan have average carbon sequestration of 225.506 t/ha the savanna patch has merely 5.06 t/ha. The land was savannized at least over 100 years ago by the shifting cultivators. Though today uninhabited the forest recovery has not taken place. Similar paradoxes exist between adjoining forest patches everywhere in the district.

Whereas in the earlier efforts towards conservation it was often the flagship species like elephant, tiger etc. and their habitats that captured major attention in the conservation priorities of the Government. Today, the Western Ghats, along with Sri Lanka constitute a hotspot of high endemism and significant threat of imminent extinctions. Therefore it has become necessary to evaluate and rank areas of high endemism, which we have attempted in this study through the application an objective method. Suggestions based on our sustained ecological research in Uttara Kannada district are:

- ☆ In the specter of climatic change that the planet is facing with its widespread implications especially on farming and biodiversity, the need has arisen to increase carbon sequestration in the forest areas. There are considerable areas of degraded forests in Uttara Kannada, the biomass of which has to be increased substantially through protection, enrichment and co-management.

- ☆ Inviolate forests should be identified range-wise for increased conservation efforts.

- ☆ Myristica swamps are among the oldest and original forest types of the Western Ghats. They have some of the highest degrees of floral and faunal endemism. Efforts should be made to make all out search for such swamps, record their locations and areas and conserve them along with their catchment area forests

- ☆ The *kan* forests and 'devarabanas' were **unique cultural identities** of bygone days. They still have portions harbouring deities and are seats of high endemism. As most of them got merged with state reserved forests they lost their pre-colonial identities as sacred groves from safety forests (except the smaller banas close to or in the middle of villages). Efforts should be made to trace them out and map and protect them.

- ☆ Conservation of Western Ghat endemism is important. High percentage of forest tree endemism even influences endemism among fishes in the streams that drain such forests.

- ☆ Biomass upgradation is an urgent necessity especially in the deciduous forest areas everywhere, especially in the maidan taluks of the district.

- ☆ Biomass and diversity are lower in the coastal minor forest tracts. Through consistent efforts involving local VFCs multiple species forests should be raised in such areas.

- ☆ Coastal lateritic hills were paid least attention so far; except for raising Acacia plantations no major activities were undertaken in them. Laterite plateaus also have great richness of monsoon herbs which flower gregariously and offer nectar for the survival of honey bees during the rainy season. Some ideal plateaus need to be conserved for their characteristic endemic flora.

☆ Regarding **scope for forestry based alternative development plan for enhancing the economic productivity of the region** we wish to state that **since** bulk of the lands in the district (over 70 per cent area) being under the control of the Forest Department there is very little scope for economic advancement of bulk of the local population beyond subsistence level. There is also not much scope for major developmental interventions due to the fragility of the terrain and the ecosystems. As economic growth gets stunted people, especially younger generation tend to migrate into the cities for better prospects. Such mass migrations from rural areas will strain the cities as well beyond their carrying capacities too- as it is happening in Bangalore. To reverse the trend as far as Uttara Kannada is concerned the following recommendations are made for creation of more of forestry based livelihoods without any major interventions into the ecosystems as such:

a. NTFP species should be widely raised.

b. Bee keeping to be promoted as an important enterprise to benefit the people and forests (through pollination). Village peripheral forests and roadsides should be planted with numerous types of nectar plants used for foraging by honey bees (separate submitted on bee keeping).

c. There is laxity among the arecanut garden owners as regards management of soppinbetta forests for fear of not getting the fruits of such improvement as the bettas are under Government ownership. It is recommended the betta owners be allowed certain tree rights if they adhere to certain norms like maintenance of the bettas to certain biomass levels, say like 30-35 sq.m of basal area/ha for trees.

d. The farmers require a helping hand from the Government in growing and marketing of medicinal plants and their primary products. Medicinal plants grown in VFC forests in home gardens or in fields, which also grow wild in the forest areas, should be procured by the Forest Department. This is to stop smuggling of medicinal plants from the forests, unauthorized exploitation by outside agencies and for betterment of local livelihoods.

e. Preparation of bio-pesticides, harmless to humans and domestic animals, may be promoted as a cottage industry using local plant resources, especially from village peripheral forests/VFC managed areas.

f. Vegetable dyes/or textiles coloured using such dyes, or for use as food colours are in increasing demand. Numerous plants in forests, mangroves and beaches are potential sources of such dyes. Village peripheral forests may be enriched using such plants to generate rural employment. Technology transfer is necessary.

g. Enormous scope for exploration of production and trade of plant based cosmetics and nutraceuticals (*e.g.*, from *Garcinias* and *Phyllanthus emblica* -amla) should be explored.

h. VFC managed sandalwood farms are recommended for the taluks of Haliyal, and Mundgod and for the eastern zone of Yellapur, Sirsi and Siddapur.

i. Being well forested district of hills and valleys, waterfalls, sea beaches and mangroves and for its cultural diversity Uttara Kannada has good scope for generating eco-friendly livelihoods through tourism promotion at grassroots level. This facet of development with the vision of upgrading livelihoods of grass root level people while also enriching forests, mangroves, sea beaches and coastal laterite plateaus has been successfully worked out by the Honavar Forest Division, at Apsarakonda, Om Beach (Gokarna), Kasarkod, Bellangi etc. The State Government should liberalise the licensing policy on home stays and community managed cottages (through VFCs) to benefit growth of decentralized ecotourism in the district, to benefit both village communities and local ecology.

j. Decentralised systems of forest nurseries for generating women's employment and providing scope for application of indigenous farming techniques for forestry purposes.

☆ Village level biodiversity hotspots should be identified and protected through the involvement VFCs/local Biodiversity Management Committees. Eventually these, through succession and vegetational enrichment will turn out to be local hotspots of biodiversity.

☆ Realizing the fact that depletion of forests of food resources and human induced vegetational changes in forests have adverse consequences on wildlife while increasing crop raids by animals enrichment of secondary forests and poor grade tree plantations with food resources for forest herbivores is highly desirable.

☆ NTFP collection, that yields only minor revenue to the state, is being carried out in many forests with gay abandon causing destruction of the resource itself. We recommend that the VFCs and other forest dwellers in respective villages be organized and trained in scientific harvesting of NTFP which also serves as medicinal plants.

☆ Rampant collection of poles, cane, fuel wood etc., has been taking a heavy toll on forest resources particularly in the village vicinities. Most of the easily accessible areas with many medicinal plants are more prone to exploitation and get converted into scrub and thickets. Even the semi-evergreen and evergreen forests higher up in more inaccessible areas are also being exploited for fuel wood, timber etc., due to which many of these forests have thorny thickets as under-growths. We recommend conduct of sustained programmes on biodiversity awareness. Also bamboo considered as 'poor man's timber' the villagers may be allowed to harvest it from designated areas for their own bonafide use, so that they will desist from pole cutting and stake removal from the forests which destroys lakhs of tree saplings and pole sized juveniles.

Recommendations

1. Forests towards Carbon Mitigation

Carbon sequestration in any given forest is related to forest biomass. Basal area/ha is an index of the forest biomass. Higher carbon sequestration in stream course/swamp forests was a significant find of this study. In Kathalekan forests 9 forest samples along the water course-swamp parts had average carbon storage of 211.87 t/ha. In the nine samples from forest away from water course areas, carbon sequestration was less at 165.54 t/ha. On a hilltop savannized part, obviously due to shifting cultivation practice in the past, the carbon sequestration was very poor at 5.03 t/ha only. Numerous hill tops and wind exposed slopes of the district are in savannized state with poor biomass, demonstrating the fact clear felling of a rain forest can bring in desertified conditions. Nevertheless, these grassy patches considering forest as a place of tree growth, often also with undergrowth, the savannised forests and several secondary forest samples subjected to ongoing human impacts, or sytems recovering from past human impacts, had some of the least basal areas, irrespective of whether they fall in high, moderate or low rainfall areas. All the taluks have such forests, which are low in biomass. Altogether 3 out of 116 samples, as studied by transect cum quadrat method, had basal areas of < 10 sq.m, 10 transects had basal areas between 10-20 sq.m/ha and 17 had between 20-30 sq.m. To such degraded forest areas also belongs bulk of the Soppinbetta or leaf manure forests allotted to arecanut garden owners of mainly the malnadu areas, for exercising the traditional privileges, importantly leaf manure collection. Thus a betta inTalekere of Siddapur had only 10.12 sq.m basal area ha and Hartebailu betta in same taluk had only 17.80 sq.m/ha basal area. Gondsur-Sampekattu betta in Sirsi taluk had just 3.74 sq.m as the basal area. Hiresara bettaland in Yellapur was exceptional in having 41.73 sq.m basal area.

2. Riparian Forest Protection

River and stream bank forests, including inland swamp area forests are to be considered as endangered ecosystems for various reasons, including for their high accumulation of biomass and higher levels of carbon sequestration. Forest rangewise river-stream-swamp protection action plans, incorporating adequate amount of inviolate vegetation growth for protection of ecology of these vital water courses along with their rare and endemic species is critical. The maps and action plans prepared for special protection of such areas should be included in the forest working plans of every forest division. If such working plans are already prepared these should be still prepared as supplements. Timber extraction, conversion into monoculture plantations, or encroachments or any developmental activities should not be allowed affecting these inviolate forests.

3. Protection of Myristica Swamps

These are remnants of the original primeval forests of the Western Ghats. The lineage of such forests could be traced to the supercontinent of Gondwanaland. The swamps, repositories of ancient and highly threatened rare biodiversity, are under various kinds of threats. They would have perished in large scale in early agricultural history of Western Ghats, being reclaimed for rice fields and betelnut gardens. Many of the last remaining fragments of swamps are also under threat from agricultural expansion. The swamps should be demarcated in the forest working plans for the relevant areas and recommended for protection through preferably co-management with the VFCs. The catchment areas for the swamps are to be protected from any kind of human disturbances being very important sources of hydrology. Kathalekan swamps in Siddapur taluk, being the most precious genepool of threatened plants and amphibains, among others, beng situated alongside the Honavar-Bangalore highway might get wiped out in case of road widening. The widening should not be permitted through any of the Myristica swamps or primary forest remnants.

4. Conservation of Unique Forest Related Cultural Identities

The district abounds in forest related unique cultural identities like sacred groves and sacred trees. Sacred groves are known by various names like kans or devarabanas (often the presiding deities' names are added to respective banas-kans- such as Jatakabana, Choudibana, Kari-kanamman-bana, Hulidevarukan, Naagarabana etc.). Numerous ancient trees, especially of genus *Ficus*, or several others like *Mimusops elengi, Mesua ferrea, Mangifera indica, Mammea suriga, Aegle marmelos* etc. are present dotting the landscapes of villages and towns signifying sacred locations of cultural value. Whereas the *kans* were traditionally large groves, of several hectares or even few sq.km in area (Kathalekan for *e.g.*), the *banas* are smaller ones, mostly within an acre in area. While the former is associated often with other forests or wilderness the latter is often found closer to or within human settlements. The *kans* were places where tree cutting was not permitted under traditional management, but NTFPs could be taken care of and harvested (*e.g.*, Wild pepper, cinnamon, toddy and starch from *Caryota urens* etc.). The *kans* while protecting wild genepool amidst secondary, human impacted landscapes, also acted as safety forests, being fireproof systems due to their evergreenness and high humidity, as sources of perennial streams and springs and as sources of NTFP. The smaller groves the *banas*, were not traditionally violated for any form of bioresources. In short both *kans* and *banas* were unique cultural identitites of the region while they preserved the region's climax vegetation. With the process of forest settlement during the British period, most of the *kans* lost their original identity as village sacred groves from safety forests, and were treated not much different from other forests. The smaller sacred groves are under shrinkage too due to erosion of conservation ethics due to changing cultural worldviews of the local communities (Chandran, 1998; Chandran and Gadgil, 1993). A detailed survey of 86 villages gave details about the presence of 241 sacred groves. We strongly recommend that the Government through the Forest Department take immediate steps to revive the system of preservation of these ancient sacred groves however small they are.

5. Identification and Recouperation Old *Kan* Forests

Kan forests were sacred forests of local rural communities of central Western Ghats. They are known as *devarakadus* in Coorg district. Devarakadus of Coorg have official recognition as sacred forests to this day. The kans of Shimoga district were demarcated in maps and their areas were already listed from early British period. But the British did not recognize the sacredness of the *kans*. In Uttara Kannada many *kans* of Sirsi and Siddapur were demarcated villagewise in forest settlement reports. At the same time many other *kans* got merged with rest of the reserved forests without any special status conferred on them and subsequently it became difficult even to locate their boundaries. Such is the case of Kathalekan in Siddapur, Karikan in Honavar and Halsollikan in Ankola which we studied in detail. All these three *kans*, despite being reserved forest areas, are associated with sacred locations within them or in their vicinity, where local people continued the worship of deities. Interestingly all these places continued to maintain their distinctness as relics of primary evergreen forests embedded in a vast matrix of secondary forests. All these forests have *Dipterocarpus indicus*, a primary evergreen forest tree of South Indian Western Ghats. This species, though commoner in more southern forests, have isolated occurrences in Uttara Kannada mostly associated with *kan* forests. The presence of this Endangered evergreen tree has enhanced the conservation value of all these forests. Asollikan is a locality where we observed also the Critically Endangered tree *Madhuca bourdillonnii*. The discovery of this rarest species in Ankola taluk, once thought to be extinct and rediscovered in southern Kerala Ghats in its original home range, is an instance of traditional, community based conservation practice. Presence of species like *Myristica magnifica* (Endangered), *Syzygium travancoricum* (Critically Endangered), *Gymnacranthera canarica* (Vulnerable) and *Semecarpus kathalekanensis* (newly described tree species from the Myristica swamps of Karikan), underscores the importance of surveying, demarcating and protecting the lost *kans* (sacred forests) of pre-colonial times, and demarcating them for more careful protection and restoration through natural regeneration. The *kan* forest areas, were considered during British period as hydrologically important areas, being associated with perennial streams and springs (Chandran and Gadgil, 1993). Even a small *kan* of just one ha, in the Mattigar village of Siddapur taluk has *Syzygium travancoricum* (Critically Endangered) and *Vateria indica* (Endangered). The *kans*, many of them in ruins, due to various reasons, should be salvaged and brought under a system of co-management involving the local VFCs, if they are closer to villages.

6. Conservation and Promotion of Forest Endemism

High rainfall areas have high biodiversity values and higher conservation values. High rainfall areas of malnadu and coastal taluks are major seats of endemic biodiversity of both plants and animals. Kathalekan studies in Siddapur taluk (by various investigators) reveal how the high endemism is associated with Myristica swamps, at least 35 species of amphibians, endemic hornbills and Imperial pigeon, Endangered primate Lion –tailed macaque etc. The very distribution of fresh water fishes is highly correlated to terrestrial landscape elements, of which quantity and quality of evergreen forests are more important. Of the 64 species of fresh water fishes reported from Sharavathi River, including in its catchment areas of Shimoga, 18 species were endemics to Western Ghats, including three new species *Batasio sharavathiensis, Schistura nagodiensis* and *S. sharavathiensis* and 24 species confined to Peninsular India (Bhat and Jayaram, 2004; Sreekantha *et al.*, 2007).

7. Upgrading Biomass in Deciduous Forests and Secondary Deciduous Forests

The quality and quantity of a deciduous forest stand is very much reflected in its total biomass of which basal area is an index. Eleven forests surveyed in the deciduous forest zone of Haliyal and Mundgod taluks reveal unsatisfactory biomass, estimated basal areas/ha being in ranges of 10-20 sq.m for three samples, 20-30 sq.m for five samples, 30-40 sq.m for just two samples and only one falls in 40-50 sq.m category (43.09 sq.m at Godnol in Mundgod). Forest fragmentation of high order, shifting cultivation practices in the past, massive conversions into monoculture plantations, clear felling and selection felling rampantly practiced in the past are some of the major causes for low basal areas. Compact stretches of forests especially in areas thinly populated by humans may be prioritised for developing ideal forests of high stature through special protection and periodical monitoring of the progress of natural succession and tree growth. The forest management should aim at developing in the deciduous forest zone of Mundgod, Haliyal, in the drier eastern parts of especially Joida, Yellapur and Sirsi compact stands with basal areas exceeding 35 sq.m/ha.

8. Increasing Biomass and Diversity in Secondary Deciduous Forests of Coastal Taluks

The secondary moist deciduous forests along the coastal taluks have been in impoverished state due to high density human impacts. Bulk of such forests constituted the 'minor forests' meant for meeting the biomass needs of coastal people, including cattle grazing. Through special protection of promising forest patches using barbed wire fencing, and closing any kind of exploitation in such protected areas, natural regeneration can be promoted, for at least five year period. Thereafter these forests can be open for free movement of wildlife and more such selected blocks can be protected, using the mode of forest working plans.

9. Demarcation of Potential Areas for Conservation of Congregation of Endemic Trees

Our survey reveals there are special areas in the forests where species like *Myristica fatua, Dipterocarpus indicus, Syzygium travancoricum* etc. congregate. More such areas should be traced out through the involvement of forest guards and village people and earmarked for special conservation efforts.

10. Importance of Conservation of the Native Flora of Coastal Laterite Hills and Plateaus

From ancient times the coastal hills and plateaus of Uttara Kannada, from Ankola to Bhatkal, presented a picture of a barren and desolate terrain with sparse growth of woody vegetation. As such these were demarcated as minor forests for meeting the biomass needs of the local population and for cattle grazing. Many have been used in the recent decades for raising monocultures of *Acacia auriculiformis*. Our studies reveal that during the rainy season, open lateritic areas get carpeted with tiny herbs, where billions of flowers bloom providing crucial off-season nectar resources for honey bees, which, especially the domesticated ones, are otherwise to be fed artificially using sugar/jaggery solutions.

11. Forest Resources for Improving Economic Conditions of Local Citizens

Regarding scope for forestry based alternative development plan for enhancing the economic productivity of the region we wish to state that since bulk of the lands in the district (over 70 per cent area) being under the control of the Forest Department there is very little scope for economic advancement of bulk of the local population beyond subsistence level unless suitable small scale enterprenureship complementary to forests and nature are nurtured in the district. This recommendation is made considering the least scope in the district for major developmental interventions due to the fragility of the terrain and the ecosystems. As economic growth gets stunted people, especially younger generation tend to migrate into the cities for better prospects. Such mass migrations from rural areas will be too exacting on the carrying capacities of cities- Bangalore, for instance is burgeoning with population and developmental activities with heavy toll on ecology the impacts far reaching even on ecology of Western Ghats. To reverse the trend as far as Uttara Kannada is concerned the following recommendations are made for creation of more of forestry based livelihoods without any major interventions into the ecosystems as such:

i. Sustainable Use of Soppinbettas

Soppinbettas are forests allotted to arecanut garden owners of mainly the malnadu areas, for exercising the traditional privileges, importantly leaf manure collection. The farmers do not have tree rights in these bettas although in most bettas we observed trees are constantly lopped for leaf manure collection, apart from collection of leaf litter from the ground. Bettas sampled were understocked in tree biomass (a betta in Talekere of Siddapur had only 10.12 sq.m basal area ha, in Hartebailu of same taluk a betta had only 17.80 sq.m/ha basal area and in Gondsur-Sampekattu betta in Sirsi taluk it was abysmally low 3.72 sq. m). Some farmers maintain bettalands in better conditions *e.g.*, Hiresara bettaland in Yellapur (basal area 41.73 sq.m/ha). One of the reasons for understocking and low biomass is that many farmers also use the bettas as tree savannas interspersed with grassy areas; as a result they are able to maintain improved cattle unlike the coastal farmers who are hard pressed for fodder grasses even to feed their diminutive indigenous cattle. The laxity in betta management is partly due to the general fear among the farmers that any improvement in the betta forests at their expenses will not be repaying for them as they do not enjoy absolute ownership over the betta lands or the trees. It is recommended here that the farmers be allowed to have rights on the trees (for timber and fuel) in the betta if they upgrade the tree biomass from present basal area indicator of less than <20 sq.m/ha to minimum of 30-35 sq.m/ha, which minimum limit the Forest Department may fix after examination of the condition of the betta on a case to case basis.

ii. Promotion of Beekeeping

Uttara Kannada has ideal district for promotion of bee keeping. Bee keeping is complementary to forestry and farming because of pollination benefits. Uttara Kannada can reap enormous benefits through especially production of forest and farming based organic honey. Even roadsides and wastelands can be planted with nectar producing plant species. Although about 7000 sq.km area is under forest cover the district has achieved only very little progress in bee keeping. One of the key reasons is the inadequacy of bee forage plant species in the village peripheral forests which are often in degraded state, with scanty attention paid to enriching them with bee forage plants. Particularly nectar producing species, groups of them flowering in different times of the year, composed of a community of site specific flowering herbs, shrubs, climbers and trees are to be promoted to support apiculture in villages. Even the landless and marginal farmers can involve in bee keeping depending on bee forage plants in forests, roadsides, mangroves and beaches. Through proper planning and implementation of 'forests for bee keeping' project, hypothetically, at the density of two bee colonies per ha of forest (not necessarily by placing bee boxes in every ha of forest, as the bees travel few km in search of forage plants; for *e.g.*, *Apis dorsata* has a foraging range of 3 km radius -Batra, 2001), at a modest estimate honey production based on 700,000 ha of forests at 40 kg/ha using native bees *Apis cerana*, and Rs.200/ - kg rate at prevailing minimum rate, can yield 28,000 tons of honey worth Rs.560 crore. Honey is a good health food in demand nationally and internationally. Proper marketing as organic forest honey can fetch much more income (for *e.g.*, Soapnut tree based honey fetches upwards of Rs.700/- kg). Surplus honey can be used in the mid-day meal programmes for school students. To achieve such ambitious target we recommend that even a wing of Forest Department be made to promote apiculture related activities.

The bettaland farmers should be assisted in bee keeping activities aiming at a minimum of one bee box for every acre of betta. They are to be guided in enriching the bettalands with bee forage plants so that the vegetation of impoverished bettas are also improved. Improved vegetation and better ground cover can also improve local hydrological conditions. A single bee colony (in a bee box) can earn for the farmer Rs.4000/- extra money, through better management and vegetational enrichment. The farmers also stand to gain from increased farm productivity due to the pollination services from bees, and NTFPs from bee forage plants. The farmers need training in bee keeping related activities. Sirsi-Siddapur taluks, which have some of the highest forest fragmentation in the district, can also substantially improve the forest wealth through betta rehabilitation.

iii. Promotion of Marketable Medicinal Plants

The farmers require a helping hand in growing and marketing of medicinal plants and their products. The farmers would look forward to the Government/ Forest Department, for acting as a purchasing agency for medicinal plants or their products. In this regard by undertaking the role of a facilitator between the producer and the purchaser (pharmaceutical companies) the Government/Forest Department would play a vital role in biodiversity conservation and enhancing the value of bettalands, minor forests, and even those who grow medicinal plants in their household gardens or private lands. The role of Forest Department as a purchasing agency while bettering local livelihoods can also stop smuggling of medicinal plants from the forests and other unauthorized exploitation by outside agencies.

iv. Biopesticides from Forest Plants

Various plant species of the district *viz.*, neem, *Pongamia, Vitex negundo* etc. are sources of biopesticides. Promotion of such plants in VFC managed forests and bettalands can further the cause of organic farming in the district while also earning extra income to the locals from production of marketable, homemade biopesticide formulations, under an assited programme from the Government. Neem based pesticide formulations are widely popular in the world. Azadirachtin, the main active principle of neem is also found in *Melia azedarach* (Hebbevu) of same family. However, use of such pesticides in India is making tardy progress, despite the fact that knowledge base for neem pesticidal properties is from India. Bark extract of *Acacia nilotica* has been found to provide complete protection to oranges from the blue mold fungus (Varma and Dubey, 1999). Leaf extract of *Clerodendron inerme*, a hedge plant and coastal shrub, is found effective against red spider mite. Use of *Lantana camara* extract to control cotton pests is a good example of a grass root level practice (Varshney, 2006). Strychnine from *Strychnos nux-vomica* is used as a rat poison. Pongamia leaves and bark are sources of traditional biopesticides, especially having insect deterrant properties (Kiruba *et al.*, 2006). Seeds of the giant forest liana *Entada pursaetha* are used to control rats in the Garo Hills of North-East India.

v. Vegetable Dyes from Forest Plants

World over, especially from developed countries, there is growing demand for textiles dyed using vegetable dyes. Total market for herbal dyes was estimated to be worth US$ one billion and growing annually at the rate of 12 per cent (Gokhale *et al.*, 2004). India has a wealth of traditional knowledge on production of plant based textile (for cotton, wool and silk) and leather dyes. The market demand for such dyes is yet unrealized in the absence of surveys. It is right time for Uttara Kannada district to capture this market using the enormous potential for growing plant sources of vegetable dyes in the VFC managed areas, including sea beaches and mangroves, under a sustained programme including training programmes for transfer of appropriate technology. Numerous plant species can be promoted for dye production in cottage industry level:

a) *Acacia catechu* (Khair): Catechin red from wood for dyeing silk, cotton and calico printing

b) *Acacia nilotica* (Jali): Catechin from wood for dyeing light yellow, dark grey, reddish brown

c) *Aegle marmelos* (Bilpatri): Marmalosin from fruit rind for yellow and gray

d) *Bauhinia purpurea* (Mandara): Chalcone and butein for dyeing and tanning purple

e) *Butea monosperma* (Muttaga): Dried flowers with several components for dyeing of silk brilliant yellow

f) *Caesalpinia sappan*: Brazilin from wood and pods for red and black

g) *Cassia fistula* (Kakkemara): Bark and sapwood for red

h) *Cassia tora* (Tagati): Rubrofusarin from seeds for tannin and dyeing blue

i) *Chukrasia tabularis* (Gnadhagarige): Leaves for red

j) *Dipterocarpus* spp. : Bark for brown and gray

k) *Madhuca indica* (Mahua): Bark for reddish yellow

l) *Mallotus phillippensis* (Kumkum): Fruits for dyeing silk red

m) *Mangifera indica* (Mango): Bark and leaves for dyeing silk yellow

n) *Morinda citrifolia* (Noni): Morindin from root and bark for dyeing silk dull red

o) *Pterocarpus marsupium* (Bet-honne): Epicatechin from bark for dyeing silk brownish red

p) *Rubia cordifolia* (Manishta): Manjistin and purpurin from stem and bark for reddish brown, light pink, light brown, gray

q) *Terminalia arjuna* (Holematti): Arjunic acid from bark for light brown

r) *Terminalia chebula* (Haritagi): Chebulinic acid from fruits for yellow and dark gray

s) *Tectona grandis* (Teak): For dyeing silk yellow

t) *Ventilago maderaspatana*: Ventilagin from root and bark for colouring cotton and tassar silk chocolate

u) *Woodfordia fruticosa:* Lawsone from leaves and flower for dyeing pink or red

v) *Zizyphus jujuba* (Bora): Fruit as modant in dyeing silk

There are many more such plant sources of dyes. The important needs before implementation are:

☆ Documentation of traditional practices, study of local and global demands

☆ Improvisation of traditional techniques

☆ Commercial cultivation of wild sources

☆ Standardisation in dyeing practices

vi. Cosmetics and Nutraceuticals from the Wild

As such lot of authorised and unauthorised extraction of NTFP used for cosmetics and nutraceuticals are happening in the district, for instance from plants like *Garcinia* spp. Kokam fat from *Garcinia* seeds has global demand as is most sought after for preparing skin creams. Following in importance is seed fat from *Madhuca indica* (mahua tree). *Garcinia cambogea* and *Phyllanthus emblica* are few among several nutraceutical plants, the multiplication and sustainable harvests of which can generate considerable rural employment. The traditional Indian cosmetic products of India came from a variety of plants like Amla, Shikakai (*Acacia concinna*), neem, soapnut (*Sapindus laurifolius*).

vii. VFC Managed Sandal Farms

Sandalwood (*Santalum album*) is perhaps the costliest of tree species in the world, Karnataka being its greatest production centre. The high cost of the wood has become baneful to the species, as the tree faces highest smuggling risks. Individual householders and farmers seldom dare to grow this valuable species due to their inability to safeguard it. Collective responsibility by village community seems to be the only course for the future of sandal. We therefore recommend the adoption of the

Biodiversity in India Vol. 8

species by VFCs in their respective jurisdiction especially in the taluks of Mundgod and Haliyal and eastern parts of Sirsi, Yellapur and Siddapur.

viii. VFC Managed Medicinal Plant Areas

Medicinal plant gardens of fast depleting and highly traded species may be promoted through VFCs for growing *Salacia chinensis, Nothopodytes foetida, Embelia* spp., *Coscinium fenestratum, Costus speciosus, Rauwolfia serpentina, Asparagus racemosus* etc. Many highly degraded forests, scrubs and thickets contain numerous medicinal plants particularly near coastal areas. These are to be mapped and brought under strict in situ conservation measures, so as to preserve the native medicinal gene pool.

ix. Forests for Ecotourism

Natural and cultural heritage are primary attractants for tourism world over. Uttara Kannada is an idyllic district of valley villages of lush greenery merging with wooded hillsides and grasslands offering tremendous scope for development of eco-tourism and study tourism. Tourism flourishes especially in areas with more than two landscape elements meet – such as sandy seashore and beach forest (*e.g.*, Kasarkod), sea shore and hillscape (*e.g.*, Apsarakonda), waterfall and forest (*e.g.*, Jog, Unchalli and Magod waterfalls), pilgrimage and picnic trail through forest to cathedral rocks (*e.g.*, Yana, or to hilltop shrine of Karikanamma in the vicinity of *Dipterocarpus* sacred grove) and so on. In all these places and in many more areas, apart from National park and sanctuary, the Forest Department has demonstrated that tourism can be conducted successfully to benefit the local communities organized into VFCs. This facet of development with the vision of upgrading livelihoods of grass root level people while also enriching forests, mangroves, sea beaches and coastal laterite plateaus has been successfully worked out by the Honavar Forest Division, at Apsarakonda, Om Beach (Gokarna), Kasarkod, Bellangi etc. The potential should be developed so as to generate income to the locals through preservation of their local environment and local cultures without the need for migration into cities in search of employment. Key elements for successful development of eco-tourism are limiting growth within sustainable limits (Jog Falls, unfortunately, is a location where ecological norms are not adhered to creating considerable negative impact on environment), generating benefits to the local community (and not to major enterprises from outside), monitoring and mitigating ecological impacts (mostly not happening in our ecotourism areas, except in PAs). Partnership with local community/VFC is of great importance of success of ecotourism. We recommend that in all areas with ongoing, potential ecotourism training be imparted to especially local youth in successful management of tourism, in running forest trails, in bird watching, familiarisation with local flora and fauna etc. Liberal issuance of licenses for home stays and community/VFC managed cottages is necessary for ecotourism to benefit grassroot level people and environment.

x. NTFP Species Raising and Utilisation

For betterment of livelihoods at local level NTFP yielding species should be raised on a larger scale in VFC areas. Auctioning of NTFP to contractors is found to be injurious to forests due to overharvests, unscientific harvesting methods and for the poor returns of revenue to the State. The local VFCs, tribal societies, self-help groups of women etc should be prioritised for NTFP harvests.

xi. Decentralised Systems of Forest Nurseries

For generating women's employment in village areas and also providing scope for application of indigenous farming techniques for forestry purposes sets of local species may be raised in household nurseries.

12. Village Level Biodiversity Hotspots

Our studies show that biodiversity conservation values are correlated to forest endemism. Although Western Ghats itself is part of a global biodiversity hotspot, the concept of village level biodiversity hotspots should be promoted through community participation. Such hotspots, which are especially centres of local level biodiversity, should be identified and special attention given to their protection through Biodiversity Management Committees/Village Forest Committees. Eventually these special patches should serve as local climax natural ecosystems also strengthening local hydrology.

13. Decentralised Systems of Forest Nurseries

Villagers in close vicinity of forest areas may be commissioned to raise small scale nurseries of selected species flowering plants for replanting in forest areas, roadsides etc. to reduce the load on the understaffed Forest Department which is required to spend considerable time and resources on large scale nurseries. This will increase rural employment, especially for women while also giving scope for application of indigenous planting techniques.

14. Promoting Food Plants for Wild Animals

Bulk of Uttara Kannada forests are of secondary nature, either old growth forests or forests, scrub and savannah in different stages of succession. As such these massive vegetational changes that have happened through centuries of human impacts, have adverse consequences on native fauna thinning the populations of many or causing their local extinctions. Leaving aside old growth forests, which should not be subjected to any kinds of tampering, the rest should be enriched with food plants for various faunal elements, particularly birds and frugivorous bats, primates and other mammals. This enrichment is also necessary to reduce crop raiding by wild animals. Care should be taken to preserve grassy blanks within forest areas, critical resources necessary for grazing wild animals. Such grassy blanks should not be subjected to afforestation.

15. VFC Based Resource Monitoring

As villages are dispersed in Uttara Kannada all over forest areas it would make much sense to adopt a system of participatory resource estimation and monitoring within their respective areas- such as estimates of *Myristica*, cinnamon, gooseberry, *Garcinias* and other NTFP plants, key medicinal plants like *Nothapodytes*, *Cosicinim*, *Salacia*, *Embelia* and so on as well as of honey bee colonies within forests. This will strengthen bonds between the Forest Department and village communities while also getting a fair idea of the worth of forests at local level for the provisional goods they contain.

16. Meeting the Fuel Needs

Fuel extraction, both legal (especially removal of dead and fallen from interior forests) and illegal by local population is instrumental in degradation of many forests. Energy efficienct stoves, biogas, solar devices, use of agricultural wastes etc. are to be promoted as fuel in rural areas. At the same time adequate fuelwood/or other alternative fuels should be granted to cottage industries run by potters, lime makers etc.

17. Selecting Appropriate Areas for Tree Plantations

Raising monocultural/mixed tree plantations has to be site specific. Planting of *Acacia auriculiformis* has to be restricted to rocky or otherwise impoverished terrain and not in lands with good soil resources where native species are to be preferred.

18. Dispensing with the Practice of Climber Cutting

Climber cutting is an archaic practice in forestry to promote tree growth. The Western Ghats harbour good diversity of climbers including endemic ones. The climber cutting practice has to be disbanded or restricted to tree plantations only as it would otherwise cause destruction of biodiversity including medicinal plants and entail adverse impacts on wildlife.

Acknowledgement

We are grateful to the NRDMS Division, The Ministry of Science and Technology (DST), Government of India and Indian Institute of Science for the financial and infrastructure support.

References

Akbar Shah, A. 1988. *Integrated development of forests in Western Ghats*. Karnataka Forest Department.

Ali, S., Rao, G. R., Divakar, K. Mesta, Sreekantha, Mukri Vishnu, Subash Chandran, M. D., Gururaja, K. V., Joshi, N. V. and Ramachandra T. V. 2007. *Ecological Status of Sharavathi Valley Wildlife Sanctuary*. Prism Books Pvt Ltd., Bangalore.

Batra, P. 2001. Spatial and temporal foraging dynamics of giant honey bees in an Indian forest. In: Ganeshaiah, K.N., Shaanker, R.U. and Bawa, K.S. 2001. *Tropical Ecosystems: Structure, Diversity and Human Welfare*. Oxford-IBH, New Delhi, pp. 196-197.

Bhat, A. and Jayaram, K.C. 2004. A new species of the genus *Batasio* Blyth (Siluriformes: Bagridae) from Sharavathi River, Uttara Kannada, Karnataka. *Zoos' Print Journal* **19**, 1339-1342

Bhat, D.M., Murali, K.S. and Ravindranath, N.H. 2001. Formation and recovery of secondary forests in India: a particular reference to Western Ghats in south India. *Journal of Tropical Forest Science*, 13(4): 601 – 620.

Bhat, D.M., Murali, K.S. and Ravindranath, N.H. 2002a. *Above ground herb layer productivity and biomass under different light gaps and varied stem density in the forests of Uttara Kannada, Western Ghats, Karnataka*. CES Technical Report no.92, Centre for Ecological Sciences, Indian Institute of Science, Bangalore.

Bhat, D.M., Murali, K.S. and Ravindranath, N.H. 2002b. *Carbon stock dynamics in tropical rainforests of Uttara Kannada district, Western Ghats, India*. CES Technical Report no.96, Centre for Ecological Sciences, Indian Institute of Science, Bangalore.

Bhat, D. M., Naik, M. B., Patagar, S. G., Hegde, G. T., Kanade, Y. G., Hegde, G. N., Shastri, C. M., Shetti, D. M. and Furtado, R. M. 2000. Forest dynamics in tropical rain forests of Uttara Kannada district in Western Ghats, India. *Curr. Sci.* 79(7): 975-985.

Bhat, D.M., Prasad, S.N., Hegde, M. and Saldanah, C.J. 1985. *Plant diversity studies in Uttara Kannada district*. CES Technical Report no.9, Centre for Ecological Sciences, Indian Institute of Science, Bangalore.

Bhat, D.M. and Ravindranath, N.H. 2011. Above – ground standing biomass and carbon stock dynamics under a varied degree of anthropogenic pressure in tropical rain forests of Uttara Kannada district, Western Ghats, India. *Taiwania*, 56(2): 85-96.

Brown, S. 1997. *Estimating biomass and biomass change of tropical forests: a Primer*. FAO Forestry Paper 134, Rome, Italy.

Buchanan, F.D. 1870. *A Journey from Madras through the countries of Mysore, Canara and Malabar, Vol. 2*. Higginbothams and Company, Madras.

Burlakova, L.E. and five co-authors. 2011. Endemic species: Contribution to community uniqueness, effect of habitat alteration, and conservation priorities. *Biological Conservation*, 144(1), 155-165.

Caldecott, J. O., Jenkins, M. D., Johnson, T. H. and Groombridge, B. 1996. Priorities for conserving global species richness and endemism. *Biodiversity and Conservation* 5(6). UNEP and WCMC.

Caratini, C.M., Fontugne, Pascal, J.P., Tisscot, C. and Bentaleb. 1991. A major change at ca. 3500 years B.P. in the vegetation of the Western Ghats In North Canara, Karnatak, *Curr. Sci.* 61: 669-672.

Champion, H.G. and Seth, S.K. 1968. *A Revised Survey of the Forest Types of India*, Government of India, New Delhi.

Chhabra, A. and Dadhwal, V. K. 2004, Assessment of major pools and fluxes of carbon in Indian forests. *Climatic Change*, 64(3): 341–360.

Chandran, M.D.S. 1995. Vegetational changes in the evergreen forest belt of Uttara Kannada district of Karnataka, India. *Ph.D. Thesis*, Karnataka University, Dharwad.

Chandran, M.D.S. 1997. On the ecological history of the Western Ghats. *Curr. Sci.* 73: 148-155.

Chandran, M.D.S. 1998. Shifting cultivation, sacred groves and conflicts in colonial forest policy in the Western Ghats. In: Grove, R.H., Damodaran, V. and Sangwan, S. (Eds) *Nature and the Orient: The Environmental Historyof South and Southeast Asia,* Oxford University Press, New Delhi, pp. 674-707.

Chandran, M.D.S. and Gadgil, M. 1993. *Kans:* safety forests of the Western Ghats. In: Brandl H (Ed) *Geschichte der Kleinprivatwaldwirtschaft Geschichte des Bauernwaldes,* Forstliche Versuchs-und Forschungsanstalt, Freiburg, pp. 49-55.

Chandran, M.D.S. and Gadgil, M. 1998. Sacred groves and sacred trees of Uttara Kannada. In: B. Saraswati (ed.) *Lifestyle and Ecology,* pp. 85-138. Indira Gandhi National Centre for Arts, New Delhi.

Chandran, M.D.S. and Mesta, D.K. 2001. On the conservation of Myristica swamps of the Western Ghats. In: Shankar, R.U., Ganeshaiah, K.N. and Bawa, K.S. (eds.). *Forest Genetic resources, status, threats and conservation strategies* pp. 1-19. Oxford and IBH, New Delhi.

Chandran, M.D.S., Mesta, D.K., Rao, G.V., Sameer, Ali, Gururaja, K.V. and Ramachandra, T.V. 2008. Discovery of two critically endangered tree species and issues related to Relic forests of the Western Ghats. *The Open Conservation Biology Journal,* 2: 1-8.

Chandran M.D.S., Rao, G.R., Gururaja, K.V. and Ramachandra, T.V. 2010. Ecology of the Swampy Relic Forests of Kathalekan from Central Western Ghats, India. *Bioremediation, Biodiversity and Bioavailability 4 (Special Issue I),* Global Science Books, 54-68.

Chokkalingam, U., De Jong, W., Smith, J. and Sabogal, C. 2000. Tropical secondary forests in Asia: Introduction and synthesis. Paper prepared for the "Tropical secondary forests in Asia: Reality and perspectives' workshop". 10–14 April 2000. Samarinda, Indonesia. Center for International Forestry Research, Bogor, Indonesia.

Conservation International, 2013. http://www.conservation.org/where/ priority_areas/ hotspots/asia-pacific/Western-Ghats-and-Sri-Lanka

Cooke, T. 1901-1908. *The Flora of the Presidency of Bombay 3 Vols.* Botanical Survey of India, Calcutta.

Daniels, R.J.R. 2003. Biodiversity of the Western Ghats: An overview. In: Gupta, A.K., Kumar, A.and Ramakantha, V. (Eds.): *ENVIS Bulletin: Wildlife and Protected Areas, Conservation of Rainforests in India.* Vol. 4, No. 1, pp. 25 – 40.

Daniels, R.J.R., Joshi, Í.V. and Gadgil, M. 1989. Changes in the bird fauna of Uttara Kannada, India, in relation to changes in land use over the past century. *Biol. Conserv.* 52 : 37-48

Daniels, R.J.R., Chandran, M.D.S. and Gadgil, M. 1993, A strategy for conserving the biodiversity of Uttara Kannada district in South India. *Environmental Conservation,* 20(2): 131-138.

Daniels, R.J.R, Gadgil, M. and Joshi, N.V. 1995. Impact of human extraction on tropical humid forests in the Western Ghats in Uttara Kannada, South India. *J. Appl. Ecol.* 32: 866-874.

Dasappa and Swaminath, M.H. 2000. A new species of *Semecarpus* (Anacardiaceae) from the *Myristica* swamps of Western Ghatsof North Kanara, Karnataka, India. *Indian Forester* 126: 78-82.

Dixon, R.K., Brown, R.A., Houghton, R.A., Solomon, A.M., Trexler, M.C. and Wisniewski, J. 1994. Carbon pools and flux of global forest ecosystems. *Science* 263: 185-190.

FAO. 2005. *State of the World's Forests*, Food and Agriculture Organisation, Rome, 168 pp.

Forest Survey of India, 1999. State of Forest Report 1999, Summary. Ministry of Environment and Forests, Govt. of India, Dehradun.

Gadgil, M. and Chandran, M.D.S. 1989, Environmental impact of forest based industries on the evergreen forests of Uttara Kannada district, A case study (Final Report). Department of Ecology and Environment, Government of Karnataka, Bangalore.

Gokhale, Y. 2001. Management of Kans in the Western Ghats of Karnataka. In: *Forest Genetic Resources: Status, Threats, and Conservation Strategies.* Shaanker, U., Ganeshaiah, K.N. and K.S. Bawa (eds), Oxford and IBH Publishing Co. Pvt. Ltd., pp. 570-573.

Gokhale, S.B., Tatiya, A.U., Balkiwal, S.R. and Fursule, R.A. 2004. Natural dye yielding plants of India. *Natural Products Radiance* 3(4): 228-234.

Gururaja, K.V. 2004. Sahyadri Mandooka, http://wgbis.ces.iisc.ernet.in/biodiversity/ sahyadri_enews/newsletter/issue6/index.htm

Haripriya, G. S. 2003. Carbon budget of the Indian forest ecosystem. *Climate Change,* 56: 291–319.

Hooker, J.D. 1872-1897. *Flora of British India, 7 vols.,* Botanical Survey of India, Calcutta.

IISc 2006. *Forest conservation and afforestation/reforestation in India: implications of forest carbon stocks and sustainable development.* Report of project No. 6/2/2006-CCC. Indian Institute of Science, Bangalore, 560 012, India.

Kiruba, S., Mishra, B.P., Stalin, S.I., Jeeva, S. and Dhaas, S.S.M. 2006. Traditional pest management practices, Kanyakumari district, southern peninsular India. *Indian Journal of Traditional Knowledge,* 5(1): 71-74.

Lele, S., Sreenivasan, V. and Bawa, K.S. 2000. Returns to investment in conservation: Disaggregated benefit-cost analysis of the creation of a wildlife sanctuary. In: Ganeshaiah, K.N., Shaanker, R.U. and Bawa, K.S. 2001. *Tropical Ecosystems: Structure, Diversity and Human Welfare.* Oxford-IBH, New Delhi, pp. 31-33.

Lugo, A. E. and Brown, S., 1992. Tropical forests as sinks of atmospheric carbon. *Forest Ecology and Management,* 48: 69-88.

Malhi, Y. and Grace, J. 2000. Tropical forests and atmospheric carbon dioxide. *Trends Ecol. Evol.* 15: 332-337.

Meadows, R. 2008. Endemism as a surrogate for bird diversity. http://www.conservationmagazine.org/2008/07/endemism-as-a-surrogate-for-biodiversity

Mesta, D.K. 2008. Regeneration status of endemic trees in the fragmented forest patches of Shravathi river basin in the Central Western Ghats. *Ph.D. Thesis*, Department of Botany, Karnatak University, Dharwad.

Murali, K.S., Bhat, D.M. and Ravindranath, N.H. 2005. Biomass estimation equations for tropical deciduous and evergreen forests. *Agricultural Resources Governance and Ecology*, 4: 81–92.

Murthy, I.K., Murali, K. S., Hegde, G. T., Bhat, P. R. and Ravindranath, N. H. 2002. A comparative analysis of regeneration in natural forests and joint forest management plantations in Uttara Kannada district, Western Ghats. *Curr. Sci.,* 83(11): 1358-1364.

Murthy, I.K., Bhat, P.R. Ravindranath, N. H. and Sukumar, R. 2005, Financial valuation of non-timber forest product flows in Uttara Kannada district, Western Ghats, Karnataka. *Curr. Sci.,* 88(10): 1573-1579.

Myers, N., Mittermeier, R.A., Mittermeier, C.G. and da Fonseca, G.A.B. 2000. Biodiversity hotspots for conservation prioritisation. *Nature* 403: 853-858.

Nelson, B.W., Ferriera, C.A.C., Da Silva, M.F. and Kawasaki, M.L. 1990. Endemism centres, refugia and botanical collection density in Brazilian Amazonia. *Nature* 345: 714-716,

Palm, C.A., Houghton, R.A., Melillo, J.M., Skole, D.L. 1986. Atmospheric carbon dioxide from deforestation in Southeast Asia. *Biotropica*, 18: 177-188

Pascal, J.P. 1982. *Vegetation Maps of South India*. Karnataka Forest Department and French Institute, Pondicherry.

Pascal, J.P. 1984. *Vegetation Maps of South India*. Karnataka Forest Department and French Institute, Pondicherry

Pascal J.P. 1986, *Explanatory booklet on the forest map of South India*, French Institute, Pondicherry, Chapter-3, pg. 19-30.

Pascal, J. P. 1988, *Wet Evergreen forests of the Western Ghats of India-Ecology, Structure, Floristic composition, and Succession.* Sri Aurobindo Ashram press, Pondicherry.

Pascal, J.P., Sunder, S.S. and Meher-Homji, M.V. 1982, *Forest Map of South-India Mercara – Mysore.* Karnataka and Kerela Forest Departments and The French Institute, Pondicherry.

Peterson, A.T., Egbert, S.L., Cordero, V.S. and Price, K.P. 2000. Geographic analysis of conservation priority: endemic birds and mammals in Veracruz, Mexico. *Biological Conservation*, 93(1): 85-94.

Prasad, S.N., Hegde, H.G., Bhat, D.M. and Hegde, M. 1987. *Estimate of standing biomass* and productivity of tropical moist forest of Uttar Kannada district of Karnataka, India, CES Technical Report no.19, Centre for Ecological Sciences, Indian Institute of Science, Bangalore.

Puri, G.S. 1960. *Indian Forest Ecology* Vol. 2. New Delhi.

Punekar, S.A. and Lakshminarasimhan, P. 2010. *Stylidium darwinii* (Stylidiaceae), a new trigger plant from Western Ghats of Karnataka, India. *J. Botanical Research Institute of Texas*, 4(1): 69-73.

Ramachandra, T.V. 2007. Vegetation status in Uttara Kannada district. *MJS*, 6(7): 1-26.

Ramachandra, T.V. and Kamakshi, G., 2005. *Bioresource Potential Of Karnataka* [Talukwise Inventory With Management Options]. CES Technical report No. 109, Centre for Ecological Sciences, Bangalore.

Ramachandra, T.V., Bharath, H. and Suja, A., 2006. Sahyadri: Western Ghats Biodiversity Information System http://ces.iisc.ernet.in/biodiversity. In: *Biodiversity in Indian Scenarios*, Ramakrishnan, N. (Ed), Daya Publishing House, New Delhi, pp. 1-22.

Ramachandra, T.V., Subash Chandran, M.D., Gururaja, K.V. and Sreekantha. 2007. *Cumulative Environmental Impact Assessment*, Nova Science Publishers, New York.

Rao, G.R., Mesta, D.K., Subhashchandran, M.D. and Ramachandra, T.V. 2008. Wetland Flora of Uttara Kannada. *Environment Education for Ecosystem Conservation*, pp. 152–159.

Ravindranath, N. H., Somashekhar, B. S. and Gadgil, M. 1997. Carbon flows in Indian forests. *Climate Change*, 35: 297–320.

Ravindranath, N.H., Rajiv Kumar Chaturvedi and Murthy, I.K. 2008, Forest conservation, afforestation and reforestation in India: Implications for forest carbon stocks. *Curr. Sci.*, 95(2): 216-222.

Ravindranath, N.H. and Ostwald, M. 2008, Estimation of carbon stocks and changes and data sources. In: *Carbon Inventory Methods: Handbook for Greenhouse Gas Inventory, Carbon Mitigation and Roundwood Production Projects*. Springer-Verlog, Berlin, pp. 237 – 270.

Richards, J. F. and Flint, E.P., 1994. *Historical land use and carbon estimates for south and South East Asia* 1880-1980, Publication Number 4174, Environmental Science Division, Oak Ridge Laboratory, USA.

Saldanha, C.J. 1984. *Flora of Karnataka*. Vol. 1. Oxford and IBH, New Delhi.

Sameer Ali, D.K. Mesta, M.D. Subash Chandran and Ramachandra, T.V. 2010. Report of *Burmannia championii* Thw. from Uttara Kannada, Central Western Ghats, Karnataka. *J.Econ. Taxon. Bot.* 34(2): 343-345.

Santapau, H. 1955. Indian Botanical Society Excursion. *J. Indian Bot. Soc.* 30: 181–191.

Shanmukhappa, G., 1966. Working Plan for the Unorganised Forests of Sirsi and

Siddapur. Karnataka Forest Department, Bangalore.

Shastri, C.M., Bhat, D.M., Nagaraja, B.C., Murali, K.S. and Ravindranath, N.H. 2002. Tree species diversity in a village ecosystem in Uttara Kannada in Western Ghats, Karnataka. *Curr. Sci.*, 82(9): 1080-1084.

Sreekantha, Chandran, M.D.S., Mesta, D.K., Rao, G.R., Gururaja, K.V. and Ramchandra, T.V. 2007. Fish diversity in relation to landscape and vegetation in central Western Ghats, India. *Curr. Sci.*, 92(11): 1592-1602.

Stattersfield, A.J., Crosby, M.J. Long, A.J. and Wege, D.C., 1998. *Endemic Bird Areas of the World: Priorities for Conservation.* Birdlife International, UK.

Stickler, C.M., Michael T. Coe, Marcos H. Costa, Daniel C. Nepstad, David G. McGrath, Livia C. P. Dias, Hermann O. Rodrigues and Britaldo S. Soares-Filh, 2010. *Dependence of hydropower energy generation on forests in the Amazon Basin at local and regional scales,* http://www.pnas.org/cgi/doi/10.1073/pn

Talbot, W.A. 1909. *Forest Flora of the Bombay Presidency and Sind (Vols. I and II).* Government Photozincographic press, Poona.

Varma, J. and Dubey, N.K. 1999. Prospectives of botanical and microbial products as pesticides of tomorrow. *Curr. Sci.*, 76(2): 172-179.

Varshney, V. 2006. Plant-based pesticide pious resolve, little action. *Down to Earth.* http://www.downtoearth.org.in/node/8689

World Water Assessment Programme. 2012. The United Nations World Water Development Report 4: Managing Water under Uncertainty and Risk (UNESCO, Paris).

Chapter 2

Diversity of Asterinaceous Fungi in Sirumalai Hills, Eastern Ghats

R. Ramasubbu[1] and A. Chandra Prabha[2]

[1]Department of Biology,
The Gandhigram Rural Institute - Deemed University,
Gandhigram, Dindigul, Tamil Nadu, India
[2]SFR College for Women,
Sivakasi, Tamil Nadu, India

ABSTRACT

Fungi are unique among the living organisms and have become omnipresent in biosphere. They represent the group of thalloid, eukaryotic, achlorophyllous, entirely heterotrophic microorganisms that are spore producing and usually reproducing through asexually, sexually or both. Black mildews are the group of fungi occurring commonly in the tropical and subtropical regions of the world. These black colonies forming organisms belong to different taxonomic groups, *viz.*, Meliolaceous fungi, Schiffnerulaceous fungi, Asterinaceous fungi, Hyphomycetous fungi, etc. Among these, Asterinaceous fungi are specialized group of fungi and widely distributed in forest areas of South India. Diversity and distribution of Asterinaceous fungi on the leaves of angiosperms of Sirumalai has been studied to document the fungal diversity. About 46 species of Asterinaceous fungi were observed on wide diversity of plant species. Among them, 34 species belong to the genus *Asterina*, 4 to *Prillieuxina*, 3 to *Lembosia* and one species each to *Asterolibertia, Asterostomella, Echidnoides, Eupelte* and *Echidnodella*. The black colonies of these fungi increase the temperature of the infected parts and cause physiological imbalance in the entire leaves. They decrease the photosynthetic

* Corresponding Author: E-mail: racprabha@yahoo.com

efficiency of the plants, affect the hormonal and phenolic compound level and in short affect the efficiency of the plants in total. The pathological control mechanism on the host plants has not been studied.

Keywords: Asterina, Follicolous, Lembosia, Prillieuxina.

Introduction

Fungi are heterotrophic, eukaryotic organisms without the capacity to produce their own food supplies and are thus completely dependent on preformed organic matters. They have neither photosynthetic nor chemosynthetic pigments nor capabilities for these processes. The Kingdom Fungi includes some of the most important organisms both in terms of their ecological and economic roles. By breaking down dead organic material, they continue the cycle of nutrients through ecosystems. In addition, most of the vascular plants could not grow without the symbiotic fungi or mycorrhizae and that inhabit their roots and supply essential nutrients. Other fungi provide numerous drugs (penicillin and other antibiotics), foods like mushrooms, truffles and morels and the bubbles in bread, champagne and beer. Fungi also cause a number of plant, animal and human diseases like ringworm, athlete's foot and several more serious diseases. Because fungi are more chemically and genetically similar to animals than other organisms and this makes fungal diseases very difficult to treat. Plant diseases caused by fungi include rusts, smuts, leaf, root and stem rots and may cause severe damage to crops.

Fungi are important components of biodiversity in tropical forest. As a major contributor to the maintenance of the earth's ecosystem, biosphere and biogeochemical cycles, fungi perform unique and indispensable activities on which organisms including human depend (Hawksworth and Colwell, 1992). Around 1,00,000 species of fungi have been formally described by taxonomists, but the global biodiversity of the fungus kingdom is not fully understood (Mueller and Schmit, 2006). On the basis of observations of the ratio of the number of fungal species to the number of plant species in selected environments, the fungal kingdom has been estimated to contain about 1.5 million species (Hawksworth, 2006). A recent estimate suggests there may be over 5 million species (Blackwell, 2011). In mycology, species have historically been distinguished by a variety of methods and concepts. Classification based on morphological characteristics such as the size and shape of spores or fruiting structures has traditionally dominated fungal taxonomy.

Depending upon the source of absorption and association with its partner or associated substratum, fungi are classified as saprophytes, parasites and symbionts. The parasites which are totally dependent on living organisms are called obligate parasites. The obligate parasites have to adjust and modify themselves with their partners for their survival. Certain parasites absorb nutrient from the living tissues, without killing them by the specialized organs like appressoria, haustoria or nutritive hyphae and these are called as biotrophs. Based on the mode of infection and symptoms produced by the parasites, they are named as rust, smut, powdery mildews, downy mildews, black mildews, sooty moulds, etc. Black mildews are obligate ectoparasites produce black colonies on the leaves of the host plant in contrast sooty moulds which grow on insect secretion or nectar produced by the plant and spread on to leaves, petiole, stem or dead bark of plants.

The leaves provide a very suitable habitat for the growth and development of fungal pathogen by providing ample surface area and nutrient supply. Such leaf inhabiting fungi are known as foliicolous or phylloplane and most of these fungi are obligate or facultative parasites. Residing on the surface of the leaves and producing special organs and opting special adaptation, they are acting as necrotrophs or biotrophs and their infection may lead to the destruction of the plant. Plant diseases have got much importance because of their direct influence on mankind. The diseases which attacked cultivated plants and caused heavy loss to the staple food and widely attracted the attention of the investigators to study them thoroughly. On the other hand, several groups of fungi which do not cause any severe symptomatic appearance on the hosts were received less attention as in the case of black mildews.

Black mildews are the group of organisms occurring commonly in the tropical and subtropical regions of the world. These black colonies forming organisms belong to different taxonomic groups, *viz.*, Meliolaceous fungi, Schiffnerulaceous fungi, Asterinaceous fungi, Hyphomycetous fungi, etc. Meliolaceous fungi are distinct from others in having brown mycelium with bi-cellular appressoria, setae, perithecial appendages and consistent brown septate ascospores (Hansford, 1961; Hosagoudar *et al.*, 1996a; Hosagoudar and Agarwal, 2008).

Of the Meliolaceous fungi, Asterinaceous fungi are ectophytic obligate biotrophs infecting wide range of flowering plants ranging from herbs to trees, weeds to economically important cultivated plants, etc. These fungi produce thin to dense black colonies on the surface of the leaves. Structurally brown superficial mycelium produces appressoria which in turn produces haustoria or nutritive hyphae in to the epidermal cells of the host plant for the nourishment. The fruiting body is flattened with radiating cells known as thyriothecium which splits radially like a star, hence they are known as Asterinaceous fungi.

The life cycle of Asterinaceous fungi start with the ascospores. Ascospores on compatible host during suitable climate initiate by producing appressoria, mostly from the terminal cells. This is the stage where parasites establish its relation with the host plant. Subsequent growth is by producing mycelium. The mycelium is dark, mostly superficial but sometimes immersed, branched with individual hyphae forming lateral appressoria. The colonies on the leaves may be epiphyllous, hyphophyllous or amphigenous. Nature of the colony may be thin, dense, crustose or velvety. Hyphae may be straight, substraight, flexuous or crooked and the branching pattern occurs as alternate, opposite or irregular. Appressoria 1-2-celled, placed in opposite, alternate or in mixed arrangement on the hyphal cells. A fine hyphal filament beneath the appressorium penetrates the host cell wall to produce a haustorium within the host cell. The apical cells vary in shape such as ovate, globose, conoid, oblong, clavate or cylindrical with entire, angular, sub-lobate or lobate margin. The fruiting body is flattened with radiating cells known as thyriothecium. Based on the shape and dehiscence of the thyriothecia, Asterinaceous fungi are placed in two families' *viz.*, Asterinaceae and Lembosiaceae. In Asterinaceae, thyriothecia are orbicular, dehisce stellately at the center. In Lembosiaceae, thyriothecia are elongated, ellipsoidal, X or Y shaped, dehisce longitudinally at the center. Species of *Asterina* and *Lembosia* are characterized by lateral appressoria where as *Asterolibertia* is characterized by intercalary appressoria. In *Ishwaramyces*, appressoria occur in clusters *i.e.*, born more than two from a same place. *Eupelte* and *Prillieuxina* are devoid of appressoria. The former differs from the latter in having conidiogenous cells. *Echidnodella* and *Echidnodes* belong to Lembosiaceae, the former differs from the latter in absence of hypostroma. The spores are born in a sac or ascus, containing eight spores (octosporous). Asci may be globose, oval, spherical or ellipsoidal. The wall of the ascospores may be smooth, echinulate, tuberculate or verrucose. The ascospores are two celled, uniseptate.

Asterinaceous fungi represent 32 genera belong to two families namely Asterinaceae and Lembosiaceae. The family Asterinaceae consists of more than 800 species, the genus *Asterina* alone is numbered more than 600 species (Hosagoudar and Abraham, 2000). No comprehensive work is available to identify Asterinaceous fungi on Sirumalai hills. Hence the present study will provide important details on Asterinaceous fungal diversity for further research.

Leveille (1845) proposed the genus *Asterina* to accommodate the fungi having orbicular thyriothecia and *Lembosia* to accommodate the fungi having elongated thyriothecia. Fries (1849) placed *Asterina* under the tribe "*Asterinaei*" and *Lembosia* under Hysteriaceae. Saccardo (1883) proposed the family Microthyriaceae to accommodate the genera having dimidiate, radiate and flattened ascomata. Theissen (1913a) proposed the order Hemisphaeriales to accommodate the genera having shield- shaped ascomata. Stevens and Ryan (1939) placed the Asterinaceous fungi under the family Microthyriaceae. Theissen (1913a) monographed the genus *Asterina* with the known species. The substantial contribution of the taxonomical analysis of *Asterina* was done by Doidge (1942) in South Africa. Works of Saccardo (1924), Stevens and Ryan (1939), Doidge (1942), Muller and Arx (1962), Crane and Jones (1997) revealed the reports of more than 500 species.

The genus *Asterina* comprises more than 578 species known on the host plants belonging to more than 106 families (Hosagoudar and Abraham, 2000) and it is the largest genus of the family Asterinaceae. The anamorphic genus of *Asterina* is *Asterostomella* Speg. and *Clasterosporium* Schwein. Asterinaceae was listed with 410 known species belonging to 37 genera (Kirk *et al.*, 2001). Thaung (2006) reported phylloplane Ascomycetes including 31 taxa of the family Asterinaceae from Burma.

Lembosia is characterized by having superficial mycelium with lateral appressoria forming intracellular haustoria; thyriothecia elongated, ellipsoidal with the layer of radiating cells that disintegrate longitudinally at the center during maturity; asci globose to ovate, bitunicate, octosporous; ascospores brown, two celled. Theissen (1913b) monographed the genus *Lembosia* with the known species and accounted for 50 species. A thorough search of works of Saccardo (1924), Stevens and Ryan (1939),

Doidge (1942) and Muller and Arx (1962) revealed the reports of more than 144 species of *Lembosia*. Mibey and Hawksworth (1997) reported 18 species from Kenya. Kirk *et al*. (2001) stated that the genus *Lembosia* contains 40 species. Four species of *Lembosia* have been recorded from China (Yamamoto, 1956; Ouyang *et al.*, 1995, 1996). Sivanesan and Shivas (2002) described two new *Lembosia* species from Australia. Song and Hosagoudar (2003) described 144 species of *Lembosia* including 5 new species. *Lembosia epidendri* was described by Silva and Pereira (2008) from Minas Gerias, Brazil.

In the recent years, substantial contribution on the taxonomical studies of foliicolous fungi including Asterinaceous fungi were done by Hosagoudar from Kerala through series of publications: Hosagoudar *et al*. (1996a,b), Hosagoudar and Abraham (1998a,b), Hosagoudar *et al*. (1999), Hosagoudar (2003a,b), Hosagoudar and Agarwal (2003). Asterinaceae was segregated and a new family Lembosiaceae was proposed to include the genera having ellipsoidal to elongated or X or Y shaped thyriothecia split or dehisce longitudinally (Hosagoudar *et al.*, 2001a). Hosagoundar and Goos (1996) reported 19 species of folicolous fungi from the Southern Western Ghats of Kanyakumari and Tirunelveli districts and the Anamalai hills of Coimbatore district. Hosagoudar *et al*. (1996a) reported *A. congesta* and *A. jambolana* from Anamalai hills of Coimbatore and they also reported *A. cipadessae* from Nilgiris. The genus *Ishwaramyces* differs from the genus *Asterina* in having axillary clusters of appressoria (Muller and Arx, 1962; Arx and Muller 1975; Hosagoudar *et al.*, 2004). In the recent years, several Asterinaceous fungi were identified and are new to science (Hosagoudar and Chandra Prabha, 2008a,b,c,d).

Study Area

Location and Physical Features

Sirumalai, locally known as small hill located in Dindigul district, Tamil Nadu and lying between 10°07'-10°18' N and 77°55'-78°12' E. Sirumalai is an isolated compact group of hills which extends between 6.5 km south of Dindigul town and 22.5 km north of Madurai city. They are rectangular in outline; 19.3 km long of north-south and 12.8 km broad of east-west, with an extended area of 317 km². The tract is bounded by the Dindigul - Madurai rail line on the west, the Dindigul - Nattam road on the east and north east and the Vadipatti-Nattam road on the south (Plate 2.1).

Geology and Soils

The Sirumalai massif is composed of acid charnockites and have the characteristic bluish grey colour and vary from coarse grained to fine recrystallised types which break with a conchoidal fracture. In some places, the rock is altered to yellow clay material. A rough gneissose banding is seen on the weathered surfaces. The charnockites carry inclusions of amphibolites and quartzite bands which are probably xenoliths (Pallithanam, 2001). The northern slopes of the Sirumalai hills are seen a complex suite of gneisses in various stages of migmatisation. The gneisses are essentially hornblendic or micaceous, but granetiferous types are seen along their contact with charnockites and granites. Fringing the Sirumalai hills along its northern

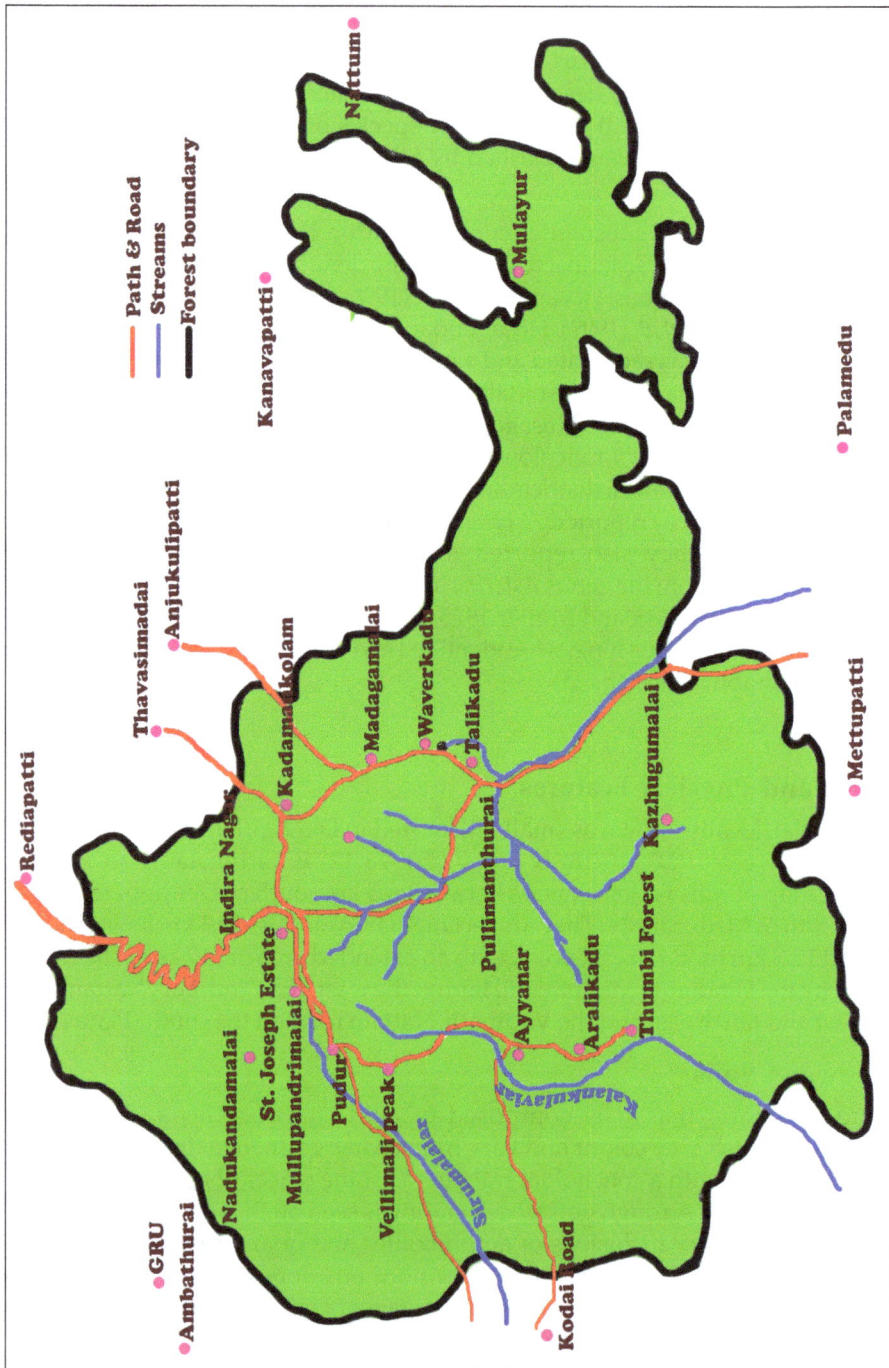

Plate 2.1: Study Area: Sirumalai Hills, Eastern Ghats.

Plate 2.3: Landscape View of Sirumalai Hills.

border are large bands of quartzites, some of them are several hundred yards wide and extending continuously over distances of 5 or 6 miles. The quartzites generally stand out as low residual ridges in barren of vegetation except in thorny scrub jungle, while residual and pebbles of quartzite extend far beyond the actual outcrops. The rock is white, massive and resembles vein quartz at places. The quartzites at a number of places carry sillimanite, magnetic and sericite (Pallithanam, 2001).

Climate and Rainfall

Sirumalai receives an annual rainfall of 132 cm on the plateau, and 120 cm in the plains during the north-east monsoon (October -December). The hottest months are April - May (40°C in the open and 37°C in the shade). The lowest relative humidity (64-4 per cent) was recorded during July- September; the highest relative humidity was recorded during south west (62-77 per cent) and north east monsoon (76-86 per cent). The average relative humidity of Sirumalai hills was recorded as 67.7 per cent.

Vegetation

The vegetation composed of two main groups, *i.e.,* the outer slopes of the hills and the plateau (the hill tops with ridges and valleys including some basins).

(A) The Outer Slopes (250 – 1000 m asl)

The area covers all the outer slopes of the hills from the foothills to 1036 m asl. The areas includes following forest types.

1. Scrub forest
2. Dry deciduous forest
3. Savannah woodlands
4. Dry evergreen forest
5. Riparian forest

(B) The Plateau: Hill Tops (>1000m asl)

This includes

1. Semi-evergreen forest
2. Dry deciduous forest
3. Savannah woodlands
4. Wet rocky slopes
5. Ponds and streamlets
6. Estates and cultivated fields.

Previous Botanical Exploration

There were 895 species from 536 genera of flowering plants and few Gymnosperms recorded, largely excluding the pantropical species of the foothills and surrounding plains (Pallithanam, 2001). In 1871, it was reported that almost all the vegetation had been cleared as far as the base, and a considerable distance up to the slopes of the Sirumalais. This was in addition to the vast areas cleared for plantations. The occasional large size trees left in the interior forests are some indication of the original vegetation. The present forest cover is secondary in origin. Next comes the illicit felling and fuel wood removal in addition to grazing and setting fire to the grasslands in summer.

In addition to the above exploration, more than hundreds of plants added and more than 1000 species were reported from the Sirumalai Hills and several ecological studies have also been conducted by several researchers (Joseph, 1999; Karuppusamy, 2007; Karuppusamy *et al.*, 1999; 2001a,b; 2002; Krishnankutty *et al.*, 2003; Kottaimuthu *et al.*, 2008; Rajasekaran, 2004). Some ethnobotanical research have been conducted by several researchers on the Paliyan tribals inhabiting at different areas of Sirumalai hills (Maruthupandian *et al.*, 2011). But no work has been conducted on diversity and distribution of foliicolous fungi in the Sirumalai hills and it is a first attempt.

Materials and Methods

Field exploration trips were conducted at different forest areas of the Sirumalai hills and collection were made in all the seasons. Infected plant parts were noticed carefully and collected from the field; field notes were prepared regarding their pathogenicity, nature of colonies, nature of infection, locality, altitude, etc. For each collection, a separate field number was given and collected separately in polythene bags along with a host twig (preferable with the reproductive parts) to facilitate the identity of the corresponding host. These collections were completely dried between the blotting paper. The host plants were identified by using the regional floras (Gamble and Fischer, 1915-35; Sharma and Sanjappa 1993; Nayar *et al.*, 2006; Ganeshaiah, 2012) and also the identity confirmed by matching them with the authentic herbarium materials and also with the help of experts.

The position of the fungal colonies on both side of the leaf surface was carefully observed (Plate 2.3). In the laboratory, the standard method (Hosagoudar and Kapoor, 1984) was used for the identification of foliicolous fungi. A drop of high quality natural coloured or well transparent nail polish was applied on the selected colonies and carefully thinned with the help of a fine brush without disturbing the colonies. Colonies with hyperparasites (wooly nature) were avoided. The treated colonies along with their host plants were kept in dust free chamber for half an hour. After the nail polish was dried, a thin colourless film or flip was formed with the colonies firmly embedded in it. For soft host parts, flip was lifted up with a slight pressure on the upper side of the leaves and just below the colonies. In case of hard host parts, the flip was eased-off with the help of a razer or scalpel. A drop of DPX was added on clean slide and the flip was spread properly on it and a coverglass was placed over it by avoiding air bubbles. These slides were labeled and placed in dust free chamber for 1-2 days for drying. The mounted specimens were observed under compound microscope for further studies. The colony character, arrangement of the hyphae, morphology of ascospores were observed and their measurements taken with the help of ocular micrometer (Plates 2.4 and 2.5). The observations were supplemented with line drawings drawn by using camera Lucida (a - Hypha, b - Thyrothecium, c - Ascus, d - Ascospores).

Plate 2.3: Infected Leaf.

A: *Syzygium* sp.; **B:** *Wrightia tinctoria*; **C:** *Santalum album*; **D:** *Flacourtia montana*;
E: *Syzygium cumini*; **F:** *Cipadessa baccifera*; **G:** *Hydnocarpus* sp.; **H:** *Ixora coccinea*.

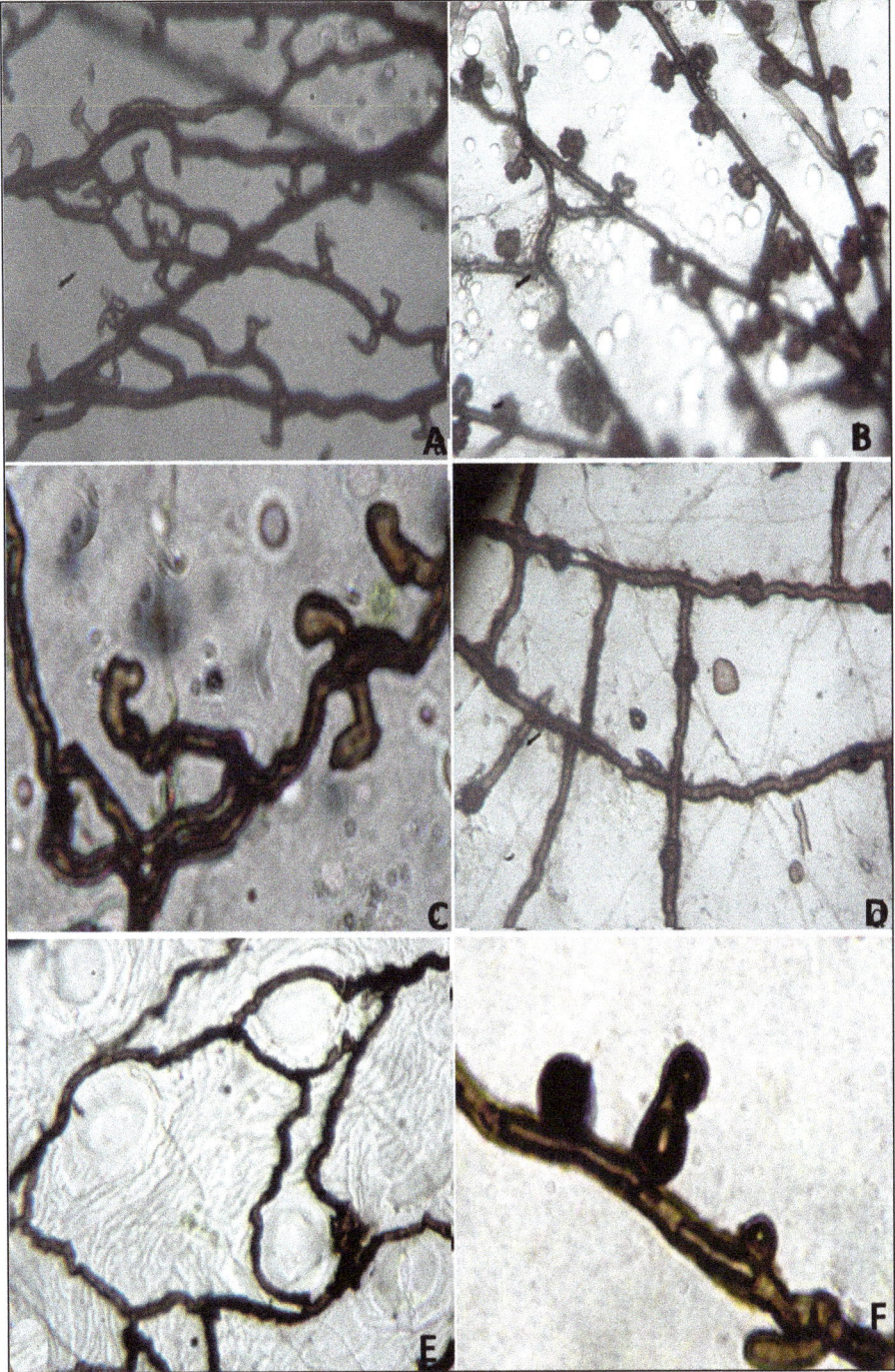

Plate 2.4: Morphology of Hyphae.

A: *Asterina* hyphae; **B:** *Asterina* hyphae; **C:** *Asterolibertia* hyphae; **D:** *Lembosia* hyphae;
E: *Prillieuxina* hyphae; **F:** *Echidnoides* hyphae.

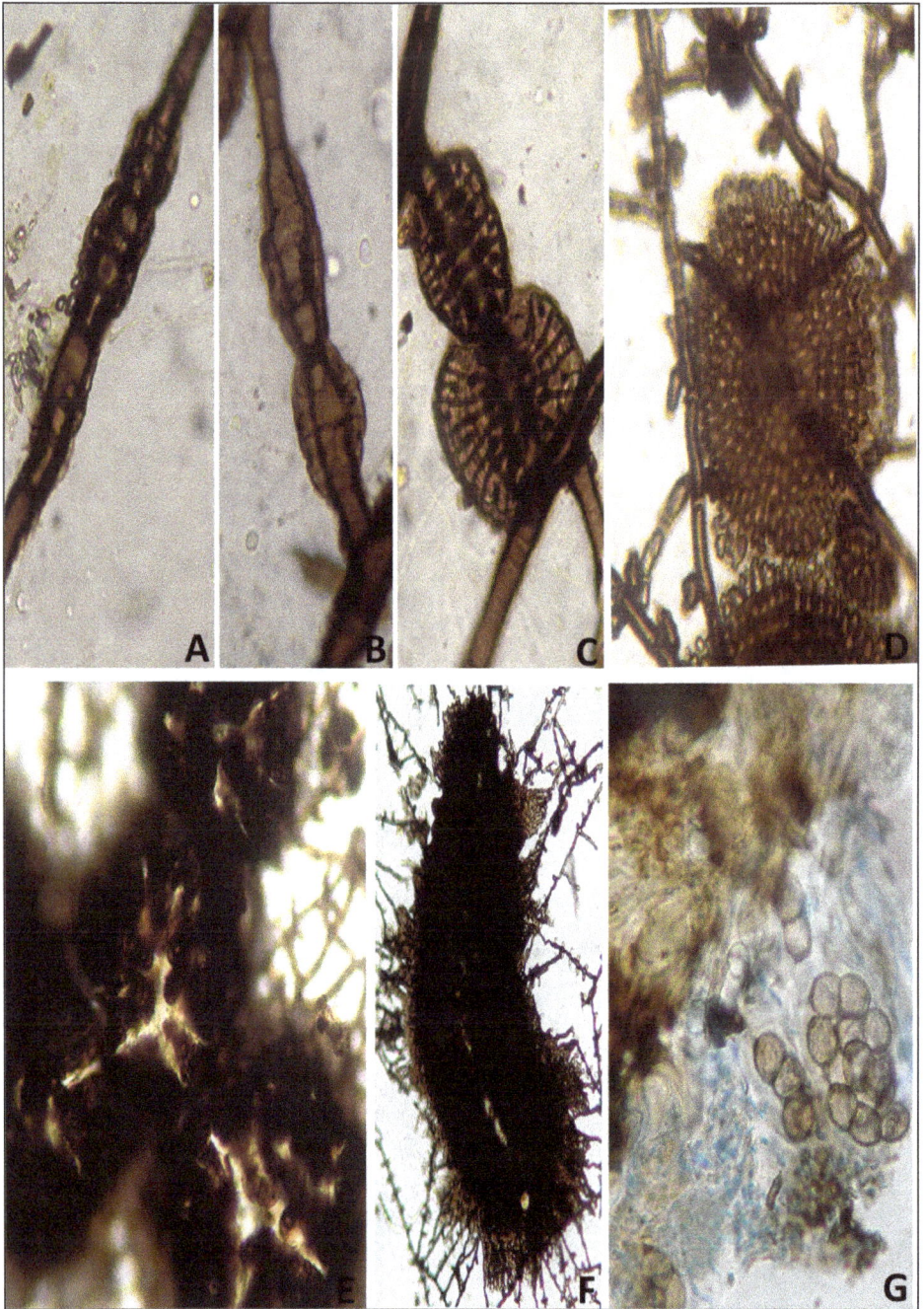

Plate 2.5: Development of Fruiting Body.

A–D: Development of fruiting body (Thyriothecia); E: Stellately dehisced thyriothecia: *Asterina*; F: Longitudinally dehisced thyriothecia: Lembosia; G: Ascospores.

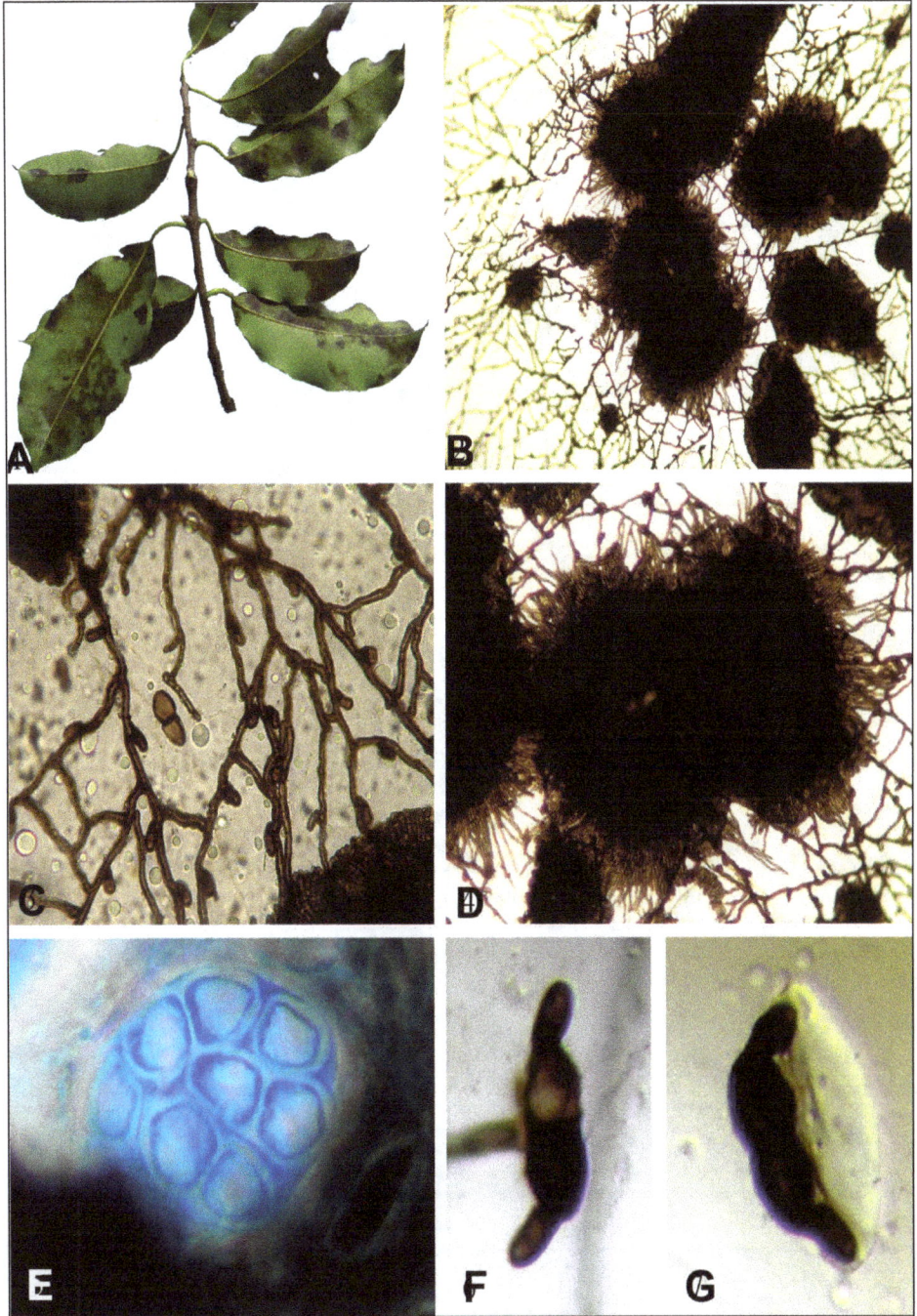

Plate 2.6: *Asterina mimusopsidicola.*

A: Infected Leaves; B: Mycelium with thyriothecia; C: Branched mycelium (Enlarged);
D: Stellately dehisced thyriothecia; E: Ascus; F-G: Germinated ascospores.

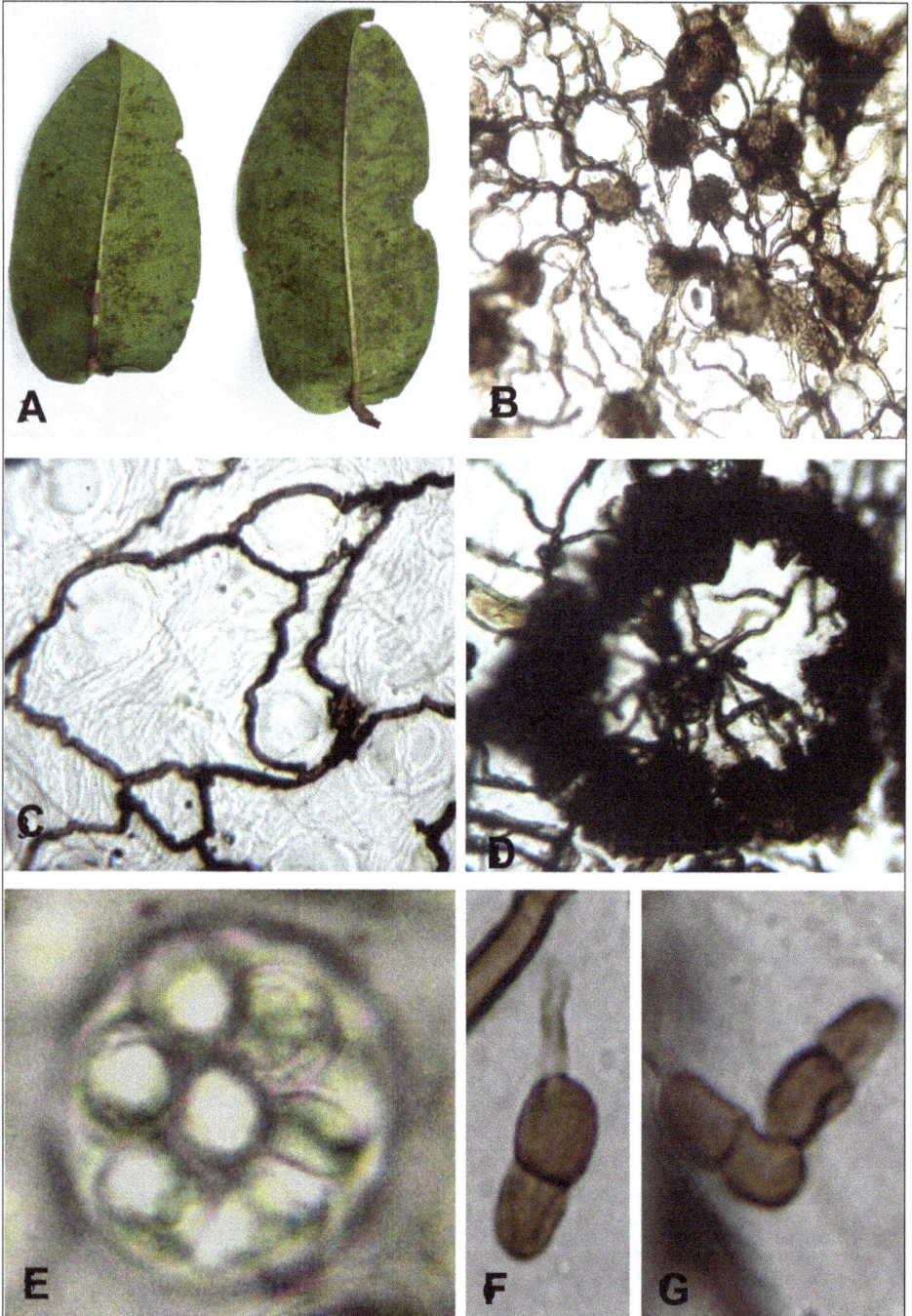

Plate 2.7: *Prillieuxina ixorigena.*

A: Infected leaves; B: Mycelium with thyriothecia; C: Crooked hyphae; D: Dehisced thyriothecia; E: Ascus; F-G: Germinated ascospores.

In some species, the septa was not visible due to heavy pigmentation, in such cases, scrape was taken directly from the infected host and mounted in 10 per cent KOH solution. After 30 minutes, KOH was replaced by lactophenol (Rangaswamy, 1975). Both the mountants worked well as clearing agents and made the septa visible.

After the detailed study of each collection, the individual material was assigned to its taxonomic rank and prepared for herbarium carrying the details of fungus name, host name, date of collection, locality, name of the collector, expert who identified the specimen and its herbarium number. The herbarium materials were deposited in the Tropical Botanic Garden Travancore herbarium (TBGT), Thiruvananthapuram, Kerala.

Results

The Order Asterinales

Leaf parasites. Mycelium ectophytic, with or without appressoria, nutrient mycelium and leaf permating stroma present. Ascomata ectophytic, dimidiate, orbicular with radiating cells, astomatous, dehisce stellately at the center; asci globose, spherical, octosporous, bitunicate; ascospores two to many septate, conglobate, hyaline to brown.

Type Family: Asterinaceae.

Key to the Families of the Order Asterinales

Thyriothecia orbicular, dehisce stellately at the center .... Asterinaceae

Thyriothecia oval to elongated, X or Y shaped,
dehisce longitudinally at the center .... Lembosiaceae

ASTERINACEAE

Asterinaceae Hansf., Mycol. Pap. 15: 189, 1946; Luttrell in Ainsworth *et al.* (eds.). The Fungi. An advanced Treatise 4: 207, 1973; Arx & Muller, Stud. Mycol. 9: 40, 1975; Hosag., Abraham & C.K. Biju, J. Mycopathol. Res. 39: 62, 2001.

Leaf parasites. Mycelium ectophytic, with or without appressoria, nutrient mycelium and leaf permating stroma present. Ascomata ectophytic, dimidiate, orbicular with radiating cells, astomatous, dehisce stellately at the center; asci globose, spherical, octosporous, bitunicate; ascospores two to many septate, conglobate, hyaline to brown.

Type Genus: *Asterina* Lev.

THE GENUS *ASTERINA*

Asterina Lev., Ann. Sci. Nat. Bot. Ser., 3 (3):57, 1845; Hansf., Mycol. Pap. 15: 189, 1946b; Luttrell in Ainsworth *et al.* (eds.). The Fungi. An advanced Treatise 4: 207, 1973; Arx & Muller, Stud. Mycol. 9: 42, 1975; Bilgrami, Jamaluddin & Rizwi, Fungi of India p. 53, 1991; Hosag., Abraham & C.K. Biju, J. Mycopathol. Res. 39: 62, 2001; Singh, Duke, Bhandari & Jain, J. Econ. Taxon. Bot. 30: 183, 2008.

Dimerosporium Fuckel, Symb. Mycol. p. 86, 1870.

Asterella (Sacc.) Speg. ex Sacc., Syll. Fung. 9: 393, 1891 *non* P. de Beauvois 1805.

Myxasterina Hohnel, Sber. Akad. Wiss. Wien 118: 870, 1909.

Englerulaster Hohnel, Sber. Akad. Wiss. Wien 119: 454, 1910.

Parasterina Theiss., H. Sydow & P. Sydow, Ann. Mycol. 15: 246, 1917.

Calothyriolum Speg., Boln Acad. nac. Cien. Cordoba 23: 498, 1919.

Opeasterina Speg., Boln Acad. nac. Cien. Cordoba 23: 498, 1919.

Englera F. Stev. in Stev. & Ryan, Illinois. Biol. Monogr. 17: 45, 1939.

Leaf parasites. Mycelium ectophytic, appressoria lateral, setae absent. Thyriothecia orbicular with radiating cells, astomatous, dehisce stellately at the center; asci globose, octosporous, bitunicate; ascospores conglobate, uniseptate, brown.

Type sp.: *A. melastomatis* Lev.

Anamorphs : *Asterostomella* Speg., *Clasterosporium* Schwein.

Asterina balakrishnanii Hosag. in Hosag., Balakr. & Goos, Mycotaxon 59: 168, 1996. (Figure 2.1)

Colonies epiphyllous, dense, crustose, up to 2 mm in diameter. Hyphae strongly appressed to the host, crooked, branching opposite to irregular at acute angles, loosely reticulate, cells 9-13 × 3-4 µm. Appressoria alternate, sessile, deep brown, globose and angularly pointed towards the apex, 7-11 × 4-7 µm. Thyriothecia closely scattered, frequently connate, orbicular to ovate, up to 120 µm in diameter, dehiscing stellately at the center, dehiscence extended up to margin, margin crenate; asci many, octosporous, ovate to globose, 30-32 × 24-26 µm; ascospores conglobate, brown, 1-septate, slightly constricted at the septum, 18-22 × 9-10 µm.

Distribution: Observed on the leaves of *Solanum torvum* Sw. (Solanaceae).

Asterina chukrasiae Hosag., H. Biju & Appaiah, Mycopathol. Res. 44: 40, 2006. (Figure 2.2)

Colonies epiphyllous, thin to subdense, up to 2 mm in diameter, rarely confluent. Hyphae substraight, branching irregular at acute to wide angles, loosely reticulate, cells 19-23 × 3-5 µm. Thyriothecia scattered, orbicular, up to 110 µm in diameter, margin crenate, stellately dehisced at the center; asci globose, octosporous, up to 26 µm in diameter; ascospores oblong, conglobate, uniseptate, constricted, brown, 20-24 × 11-13 µm, wall smooth.

Distribution: Observed on the leaves of *Chukrasia tabularis* A. Juss. (Meliaceae).

Asterina cipadessae Yates, Philippin J. Sci. 12: 371, 1917. (Figure 2.3)

Parasterina cipadessae (Yates) Mendoza, Philippin J. Sci. 49: 446, 1932.

Colonies amphigenous, mostly epiphyllous, dense, up to 2 mm in diameter, confluent. Hyphae flexuous, crooked, branching opposite at acute to wide angles, loosely to closely reticulate, cells 12-33 × 4-7 µm. Appressoria alternate (80 per cent), opposite (20 per cent), sessile, lobate, 7-11 × 4-7 µm. Thyriothecia scattered to grouped,

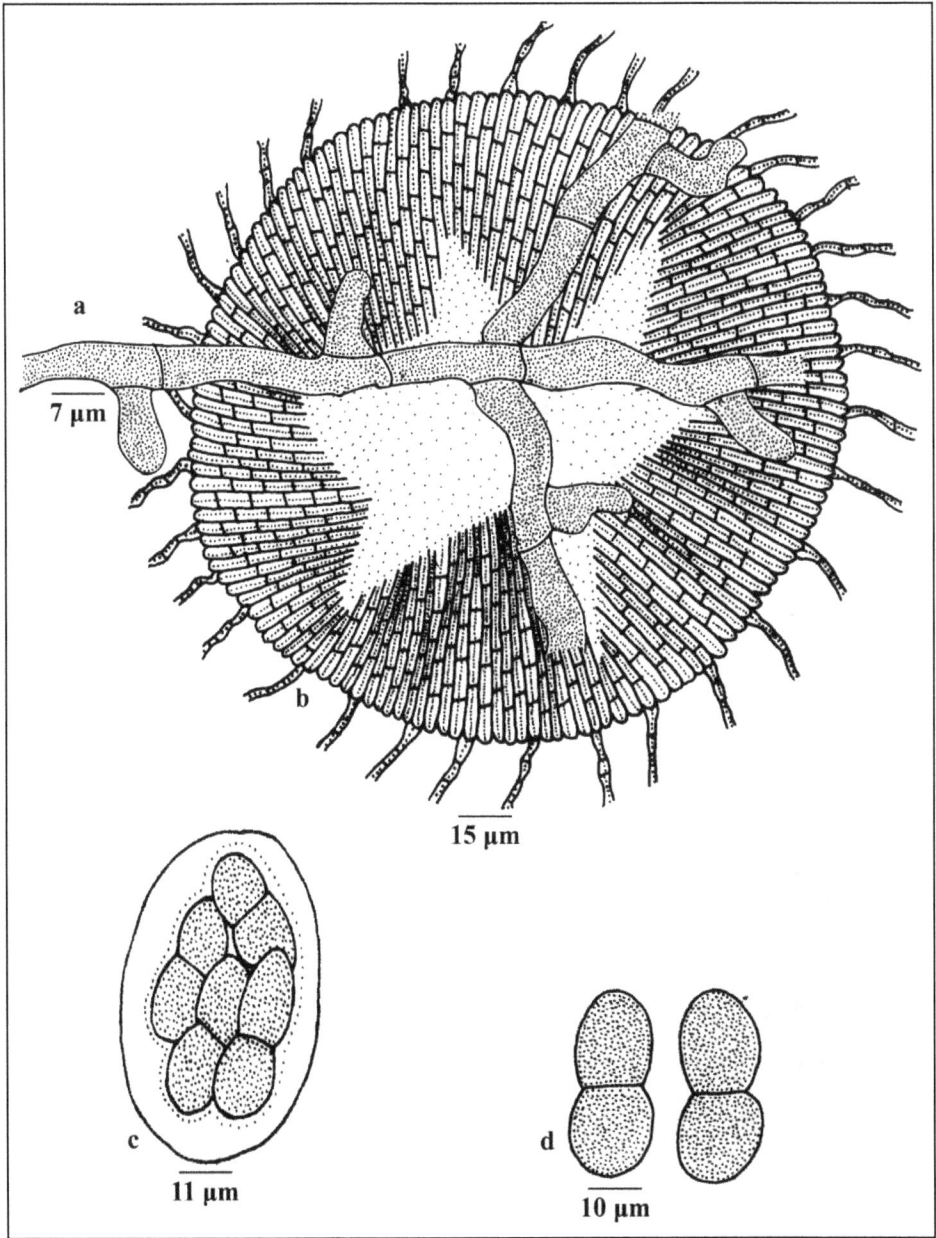

Figure 2.1: *Asterina balakrishnanii* **Hosag.**

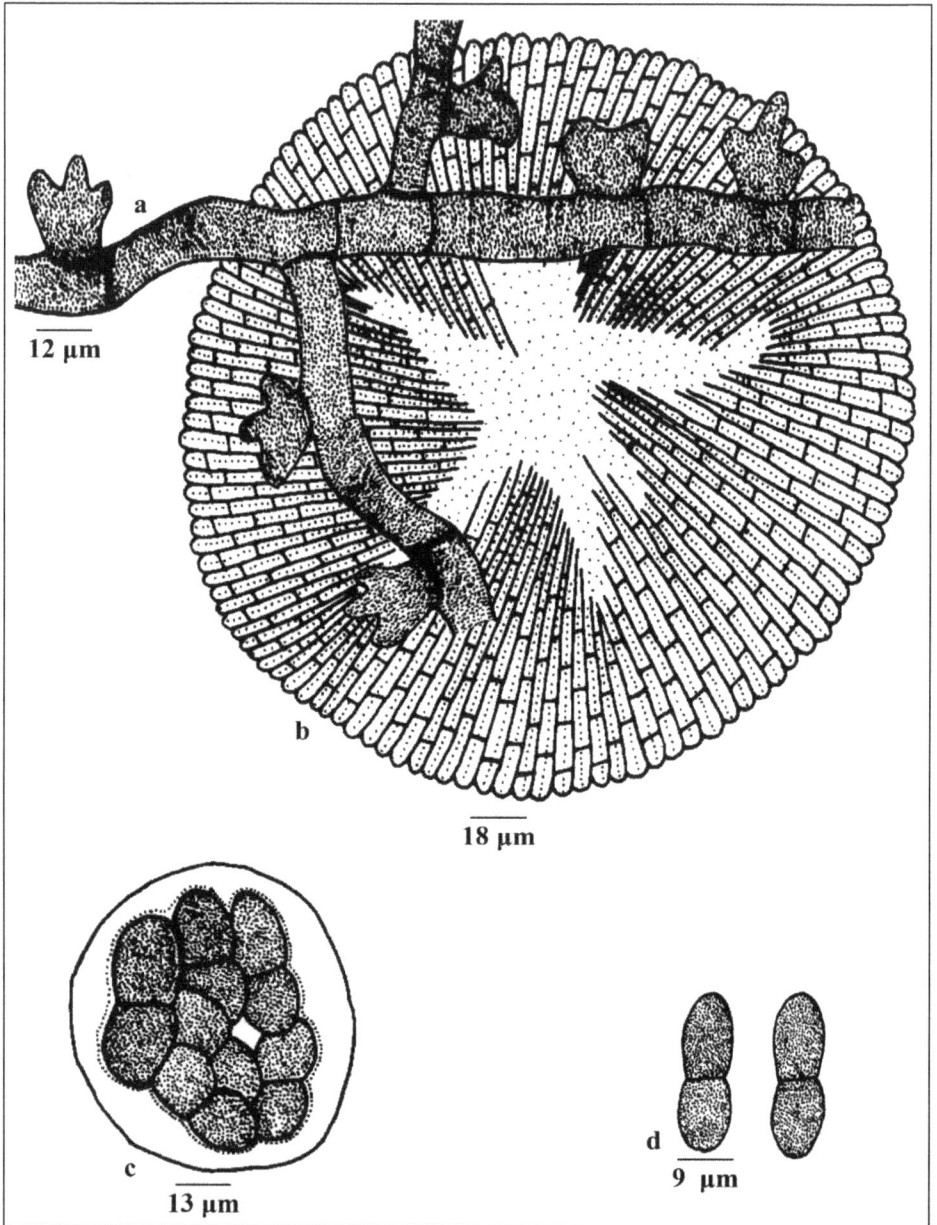

Figure 2.2: *Asterina chukrasiae* Hosag. *et al.*

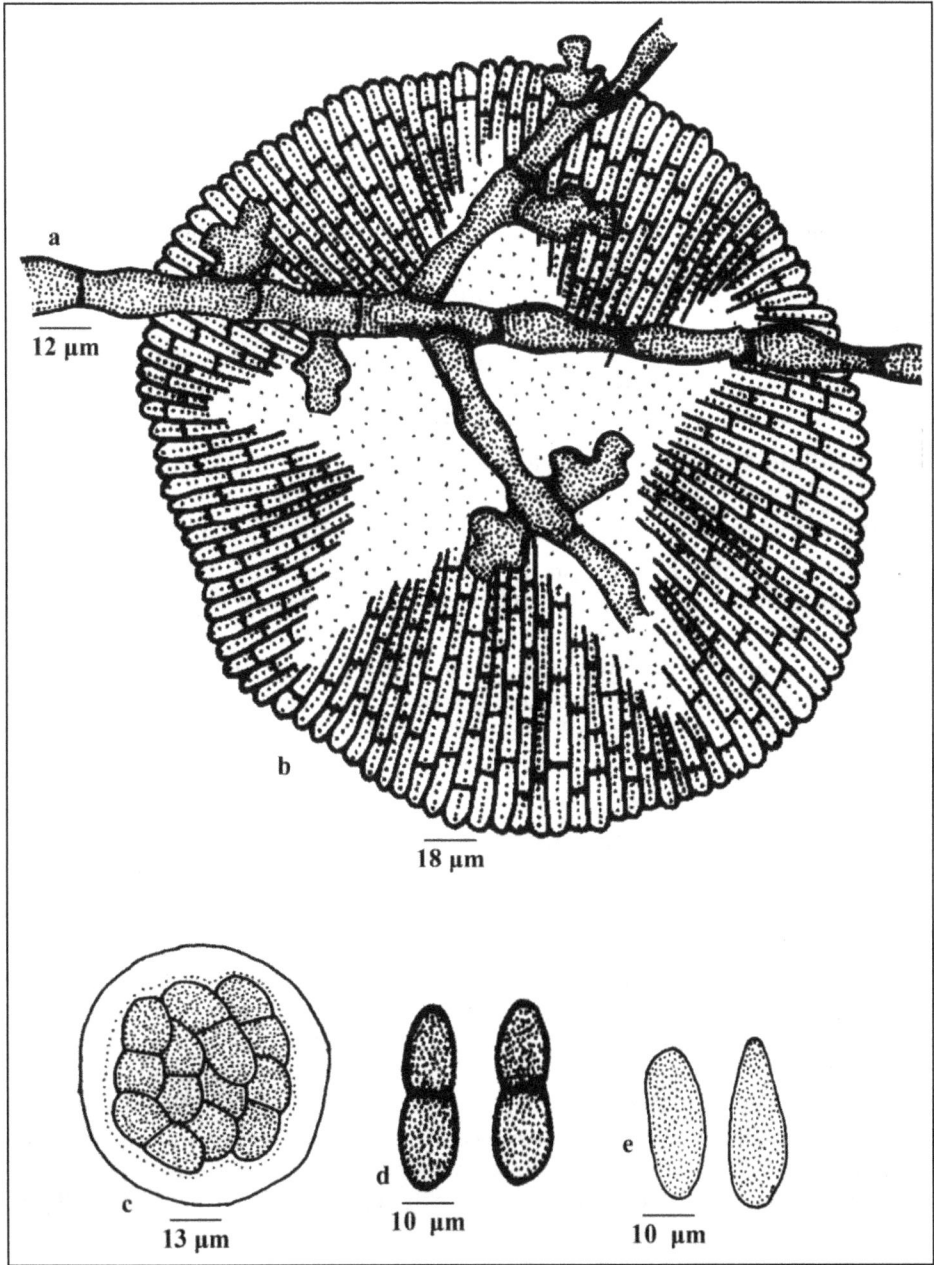

Figure 2.3: *Asterina cipadessae* **Yates.**

orbicular, up to 143 μm in diameter, dehisce stellately at the center, margin crenate; asci globose, octosporous up to 30 μm in diameter; ascospores conglobate, uniseptate, constricted at the septum, brown, 11-26 × 4-11 μm, wall smooth. Pycnothyriospores ovate, pyriform, 11-20 × 4-11 μm, wall smooth.

Distribution: Observed on the leaves of *Cipadessa baccifera* (Roth.) Miq. (Meliaceae).

Asterina cissi Hughes, Mycol. Pap. 48: 10, 1952. (Figure 2.4)

Colonies epiphyllous, subdense, up to 3 mm in diameter, confluent. Hyphae flexuous, branching irregular at acute angles, loosely reticulate, cells 15-22 x 4-7 μm. Appressoria alternate to unilateral, ovate to globose, entire, angular, slightly sublobate, 7-9 x 4-7 μm. Thyriothecia scattered, rarely connate, orbicular, up to 120 μm in diameter, stellately dehisced at center, margin crenate to fimbriate, fringed hyphae substraight; asci globose, octosporous, up to 33 μm in diameter; ascospores conglobate, brown, 1-septate, constricted at the septum, 18-20 x 9-11 μm, wall smooth.

Distribution: Observed on the leaves of *Cissus* sp. (*Vitis* sp.) (Vitaceae).

Asterina clausenicola Doidge, Trans. Royal Soc. South Africa 8: 263, 1920. (Figure 2.5)

Colonies epiphyllous, dense up to 3 mm in diameter, confluent. Hyphae substraight to flexuous, branching mostly opposite at acute angles, loosely reticulate, cells 15-26 × 4-7 μm. Appressoria alternate and about 30 per cent opposite, unicellular, uni to multilobate, conoid towards the apex, 7-11 × 4-7 μm. Thyriothecia scattered, orbicular, up to 132 μm in diameter, margin crenate, stellately dehisced at the center, asci globose, octosporous, up to 30 μm in diameter, ascospores brown, conglobate, oblong, 1-septate, strongly constricted at the septum, rounded at both ends, 17-22 × 7-11 μm, wall smooth to echinulate.

Distribution: Observed on the leaves of *Melicope lunu-ankenda* (Gaertn.) T. Hartley (*Euodia lunu-ankeda* (Gaetrn.) Merr. (Rutaceae).

Asterina combreti Sydow, Engl. Bot. Jahrb. 44: 264, 1920. (Figure 2.6)

Colonies epiphyllous, dense, crustose, up to 4 mm in diameter, confluent. Hyphae flexuous, branching irregular at acute to wide angles, loosely reticulate, cells 9-15 × 2-4 μm. Appressoria alternate, two celled, distantly placed, mostly perpendicular to the hyphae, 9-11 μm long; stalk cells cylindrical to cuneate, 4-7 μm long; head cells ovate, cylindric, entire to slightly lobate, 4-9 × 7-9 μm. Thyriothecia scattered, orbicular, up to 110 μm in diameter, stellately dehisced at the center, margin crenate to fimbrate, fringed hyphae flexuous, asci globose, octosporous, up to 30 μm in diameter; ascospores brown, conglobate, 1-septate, constricted at the septum, 11-15 × 7-9 μm, wall smooth.

Distribution: Observed on the leaves of *Calycopteris floribunda* (Roxb.) Poiret (Combretaceae).

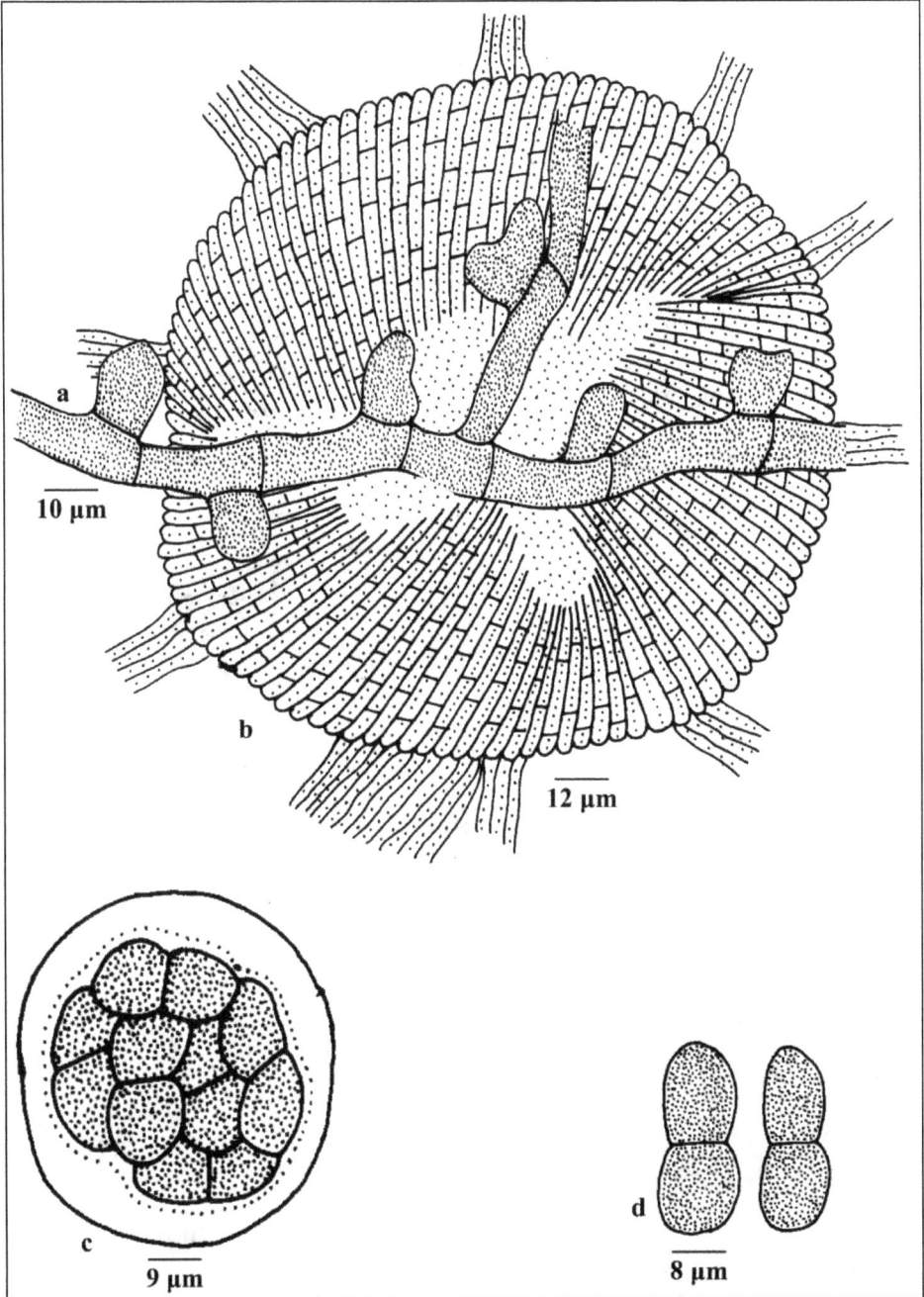

Figure 2.4: *Asterina cissi* Hughes.

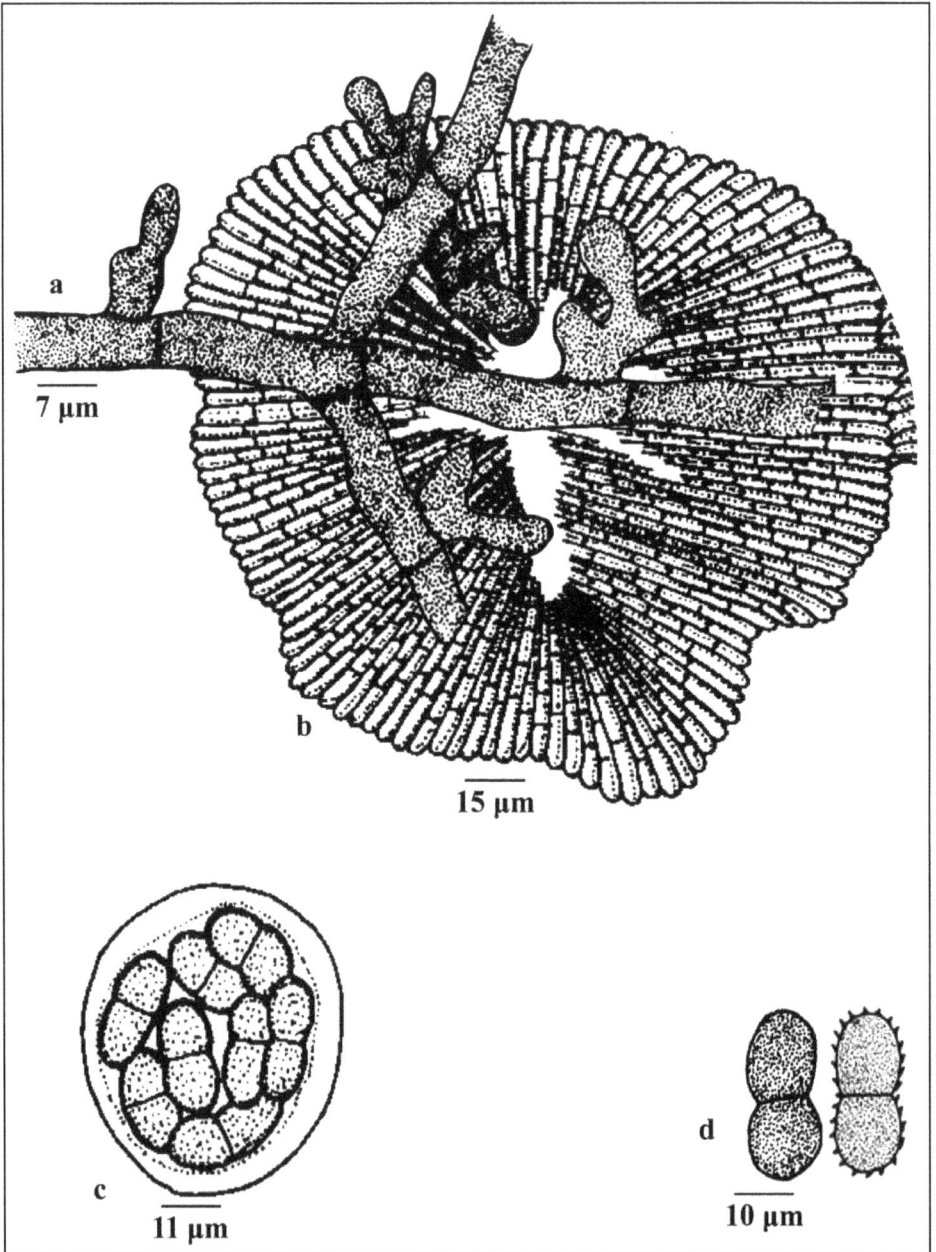

Figure 2.5: *Asterina clausenicola* **Doidge.**

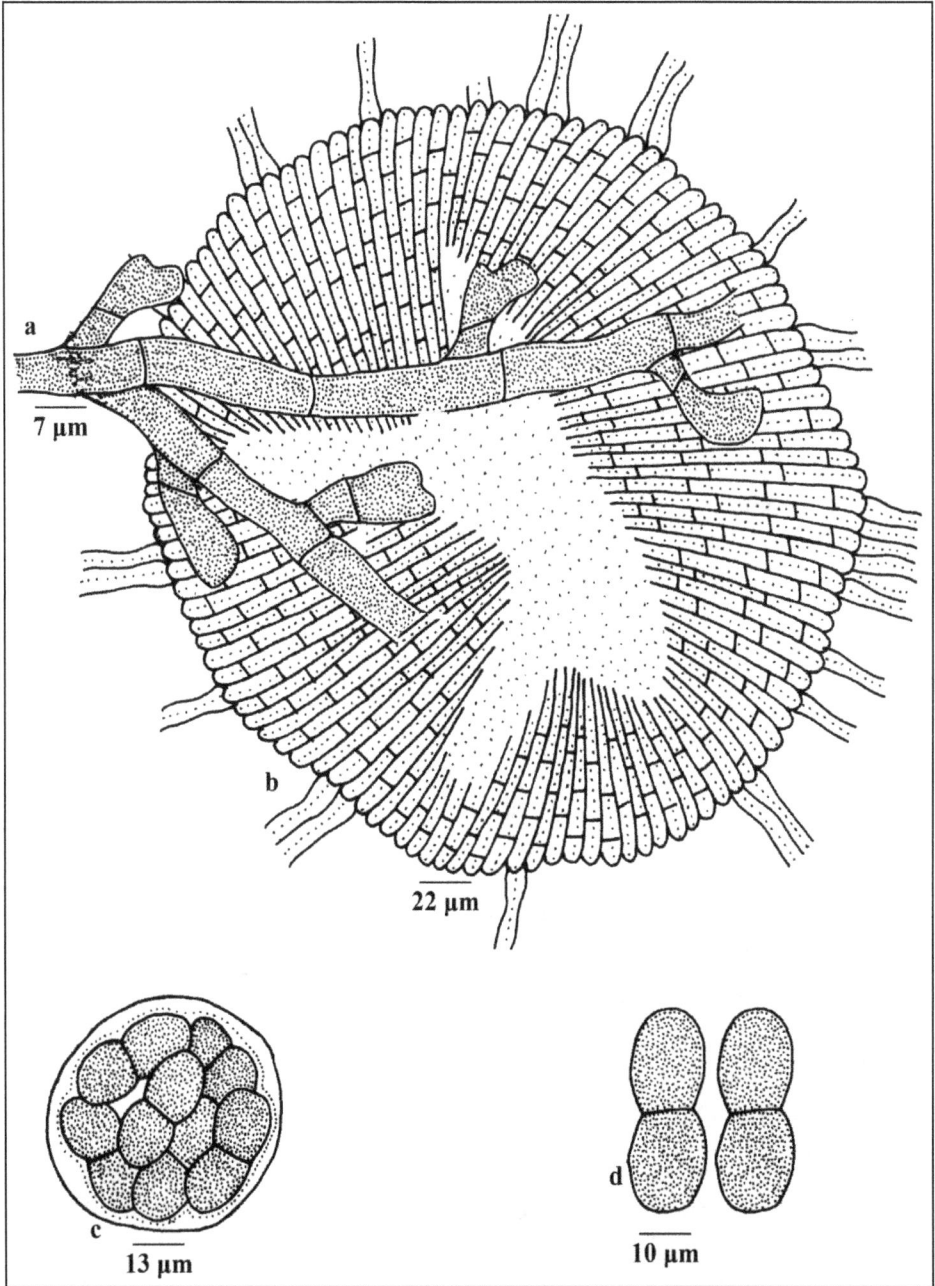

Figure 2.6: *Asterina combreti* **Sydow.**

Asterina congesta Cooke, Grevillea 7: 95, 1879. (Figure 2.7)

Colonies amphigenous, dense, up to 2 mm in diameter. Hyphae flexuous, branching mostly alternate at acute to wide angles, closely reticulate, cells 11-24 × 4-7 µm. Appressoria unicellular, unilateral, alternate, straight to slightly curved, oblong, cylindric, entire, sublobate, 9-15 × 4-7 µm. Thyriothecia scattered to grouped, orbicular, up to 154 µm in diameter, stellately dehisced at the center, margin crenate; asci many, globose, octosporous, up to 35 µm in diameter; ascospores oblong, conglobate, brown, uniseptate, constricted at the septum, 13-26 × 9-11 µm, wall smooth. Pycnothyria many, similar to thyriothecia; pycnothyriospores ovate, pyriform, brown, 9-13 × 4-9 µm, wall smooth.

Distribution: Observed on the leaves of *Santalum album* L. (Santalaceae).

Asterina dallasica Petrak, Sydowia 8: 14, 1954. (Figure 2.8)

Colonies epiphyllous, dense, up to 3 mm in diameter. Hyphae flexuous to crooked, branching irregular at acute to wide angles, loosely reticulate, cells 11-26 × 4-7 µm. Appressoria scattered, unicellular, alternate, unilateral, about 2 per cent opposite, antrorse to subantrorse, globose, mammiform, entire, 7-9 × 4-7 µm. Thyriothecia closely scattered, orbicular, up to 115 µm in diameter, stellately dehisced at the center, margin crenate, rarely fimbriate; asci globose, octosporous, bitunicate, up to 26 µm in diameter; ascospores brown, conglobate, uniseptate, constricted at the septum, 17-22 × 6-11 µm, wall smooth.

Distribution: Observed on the leaves of *Trema orientalis* (L.) Blume (Ulmaceae).

Aterina deightonii Sydow, Ann. Mycol. 36: 172, 1938. (Figure 2.9)

Colonies amphigenous, mostly epiphyllous, thin to subdense, up to 2 mm in diameter, rarely confluent. Hyphae flexuous, branching irregular at acute to wide angles, loosely reticulate, cells 13-20 × 4-7 µm. Appressoria unicellular, alternate, globose to ovate, entire, 7-11 × 7-9 µm. Thyriothecia scattered, often loosely grouped, orbicular, up to 140 µm in diameter, margin crenate to fimbriate, fringed hyphae flexuous, stellately dehisced at the center; asci few to many, globose, octosporous, up to 40 µm in diameter; ascospores brown, oblong, conglobate, uniseptate, constricted, 20-22 × 11-13 µm, wall minutely echinulate. Pycnothyria similar to thyriothecia, smaller; pycnothyriospores few, globose, pyriform, brown, 15-19 × 11-15 µm, wall smooth.

Distribution: Observed on the leaves of *Dendrophthoe* sp. (*Loranthus* sp.), (Loranthaceae).

Asterina diospyri Hosag. & C.K. Pradeep in Hosag., C.K. Biju, Abraham & C.K. Pradeep, J. Econ. Taxon. Bot. 25: 279, 2001 (Figure 2.10).

Colonies amphigenous, mostly hypophyllous, dense, up to 5 mm in diameter, confluent. Hyphae straight to substraight, branching opposite to irregular at acute to wide angles, closely reticulate, cells 13-24 × 4-7 µm. Appressoria alternate, about 20 per cent opposite, unicellular, cylindrical to ampulliform, subantrorse, entire, broadly

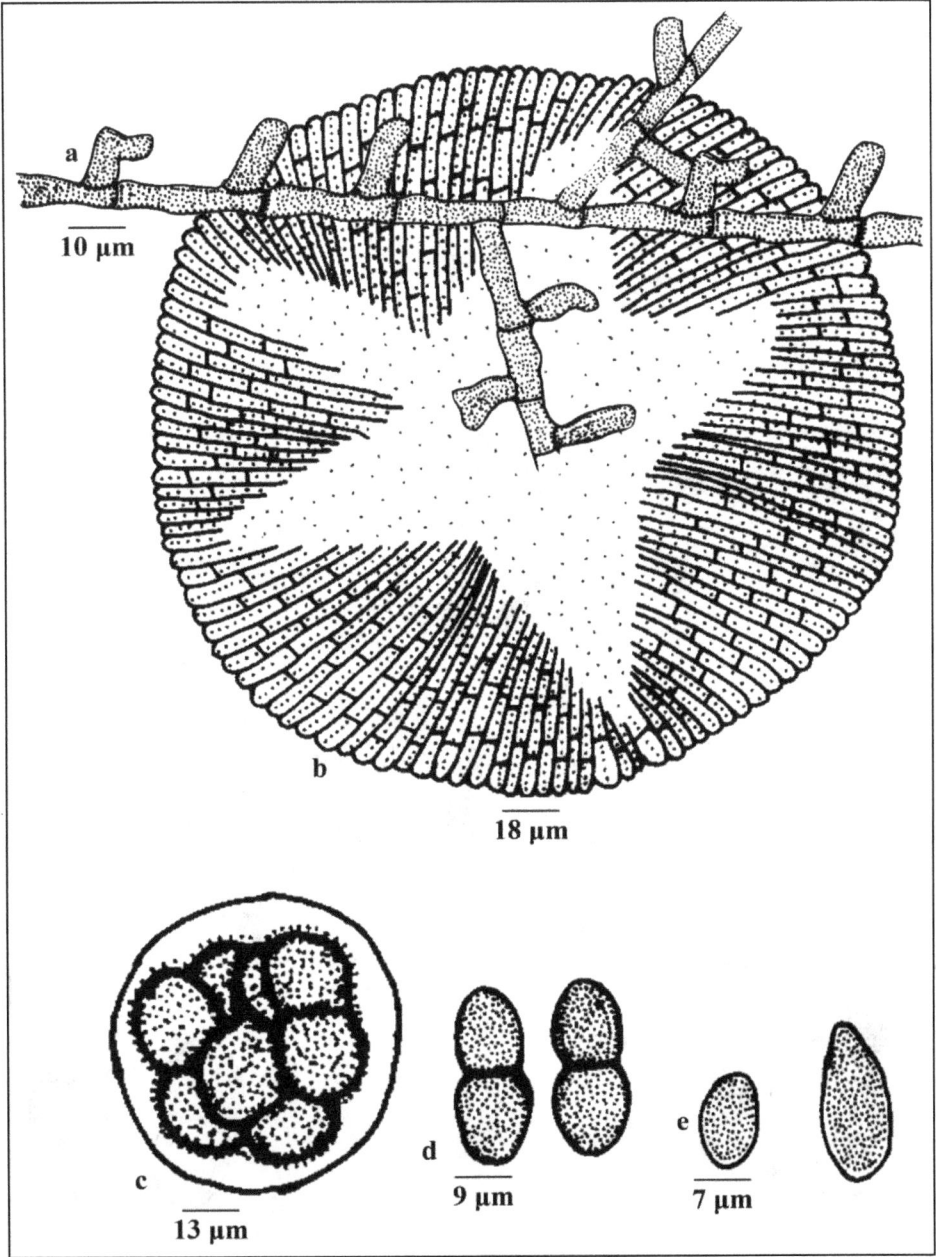

Figure 2.7: *Asterina congesta* **Cooke.**

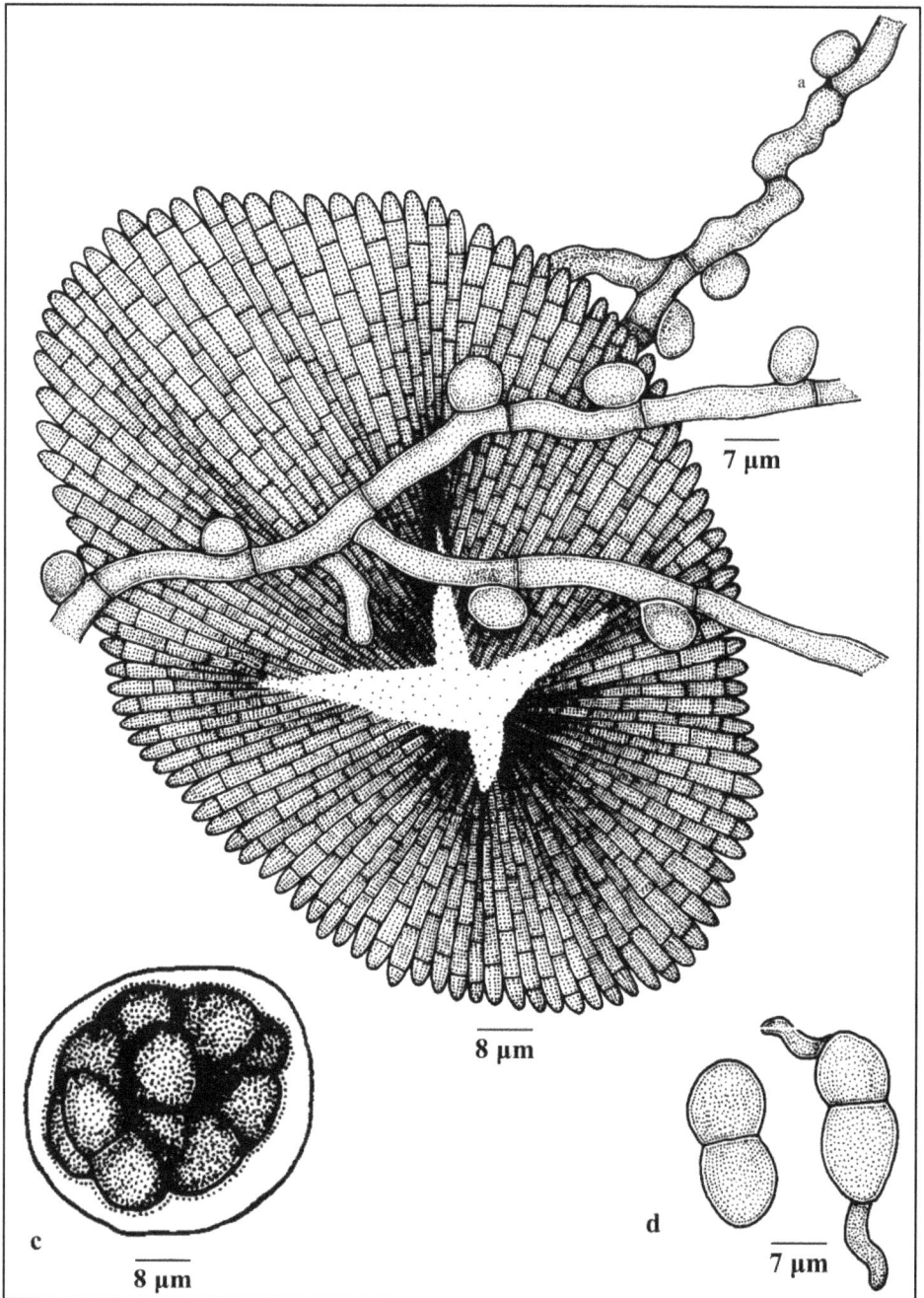

Figure 2.8: *Asterina dallasica* **Petrak.**

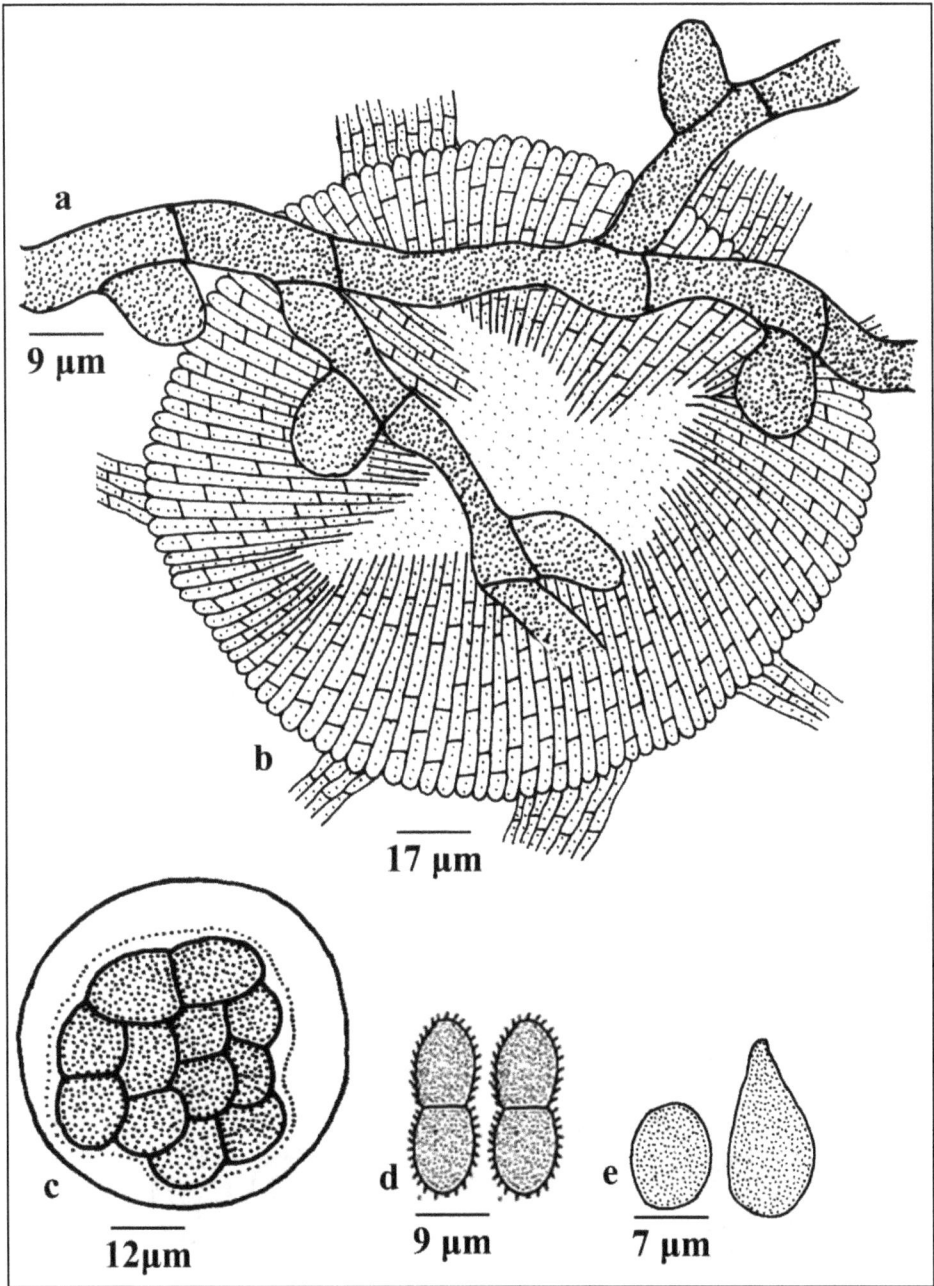

Figure 2.9: *Aterina deightonii* Sydow.

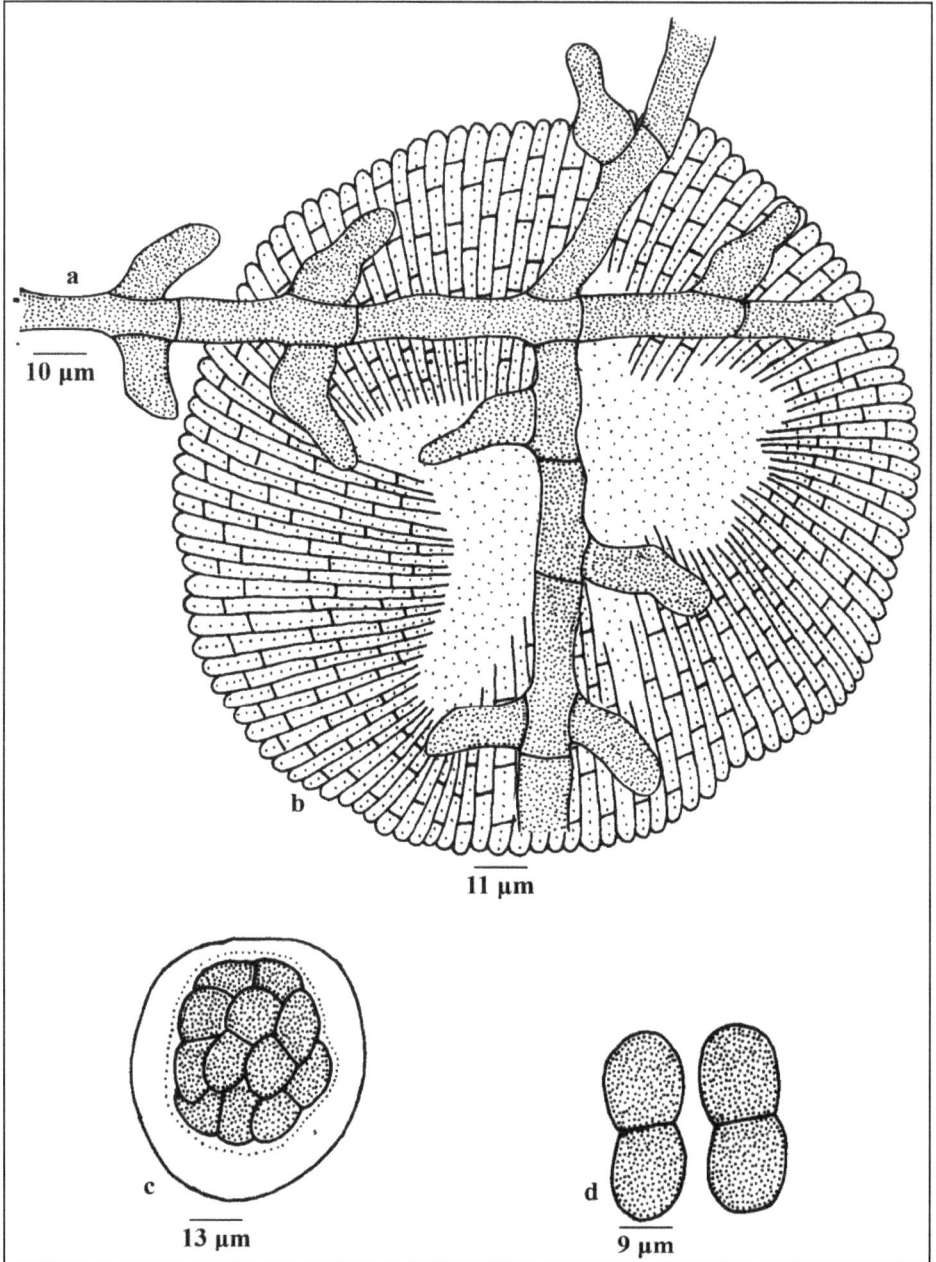

Figure 2.10: *Asterina diospyri* Hosag. & C.K. Biju.

rounded at the apex, 7-11× 4-7 μm. Thyriothecia scattered to grouped in the center of the colonies, orbicular, up to 60 μm in diameter, stellately dehisced and later widely opened at the center, margin crenate; asci few, globose, octosporous, 35-40 μm in diameter; ascospores oblong, conglobate, brown, uniseptate, constricted at the septum, 20-26 ×11-13 μm, wall smooth.

Distribution: Observed on the leaves of *Diospyros* sp. (Ebenaceae).

Asterina diplocarpa Cooke, Grevillea 10: 129, 1882 (Figure 2.11).

Colonies amphigenous, mostly epiphyllous, subdense, up to 3 mm in diameter, confluent. Hyphae crooked, branching irregular at acute to wide angles, loosely to closely reticulate, cells 9-22 × 4-7 μm. Appressoria scattered, alternate to unilateral, antrorse to retrorse, unicellular, broad based to slightly stipitate, ovate to globose, sublobate to deeply lobate, 7-11 × 4-7 μm. Thyriothecia scattered to grouped, orbicular, up to 110 μm in diameter, margin crenate, stellately dehisced at the center; asci globose, octosporous, up to 30 μm in diameter, ascospores brown, conglobate, uniseptate, constricted at the septum, 15-18 × 7-9 μm, wall slightly crenulated. Pycnothyria many, similar to thyriothecia, smaller; pycnothyriospores pyriform, brown, 13-20 × 7-11 μm.

Distribution: Observed on the leaves of *Sida cordata* (Burm.f.) Borssum (Malvaceae).

Asterina elaeocarpi Sydow var. *ovalis* Kar & Ghosh, Indian phytopathol. 39: 218, 1986 (Figure 2.12).

Colonies epiphyllous, subdense, up to 2 mm in diameter, confluent. Hyphae substraight to flexuous, branching opposite to alternate at acute to wide angles, loosely reticulate, cells 11-22 × 4-7 μm. Appressoria unicellular, alternate, opposite to subopposite, ovate, oblong, entire, 9-22 × 4-7 μm. Thyriothecia scattered, orbicular, up to 132 μm in diameter, stellately dehisced at the center, margin crenate to fimbriate, fringed hyphae flexuous; asci few to many, globose, octosporous, 30-40 μm in diameter, ascospores oblong, conglobate, brown, uniseptate, constricted at the septum, 22-26 × 9-13 μm, wall echinulate.

Distribution: Observed on the leaves of *Elaeocarpus tuberculatus* Roxb. (Elaeocarpaceae).

Asterina girardiniae Hosag., H.Biju & A.Manojkumar, Zoos' Print J. 21: 2304, 2006 (Figure 2.13).

Colonies epiphyllous, thin, up to 2 mm in diameter, rarely confluent. Hyphae flexuous, branching alternate to irregular at acute to wide angles, loosely reticulate, cells 9-22 x 4-7 μm. Appressoria 2-celled, alternate, about 10 per cent opposite, antrorse, subantrorse, straight to curved, 9-15 μm long; stalk cells cylindrical to cuneate, 2-4 μm long; head cells ovate, cylindric, entire to rarely sublobate, 7-11 x 4-7 μm. Thyriothecia scattered to connate, orbicular, up to 112 μm in diameter, crenate at the margin, stellately dehisced at the center; asci globose, octosporous, up to 26 μm in

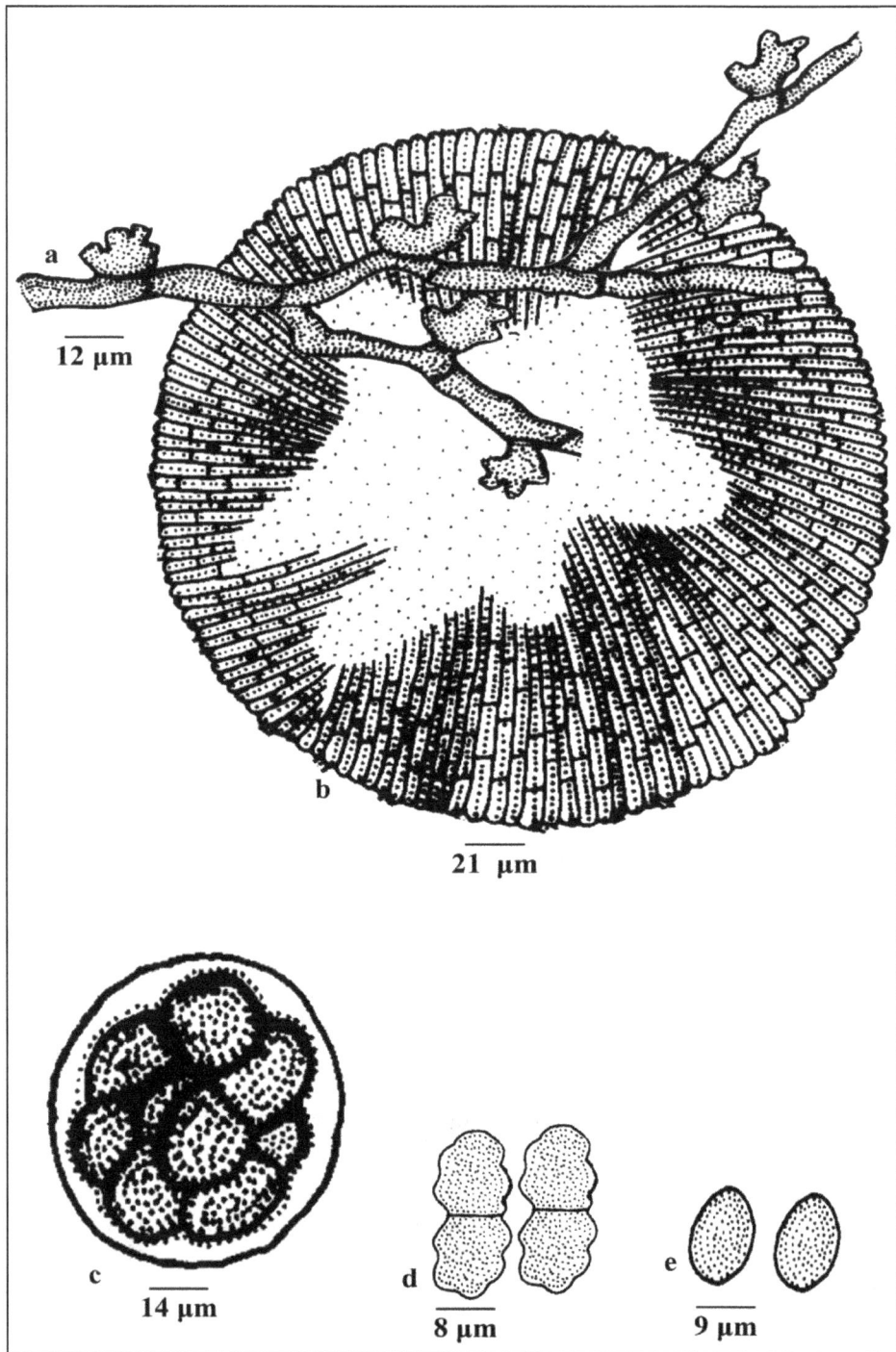

Figure 2.11: *Asterina diplocarpa* Cooke.

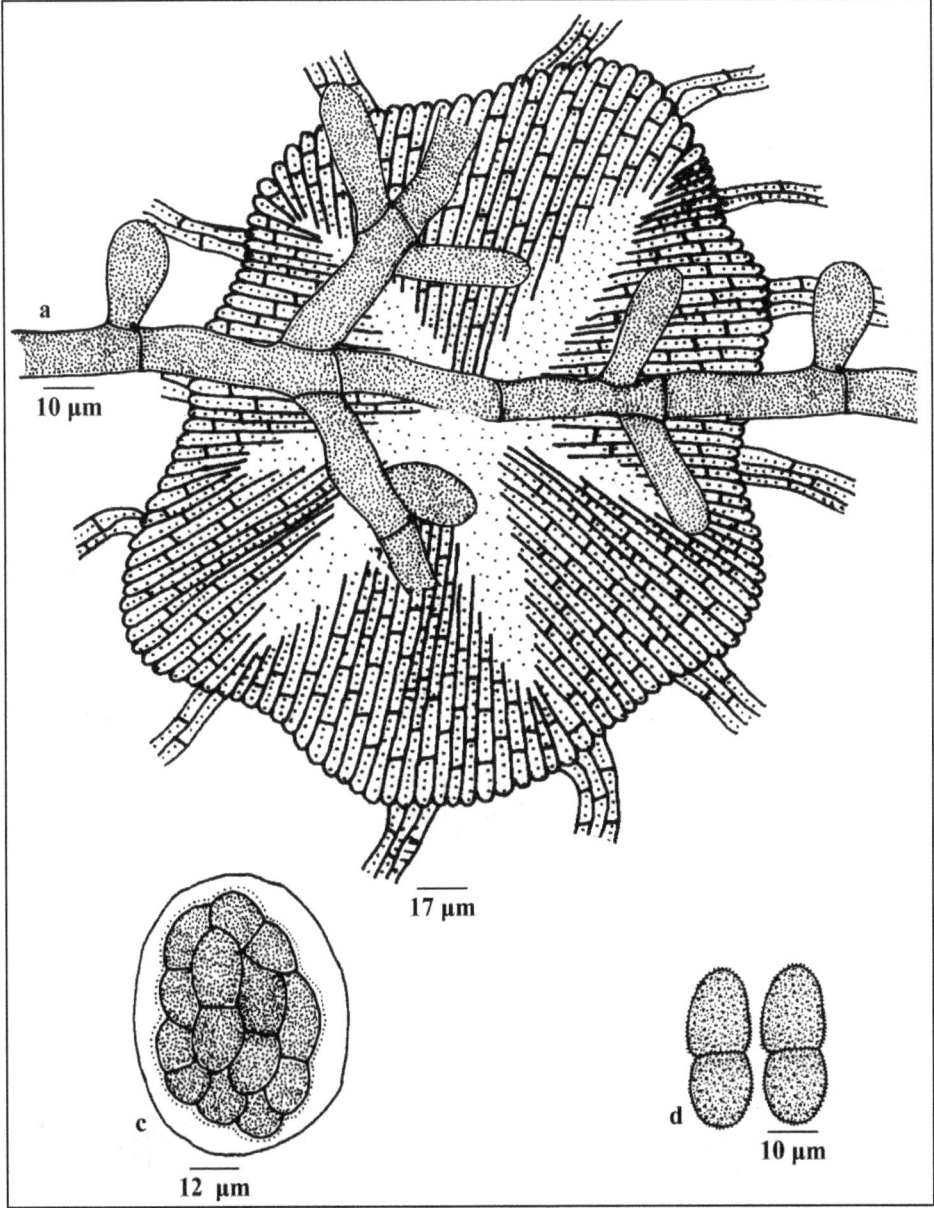

Figure 2.12: *Asterina elaeocarpi* Sydow var. *ovalis* Kar & Ghosh.

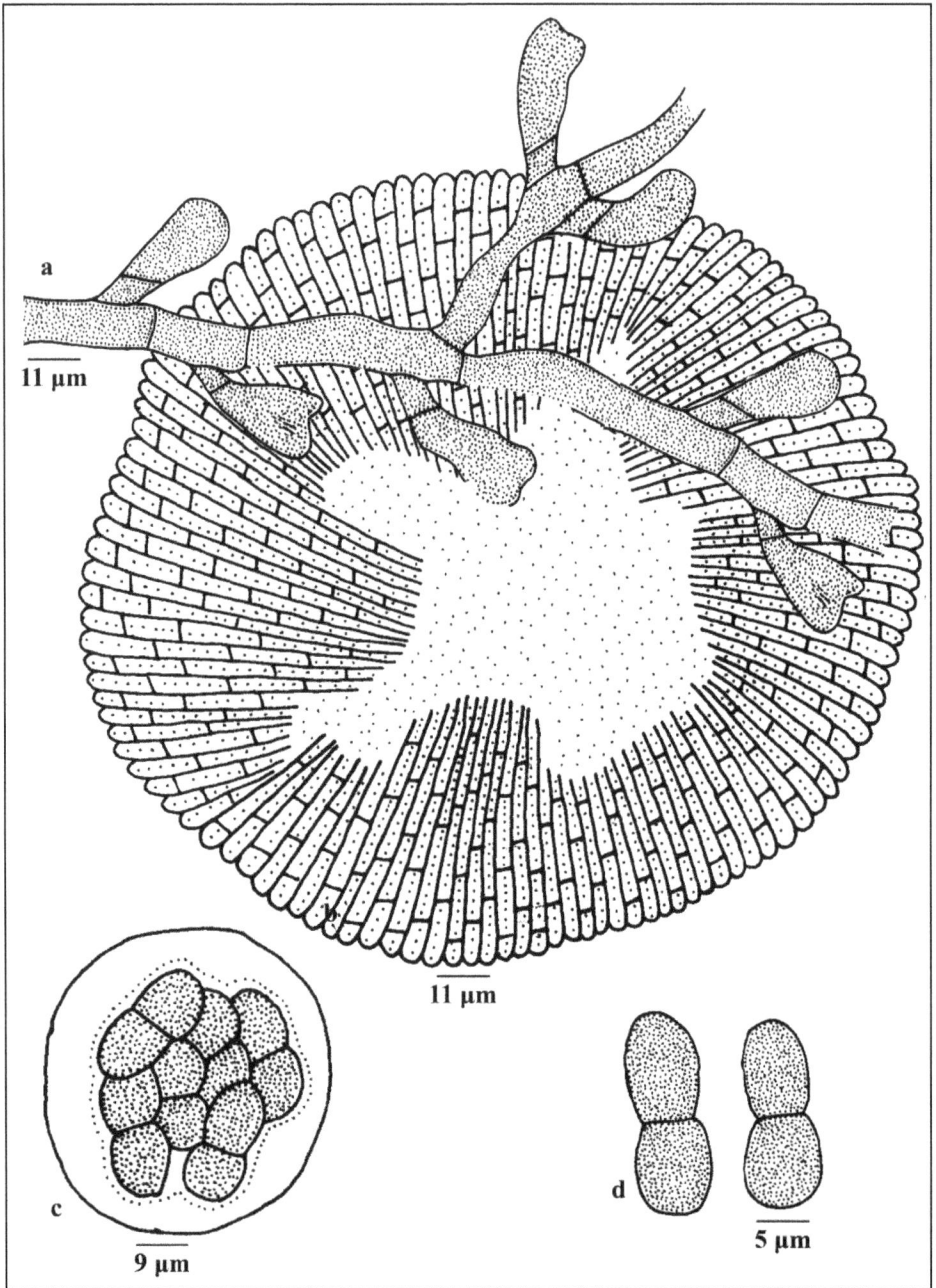

Figure 2.13: *Asterina girardiniae* **Hosag.** *et al.*

diameter; ascospores conglobate, uniseptate, constricted, brown, 13-15 x 7-9 μm, wall smooth.

Distribution: Observed on the leaves of *Girardinia diversifolia* (Link) Fries (Urticaceae).

Asterina helicteridis Ouyang & Hu in Yousheng, Song & Hu, Acta Mycol. Sinica 15: 88, 1996 (Figure 2.14).

Colonies epiphyllous, dense, up to 3 mm in diameter. Hyphae flexuous to crooked, branching irregular at acute to wide angles, loosely reticulate, cells 15-26 × 2-4 μm. Appressoria unicellular, alternate, sessile, ovate, lobate, hamate, 7-9 × 4-7 μm. Thyriothecia scattered, orbicular, up to 80 μm in diameter, stellately dehisced at the center, crenate at the margin; asci globose, octosporous, up to 25 μm in diameter; ascospores conglobate, brown, uniseptate, constricted at the septum, 13-20 × 7-9 μm, wall smooth. Pycnothyria many, similar but smaller than the thyriothecia; pycnothyriospores ovate, brown, 11-18 × 4-7 μm, wall smooth.

Distribution: Observed on the leaves of *Helicteres isora* L. (Sterculiaceae).

Asterina hibisci (Doidge) Hosag. in Hosag., C.K. Biju & Abraham, J. Econ. Taxon. Bot. 28: 175, 2004 (Figure 2.15).

Colonies mostly epiphyllous, thin to subdense, up to 5 mm in diameter, confluent and cover an entire upper surface of the leaves. Hyphae substraight to undulate, branching alternate to opposite at acute angles, loosely reticulate, cells 13-22 × 4-7 μm. Appressoria unicellular, alternate, ovate, globlose, entire to sublobate, 9-13 × 6-7 μm. Thyriothecia scattered, orbicular, up to100 μm in diameter, stellately dehisced at the center, margin crenate; asci many, globose, octosporous, up to 33 μm in diameter; ascospores brown, conglobate, uniseptate, constricted, 20-24×9-11 μm, wall verrucose.

Distribution: Observed on the leaves of *Hibiscus* sp. (Malvaceae).

Asterina hydnocarpi Hosag. & Abraham, Indian phytopathol. 51: 389, 1998 (Figure 2.16).

Colonies hypophyllous, subdense, up to 3 mm in diameter, confluent. Hyphae straight to substraight, branching mostly opposite at wide angles, loosely reticulate, cells 11-18 × 4-7 μm. Appressoria alternate, about 3 per cent opposite, 2-celled, straight, curved, 9-13 μm long; stalk cells cylindrical to cuneate, 4-7 μm long; head cells ovate, globose, hamate, straight to curved, entire, bifid to 3-4 times sublobate, 7-9 × 4-7 μm. Thyriothecia scattered, orbicular, up to 120 μm in diameter, stellately dehisced at the center, margin crenate; asci few, globose, octosporous, up to 40 μm in diameter; ascospores brown, conglobate, 1-septate, constricted at the septum 20-24 × 9-11 μm, wall tubercled.

Distribution: Observed on the leaves of *Hydnocarpus* sp. (Flacourtiaceae).

Asterina indica Sydow in Sydow & Butler, Ann. Mycol. 9: 390, 1911 (Figure 2.17).

Colonies epiphyllous, thin up to 2 mm in diameter. Hyphae straight to substraight, branching mostly opposite at acute to wide angles, loosely reticulate, cells 15-26 x 4-

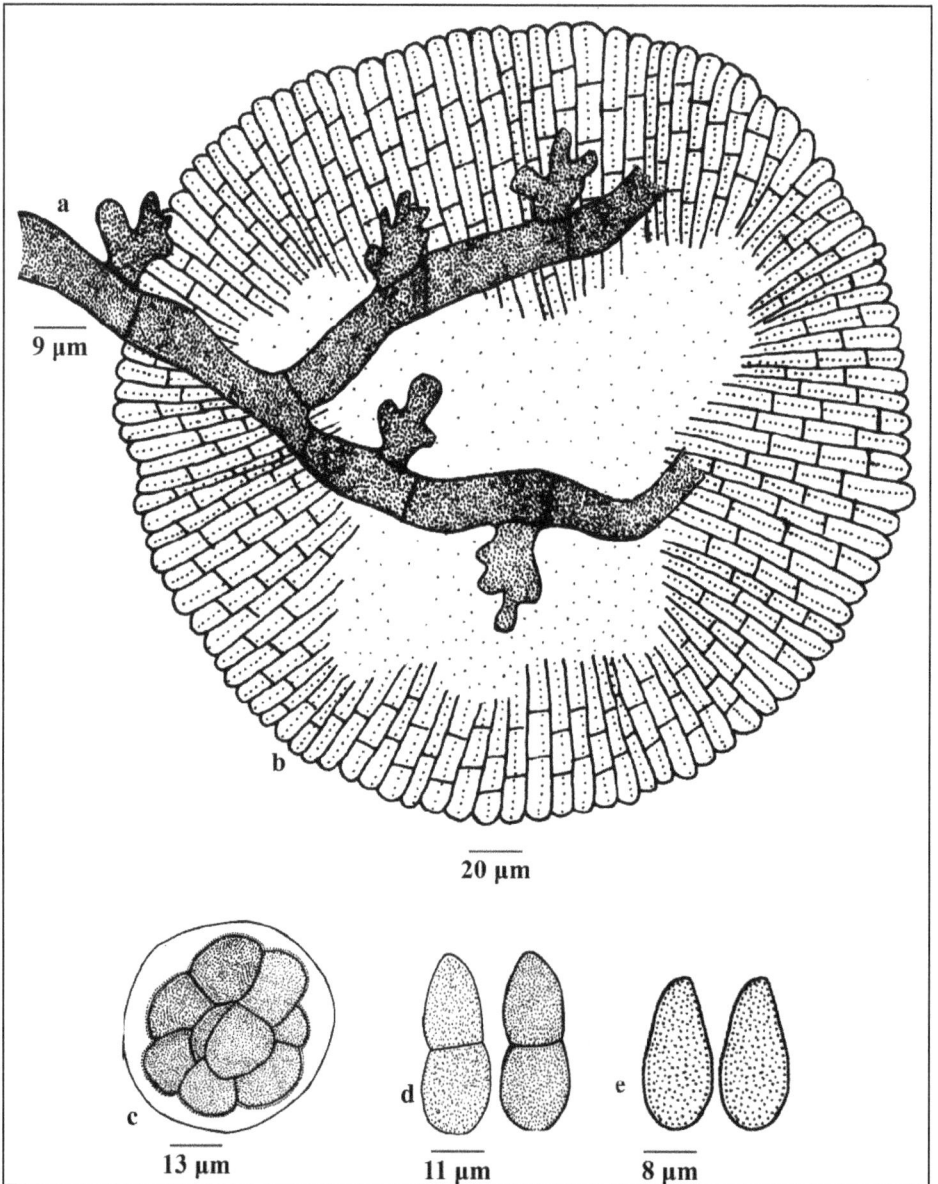

Figure 2.14: *Asterina helicteridis* **Ouyang & Hu.**

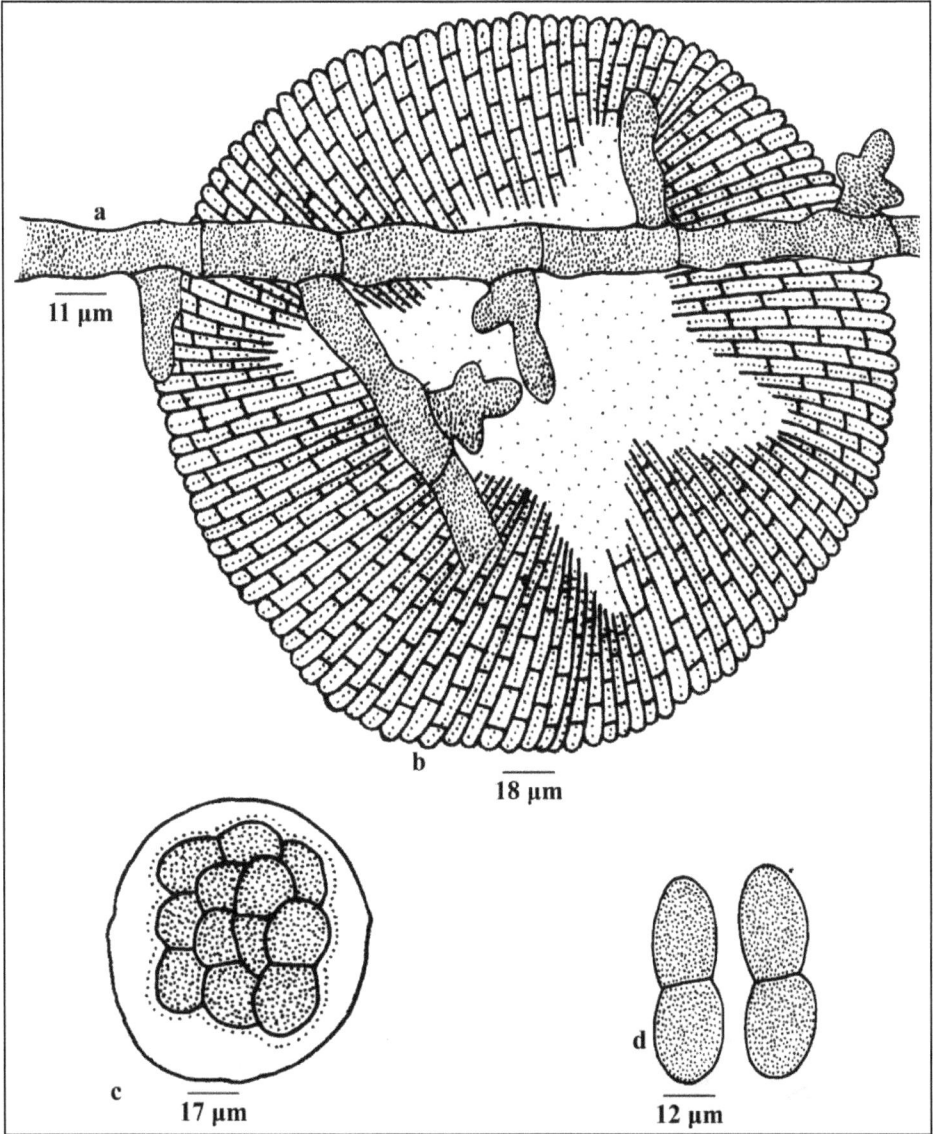

Figure 2.15: *Asterina hibisci* **(Doidge) Hosag.**

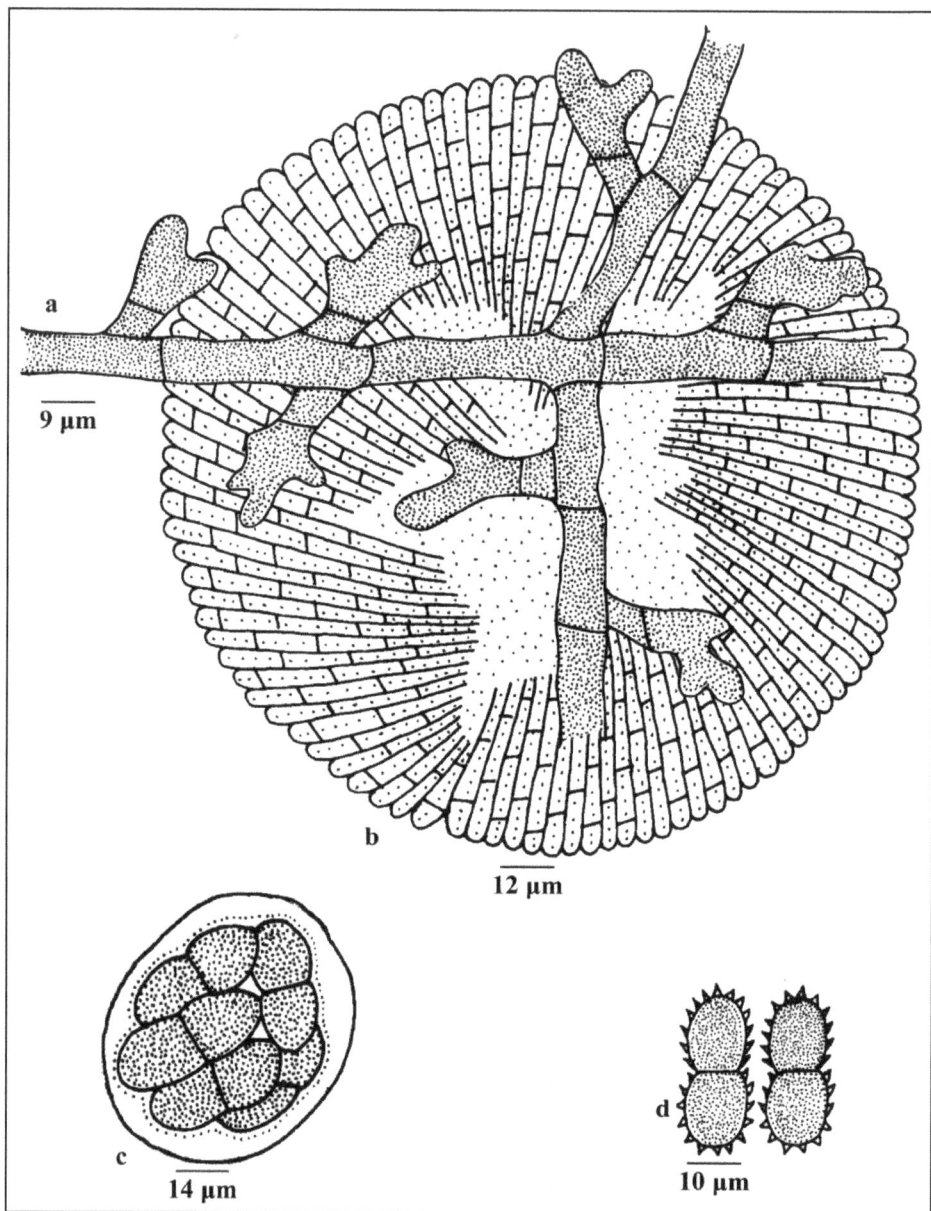

Figure 2.16: *Asterina hydnocarpi* **Hosag. & Abraham.**

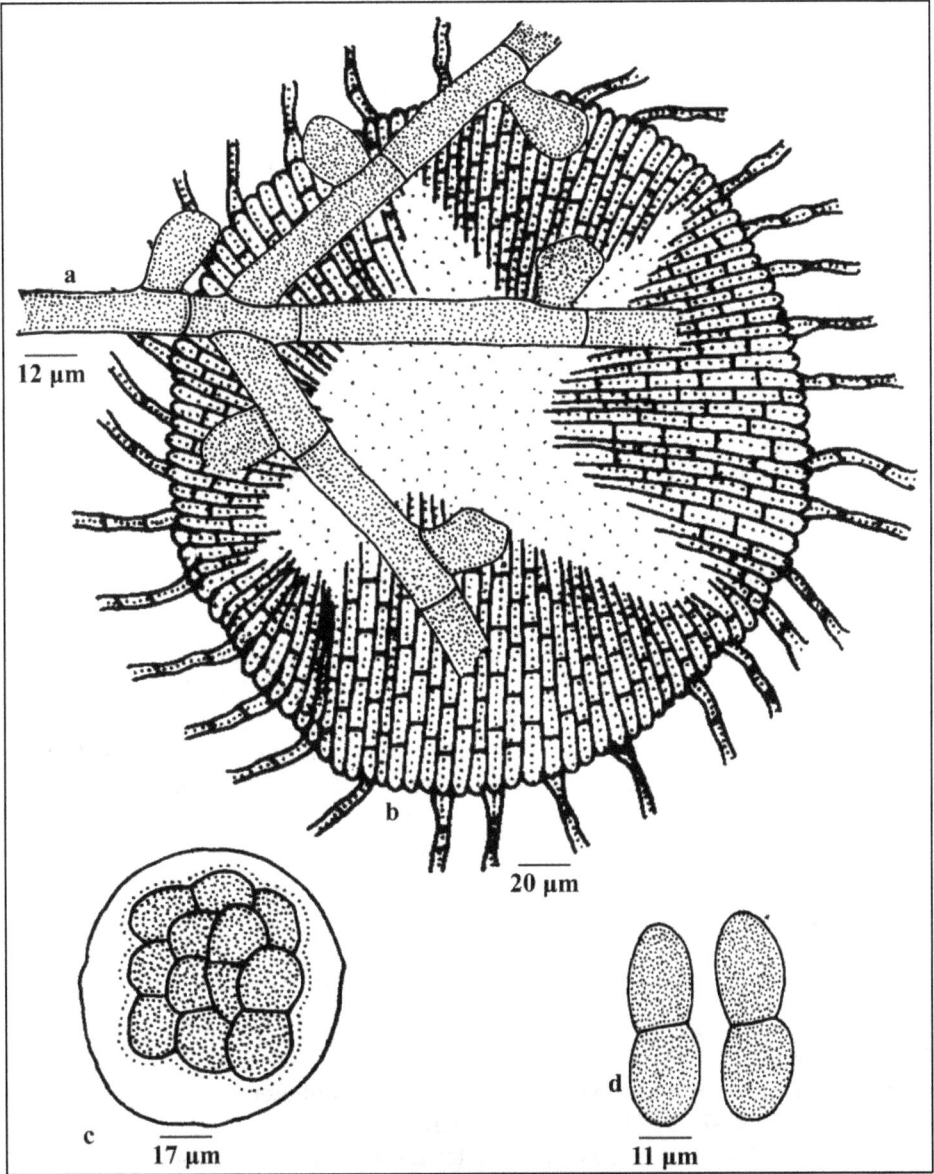

Figure 2.17: *Asterina indica* Sydow.

9 μm. Appressoria alternate, unilateral, unicellular, ovate, globose, angular, slightly sublobate, 9-11 x 7-11 μm. Thyriothecia scattered, orbicular, up to 300 μm in diameter, margin crenate to fimbriate, stellately dehisced at the center; asci few, globose, octosporous, up to 35 μm diameter; ascospores oblong, conglobate, brown, uniseptate, constricted at the septum, 15-24 x 7-11 μm wall smooth.

Distribution: Observed on the leaves of *Symplocos cochinchinensis* subsp. *laurina* (Rets) Nooteb. (Symplocaceae).

Asterina jambolana Kar & Maity, Trans. Brit. Mycol. Soc. 54: 438, 1970 (Figure 2.18).

Colonies amphigenous, mostly epiphyllous, dense, velvety, up to 3 mm in diameter. Hyphae substraight to flexuous, branching irregular at acute angles, loosely to closely reticulate, cells 22-33 × 4-9 μm. Appressoria two celled, alternate, antrorse, subantrorse to spreading, appressed to the hyphae, often curved, entire, 13-20 μm long; stalk cells cylindrical to cuneate, 4-9 μm long; head cells ovate, globose, straight to slightly curved, entire, 7-11 × 7-9 μm. Thyriothecia scattered, orbicular, up to 209 μm in diameter, stellately dehisced at the center, margin crenate to fimbriate; asci globose, octosporous, up to 45 μm in diameter; ascospores oblong, conglobate, brown, uniseptate, constricted at the septum, 24-35 × 11-15 μm, wall smooth.

Distribution: Observed on the leaves of *Syzygium* sp. (Myrtaceae).

Asterina lanneae Hosag. & Manoj in Hosag., Zoos' Print J. 18: 1037, 2003 (Figure 2.19).

Colonies epiphyllous, dense, crustose, up to 3 mm in diameter, confluent. Hyphae flexuous, branching mostly opposite at acute angles, loosely to closely reticulate, cells 15-19 × 4-7 μm. Appressoria unicellular, unilateral, opposite, ovate, conoid, entire to variously lobed, 9-11 × 4-7μm. Thyriothecia scattered to grouped, often connate, orbicular, up to 190 μm in diameter, margin fimbriate, fringed hyphae few, flexuous, stellately dehisced at the center; asci globose, octosporous, up to 35μm in diameter; ascospores conglobate, brown, uniseptate, 19-22 × 9-11μm, wall punctate.

Distribution: Observed on the leaves of *Lannea coromandelica* (Houtt.) Merr. (Anacardiaceae).

Asterina lawsoniae Henn. & Nyn., Monsumia 1:159, 1899 (Figure 2.20).

Colonies amphigenous, dense, up to 3 mm in diameter, confluent, Hyphae flexuous, branching irregular at acute to wide angles, loosely to closely reticulate, cells 9-7 × 4-7 μm. Appressoria unicellular, alternate, unilateral, unicellular, sessile, ovate, lobate, 4-9 × 4-7 μm. Thyriothecia scattered, orbicular, up to 120 μm in diameter, margin crenate, stellately dehisced at the center; asci few to many, octosporous, globose, up to 30 μm in diameter; ascospores, conglobate, brown, uniseptate, constricted at the septum, 20-24 × 9-11 μm, wall smooth. Pycnothyria many, similar but smaller than the thyriothecia; pycnothyriospores pyriform, brown, 11-15 × 4-7 μm, wall smooth.

Distribution: Observed on the leaves of *Lawsonia inermis* (Lythraceae).

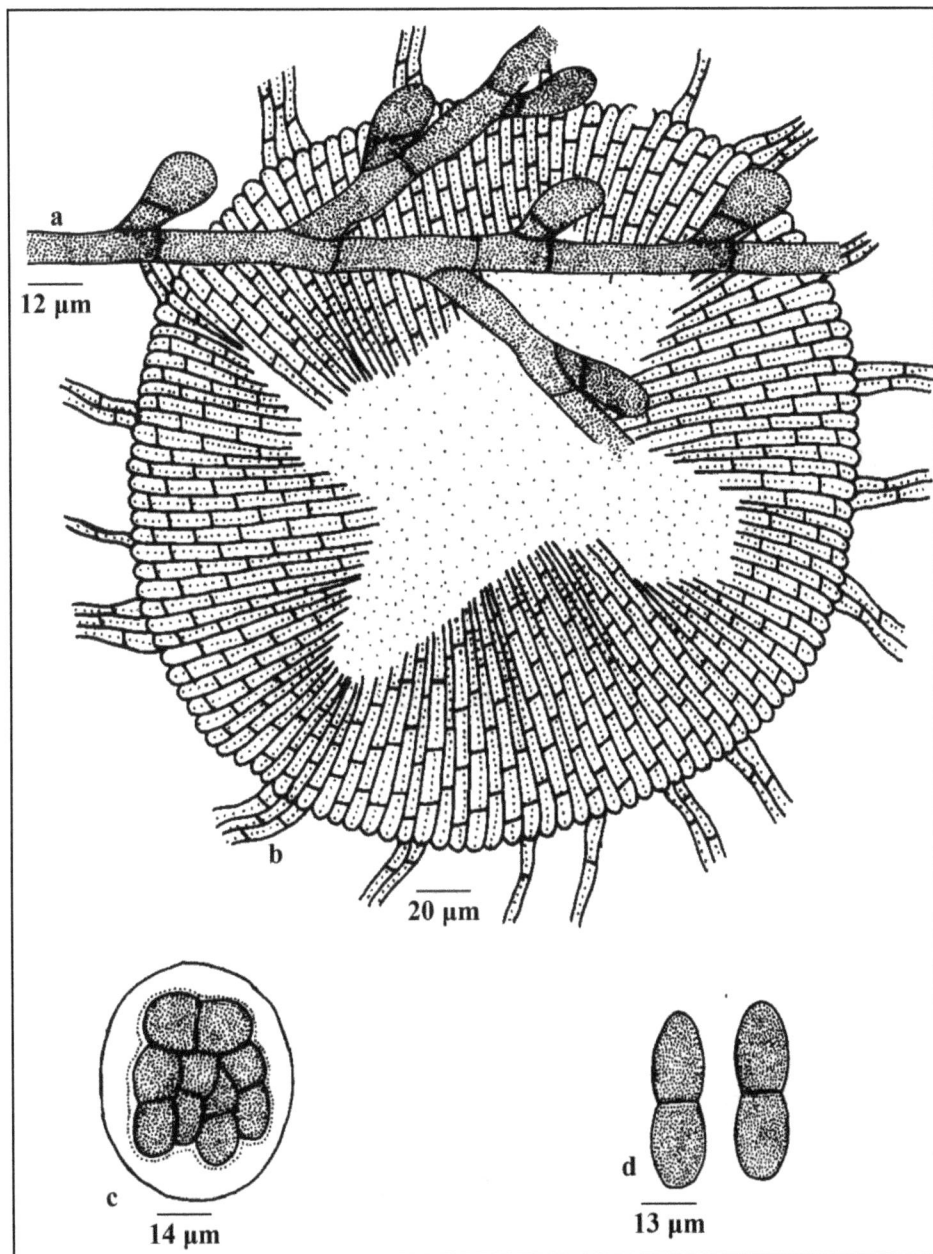

Figure 2.18: *Asterina jambolana* **Kar & Maity.**

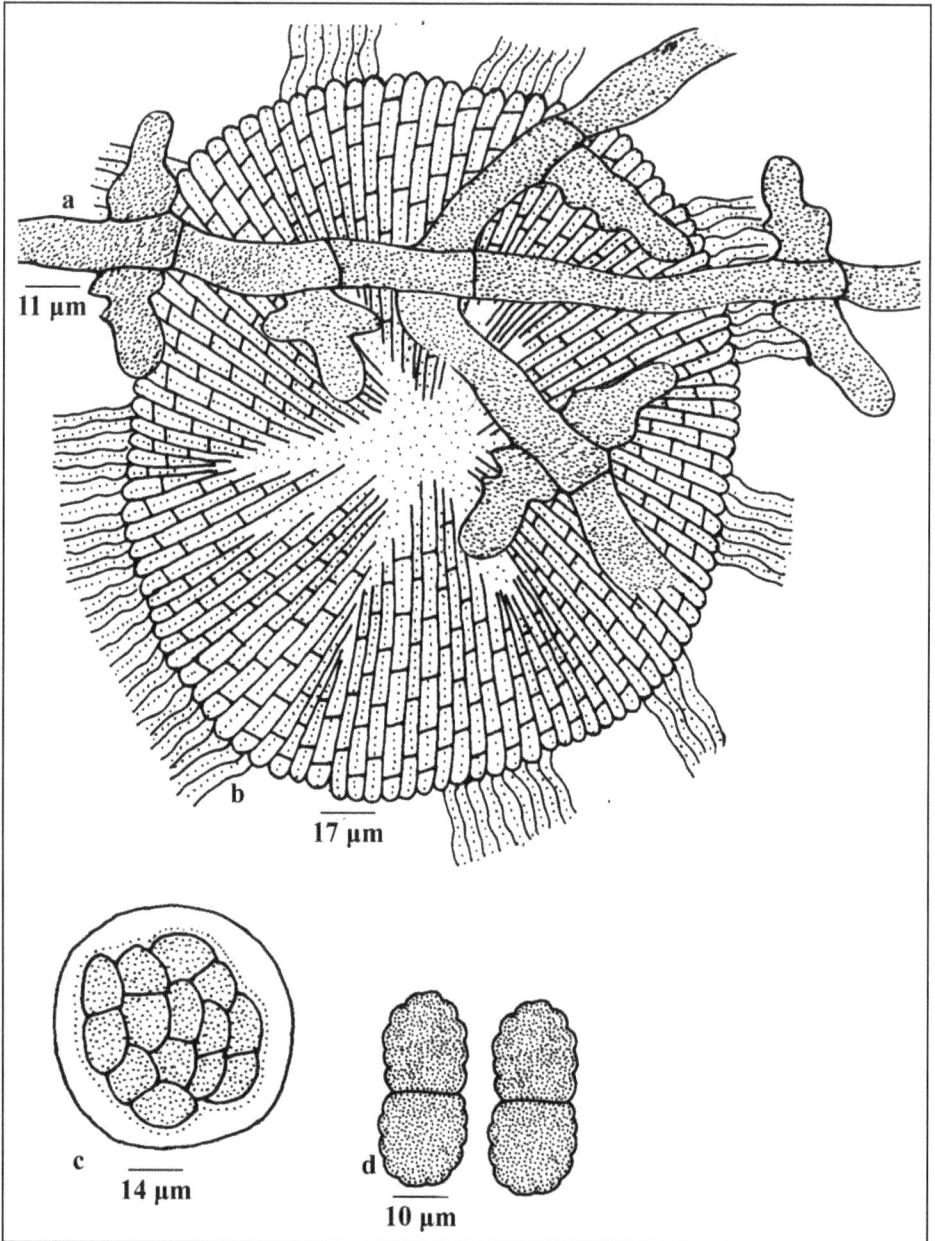

Figure 2.19: *Asterina lanneae* **Hosag. & Manoj.**

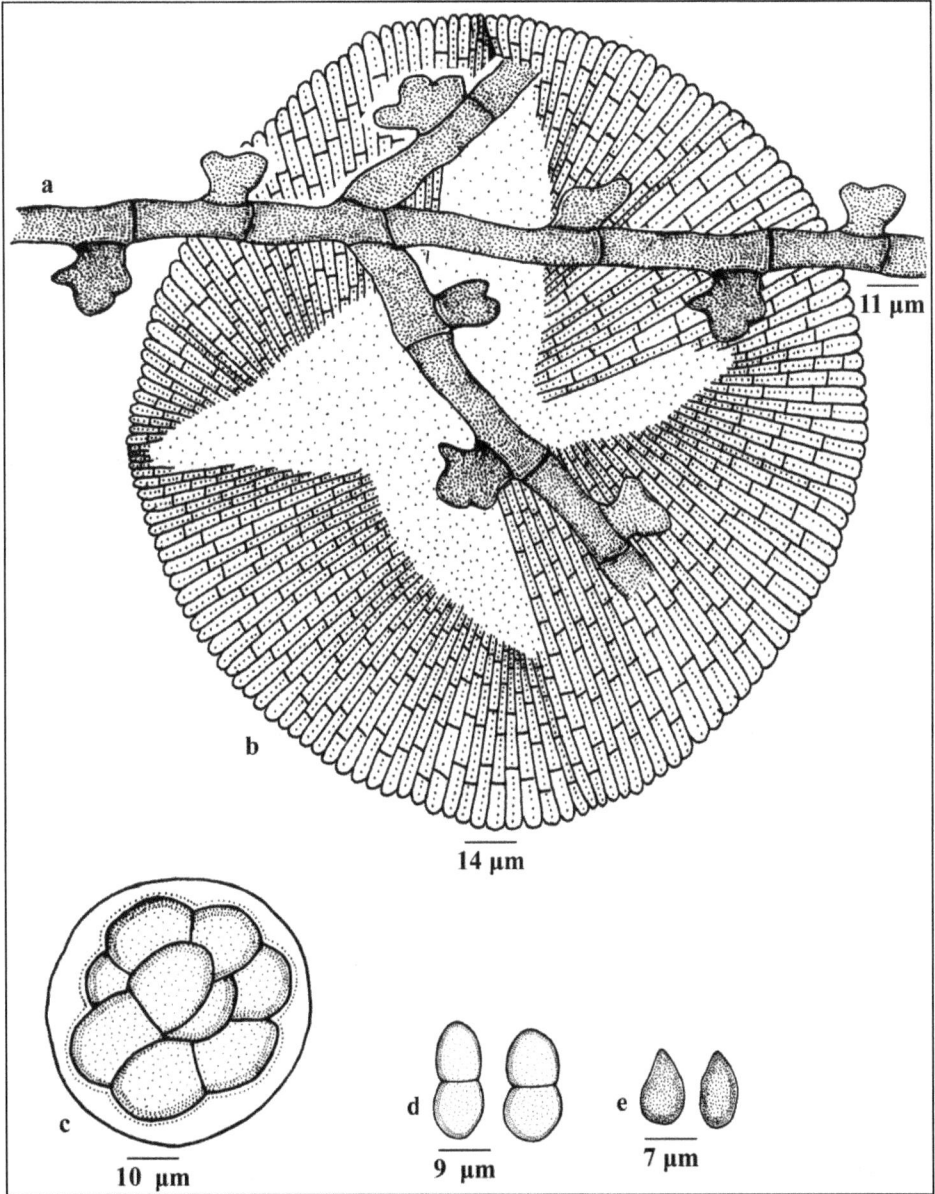

Figure 2.20: *Asterina lawsoniae* **Henn. & Nyn.**

Asterina lobulifera Sydow, Phillippine J. Sci. 9: 181, 1914 (Figure 2.21).

Colonies epiphyllous, thin, subdense, up to 3 mm in diameter, rarely confluent. Hyphae flexuous, branching irregular at acute angles, loosely retriculate, cells 15-24 × 4-7 µm. Appressoria two celled, alternate, antrorse, subantrorse, 9-15 µm long; stalk cells cylindrical to cuneate, 4-7 µm long; head cells ovate, globose, entire, sublobate to lobate, 7-9 × 7-11 µm. Thyriothecia scattered to loosely grouped, often connate, orbicular, up to 150 µm in diameter, stellately dehisced at the center, margin crenate to fimbriate, fringed hyphae straight to flexuous; asci globose, octosporous, up to 30 µm in diameter; ascospores conglobate, brown, 1-septate, strongly constricted at the septum, 18-22 × 11-13 µm, wall smooth. Pycnothyria smaller and similar to thyriothecia; pycnothyriospores brown, pyriform, 15-18 × 9-11 µm, wall smooth.

Distribution: Observed on the leaves of *Glochidion* sp. (Euphorbiaceae).

Asterina lobulifera Sydow var. *indica* Hosag. & Chandra., Indian J. Sci. Techn. 2(6): 15, 2009 (Figure 2.22).

Colonies amphigenous, dense, up to 2 mm in diameter. Hyphae flexuous to crooked, branching opposite to irregular at acute to wide angles, loosely to closely reticulate, cells 15-26 × 4-7 µm. Appressoria 2-celled, alternate, opposite (5-10 per cent), subantrorse, straight to curved, 11-15 µm long; stalk cells cylindrical to cuneate, 4-7 µm long; head cells ovate, lobate, 7-11 × 4-7 µm. Thyriothecia scattered to grouped, orbicular, up to 121 µm in diam., margin crenate to fimbriate, stellately dehisced at the center; ascospores oblong, conglobate, uniseptate, constricted at the septum, 13-22 × 7-9 µm, wall smooth. Pycnothyria smaller, similar to thyriothecia; pycnothyriospores ovate, pyriform, brown, 11-22 × 7-11 µm, wall smooth.

Distribution: Observed on the leaves of *Glochidion* sp. (Euphorbiaceae).

Asterina malloticola Hosag., Kamar. & Rajkumar, Indian phytopathol. 56: 99, 2003 (Figure 2.23).

Colonies epiphyllous, subdense up to 3 mm in diameter, confluent. Hyphae straight to flexuous, branching alternate, irregular at acute angles, loosely to closely reticulate, cells 13-20 × 4-7 µm. Appressoria alternate, about 1 per cent opposite, two celled, subantrorse, 9-13 µm long; stalk cells cylindrical to cuneate, 2-4 µm long; head cells ovate, globose, irregular lobate, 7-9 × 4-9 µm. Thyriothecia scattered, orbicular, up to 120 µm in diameter, stellately dehisced at the center, margin crenate, fringed hyphae very small; asci globose, octosporous, up to 30 µm in diameter; ascospores, conglobate, brown, uniseptate, constricted at the septum, 19-22 × 11-13 µm, wall minutely echinulate.

Distribution: Observed on the leaves of *Mallotus philippensis* Muel-Arg. (Euphorbiaceae).

Asterina melicopecola Hosag. & Abraham, Indian Phytopathol. 50: 216, 1997 (Figure 2.24).

Clolonies amphigenous, mostly epiphyllous, dense, up to 1 mm in diameter, confluent. Hyphae flexuous, branching opposite to irregular, at acute angles, loosely

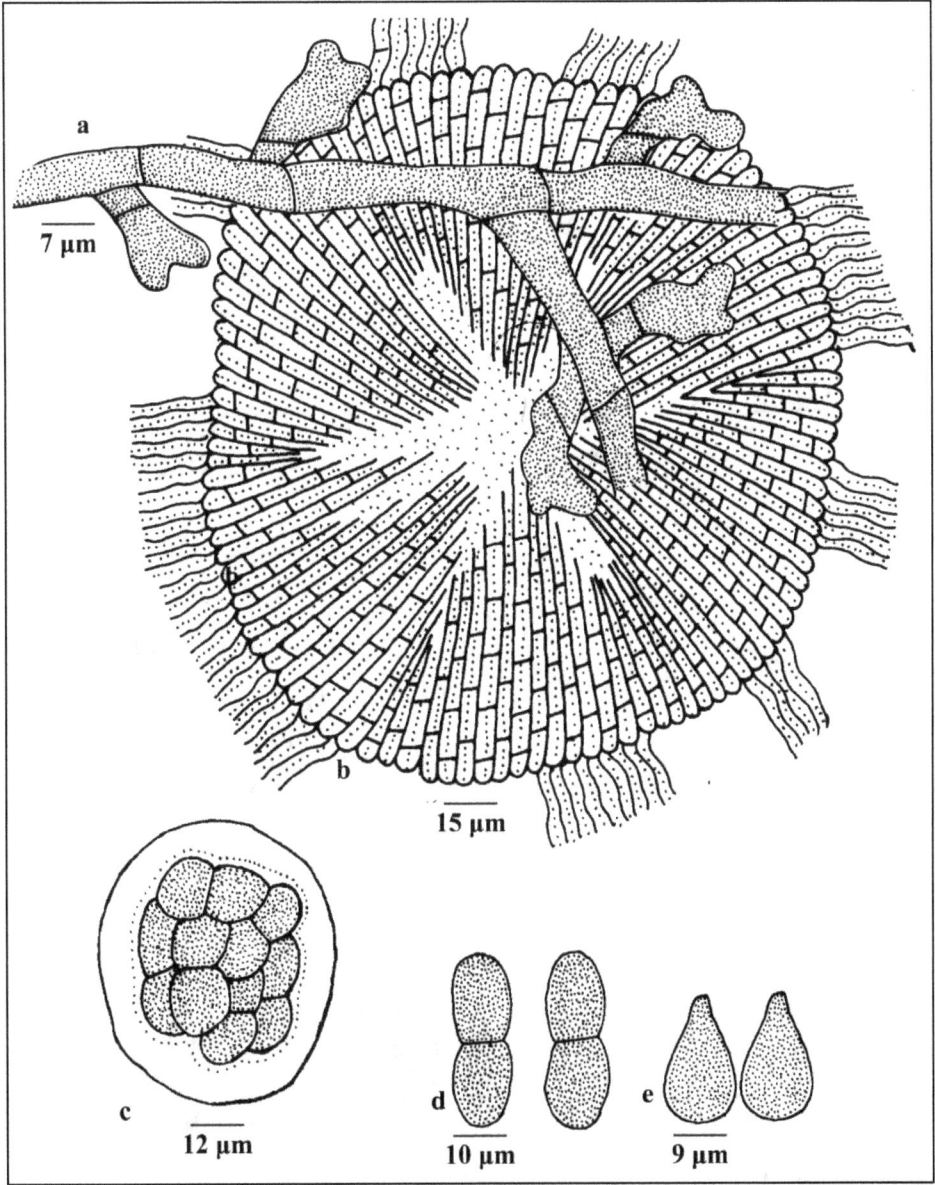

Figure 2.21: *Asterina lobulifera* **Sydow.**

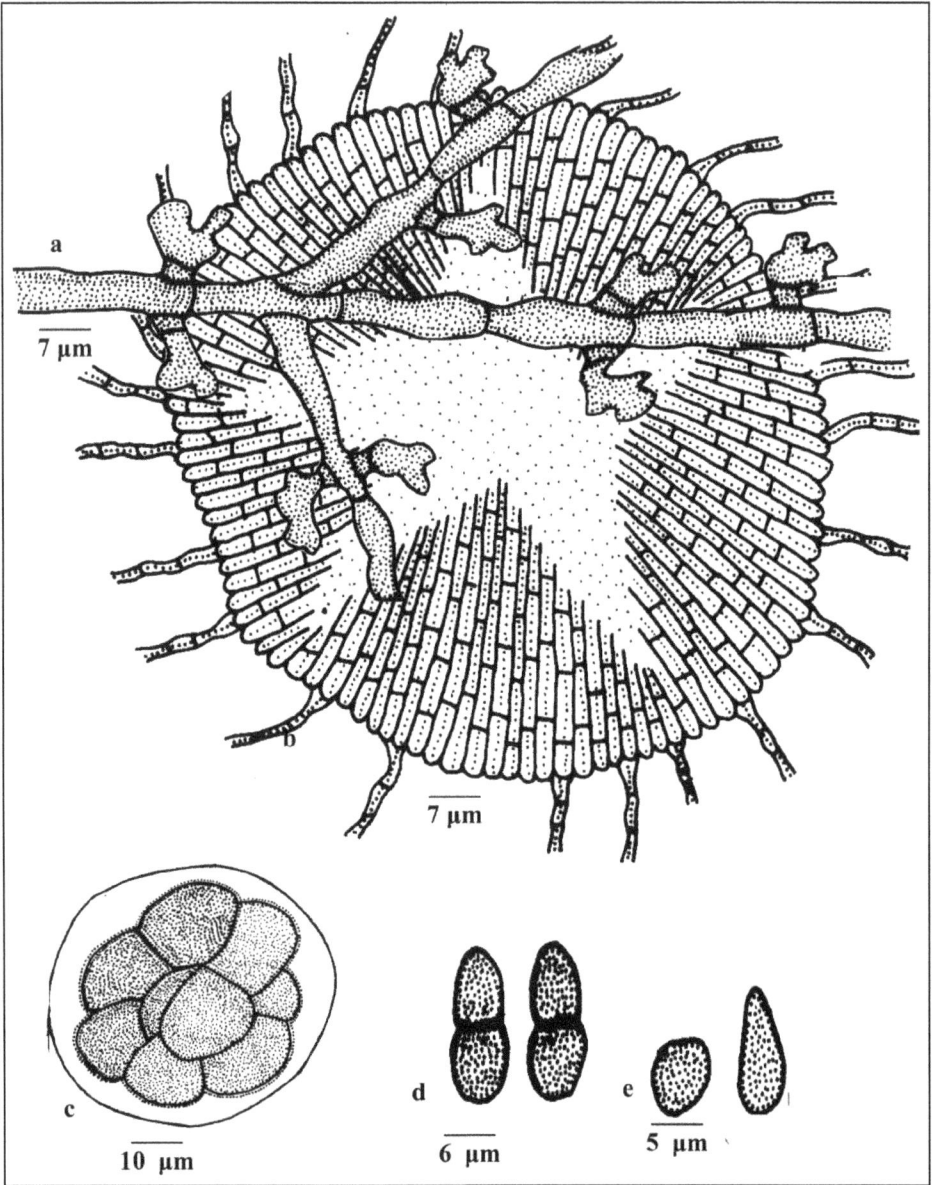

Figure 2.22: *Asterina lobulifera* Sydow var. *indica* Hosag. & Chandra.

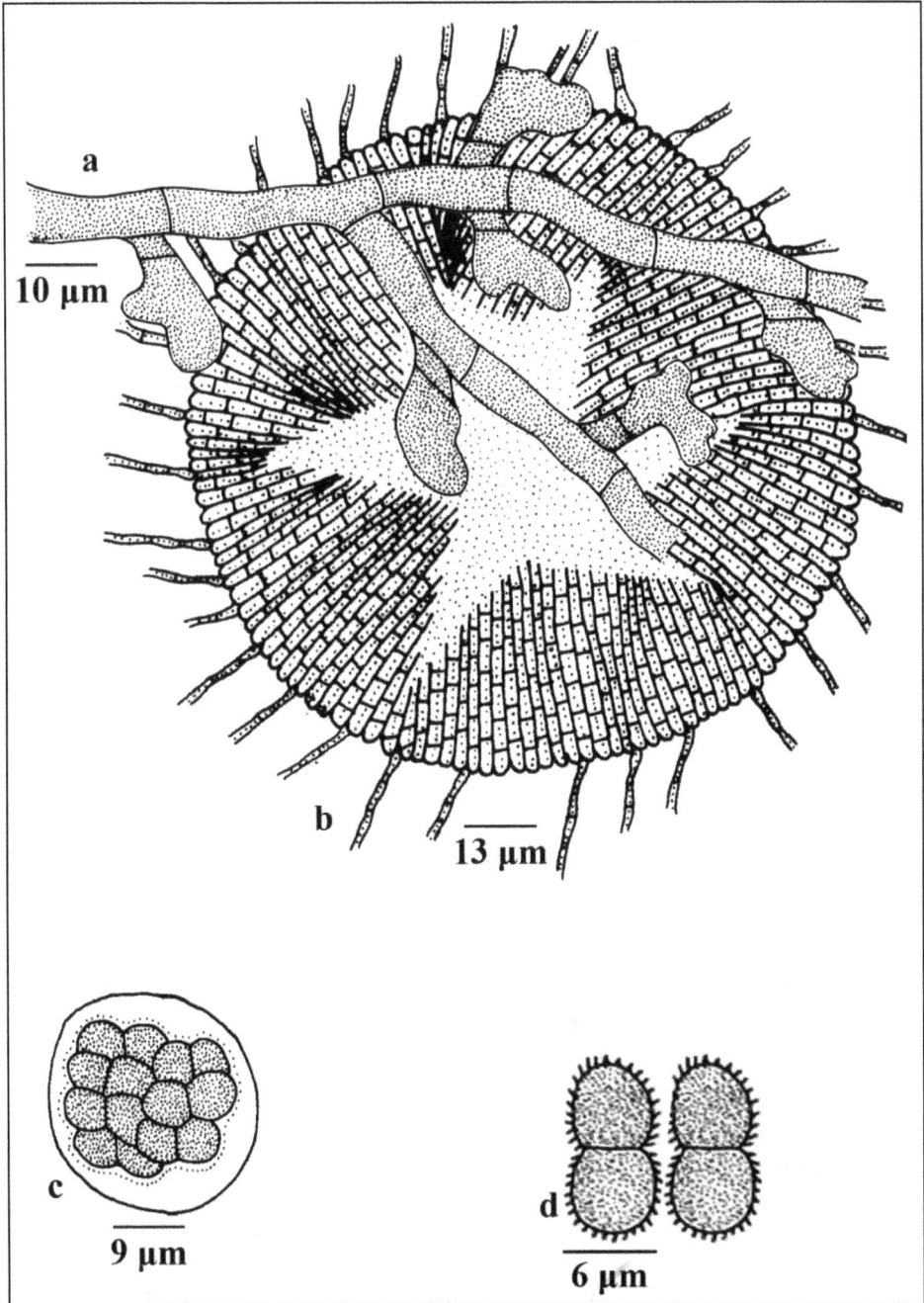

Figure 2.23: *Asterina malloticola* Hosag. *et al.*

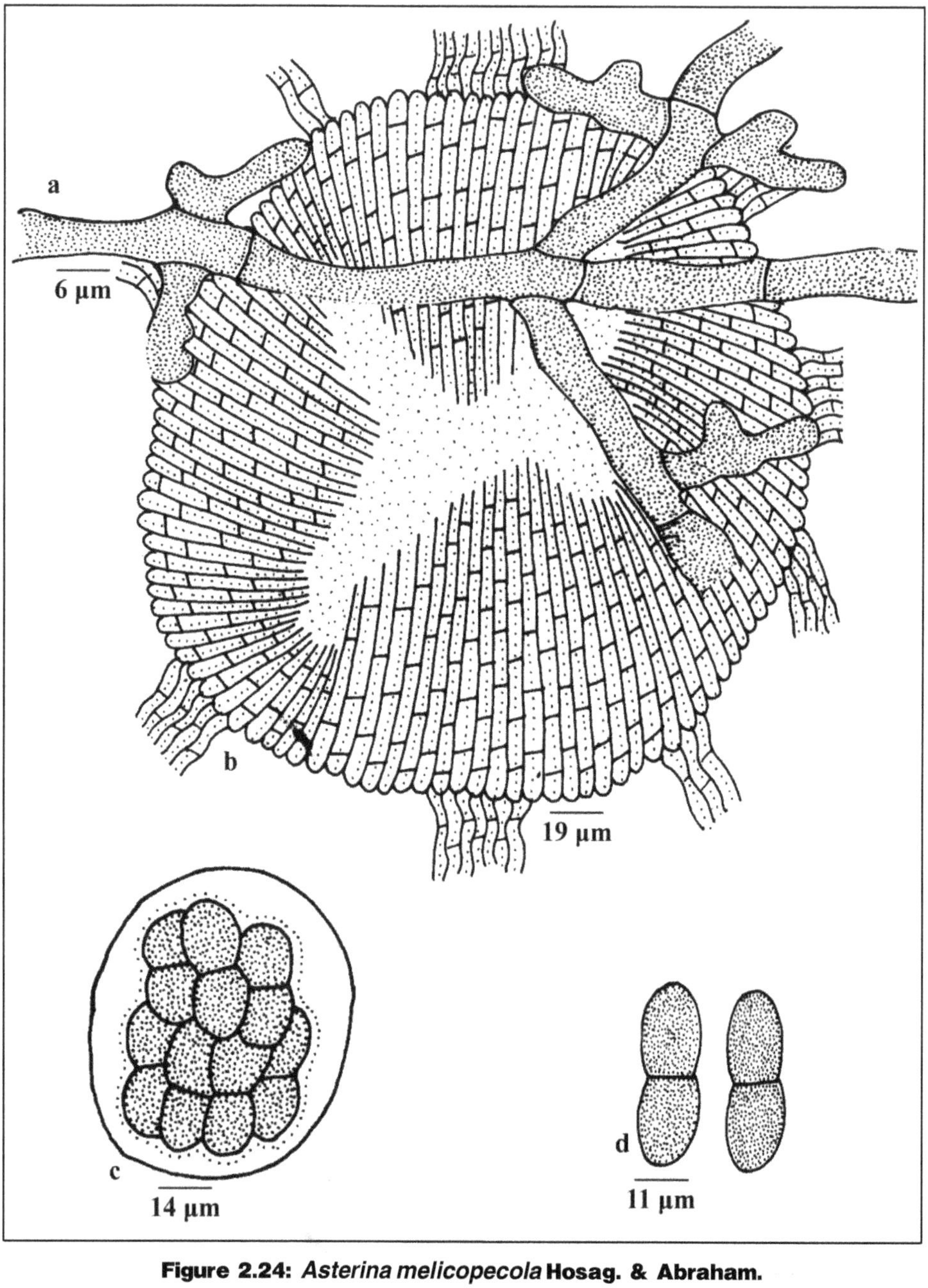

Figure 2.24: *Asterina melicopecola* **Hosag. & Abraham.**

reticulate, cells 13-20 x 2-4 µm. Appressoria opposite, about 20 per cent alternate, unicellular, ovate, conoid towards apex, sublobate, 9-13 x 7-9 µm. Thyriothecia orbicular, scattered to connate, fringed hyphae small; asci globose, octosporous, up to 37 µm in diameter; ascospores conglobate, uniseptate, brown, deeply constricted at the septum, 20-26 x 11-13 µm, wall smooth.

Distribution: Observed on the leaves of *Melicope lunu-ankenda* (Gaertn.) T. Hartley (*Euodia lunu-ankenda* Gaertn.) (Rutaceae).

Asterina mimusopsidicola Hosag., Sabeena & Agarwal, Indian Phytopath. 62: 729, 2009 (Figure 2.25).

Colonies mostly hypophyllous, subdense, up to 3 mm in diameter, confluent. Hyphae flexuous, branching opposite, alternate to unilateral at acute angles, loosely to closely reticulate, cells 22-35 x 4-6 µm. Appressoria alternate, unilateral, antrorse, subantrorse to closely appressed to the hyphae, two celled, 11-22 µm long; stalk cells cylindrical to cuneate, 4-9 µm long; head cells ovoid, oblong, cylindrical to linear, entire, 6-13 x 4-9 µm. Thyriothecia scattered to connate, orbicular, up to 500 µm in diameter, stellately dehisced or dissolved at the center; margin crenate to fimbricate, fringed hyphae flexuous, devoid of appressoria; asci globose, octosporous, up to 55 µm in diameter; ascospores conglobate, brown, uniseptate, constricted at the septum, 20-22 x 11-13 µm, wall smooth.

Distribution: Observed on the leaves of *Mimusops elengi* L. (Sapotaceae).

Asterina murrayae Hansf., Proc. Linn. Soc. London 158: 45, 1947 (Figure 2.26).

Colonies amphigenous, mostly epiphyllous, dense, up to 2 mm in diameter, confluent. Hyphae flexuous, branching alternate to irregular at acute angles, loosely to closely reticulate, cells 9-15 × 4-7 µm. Appressoria alternate, about 40 per cent opposite, unicellular, ovate, oblong, lobate, 7-13 × 4-7 µm. Thyriothecia scattered to loosely grouped, orbicular, up to 160 µm in diameter, stellately dehisced at the center, margin crenate; asci few, globose, octosporous, up to 33 µm in diameter; ascospores oblong, conglobate, brown, uniseptate, slightly constricted at the septum, 20-22 × 9-11 µm, wall smooth.

Distribution: Observed on the leaves of *Murraya koenigii* (L.) Sprengel (Rutaceae).

Asterina naraveliae Hosag., C.K. Biju & Agarwal, Indian Phytopathol. 55: 499, 2002 (Figure 2.27).

Colonies amphigenous, thin to subdense, up to 2 mm in diameter. Hyphae flexuous to crooked, branching irregular at acute angles, loosely reticulate, cells 13-22 × 4-7 µm. Appressoria two celled, scattered, alternate, 9-15 µm long; stalk cells cylindrical to cuneate, 2-4 µm long; head cells ovate, 3-4 times lobate, 9-11 × 7-9 µm. Thyriothecia scattered, orbicular, up to 80 µm in diameter, stellately dehisced at the center, crenate at the margin; asci few to many, globose, octosporous, up to 28 µm in diameter; ascospores conglobate, 1-septate, brown, 15-19 × 7-9 µm, wall smooth.

Distribution: Observed on the leaves of *Naravelia zeylanica* (L.) DC. (Ranunculaceae).

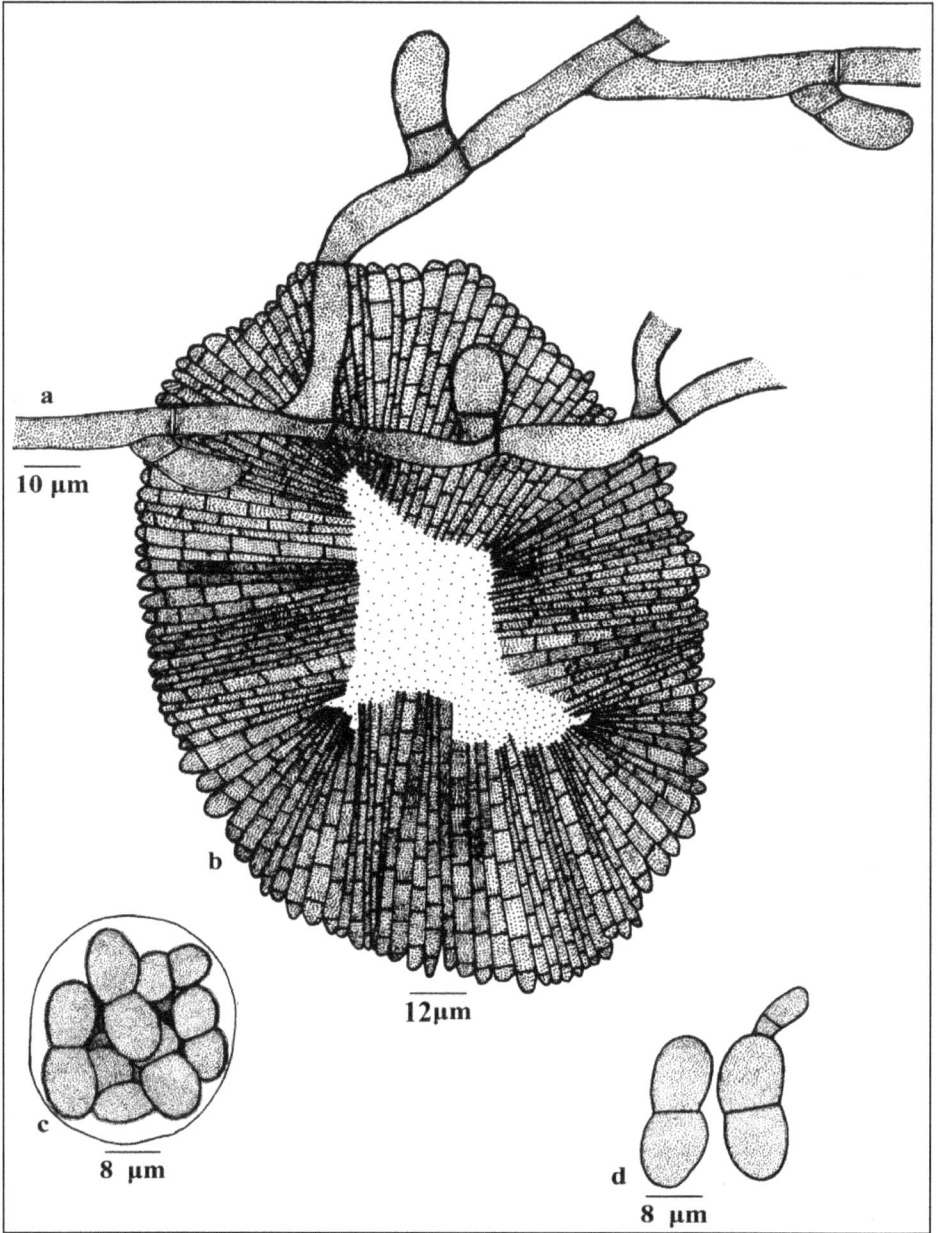

Figure 2.25: *Asterina mimusopsidicola* **Hosag.** *et al.*

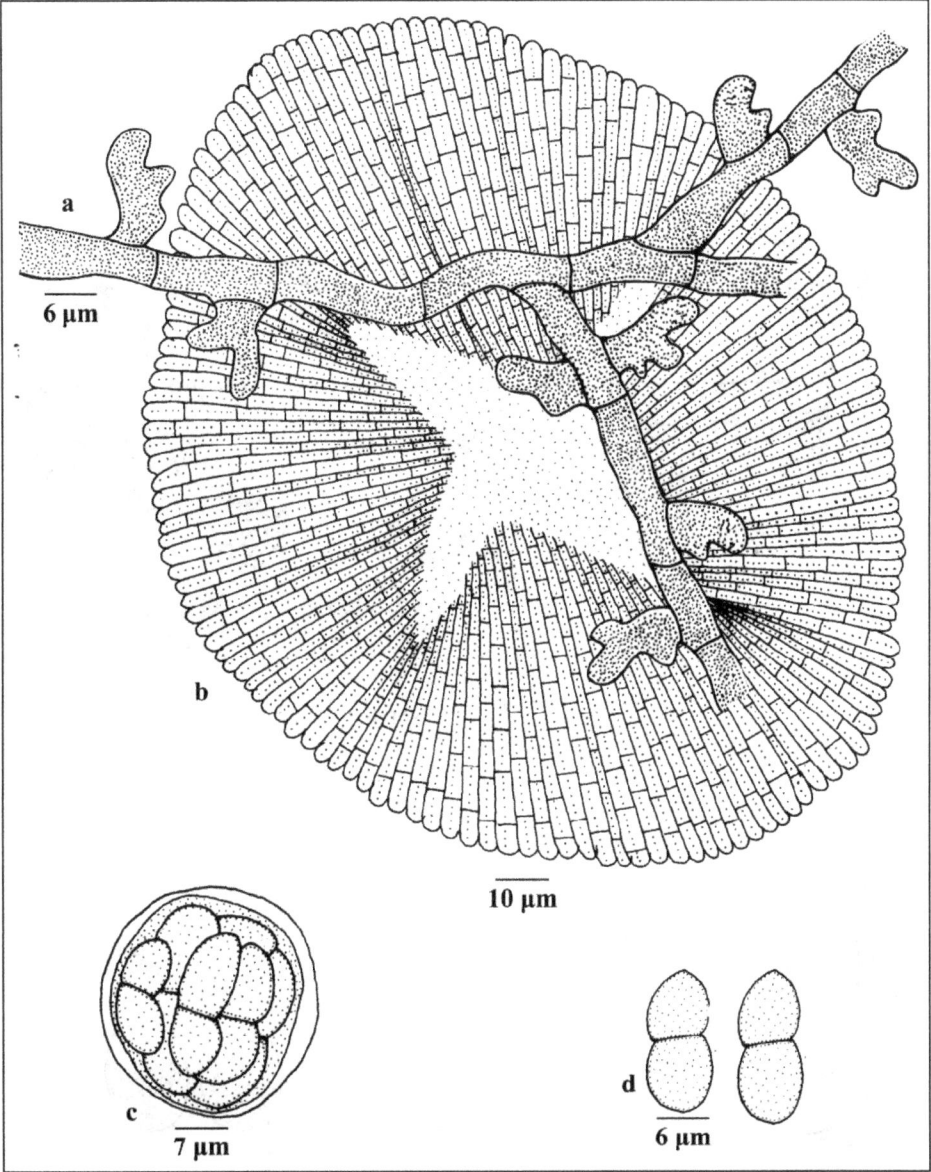

a

6 μm

b

10 μm

c

7 μm

d

6 μm

Figure 2.26: *Asterina murrayae* **Hansf.**

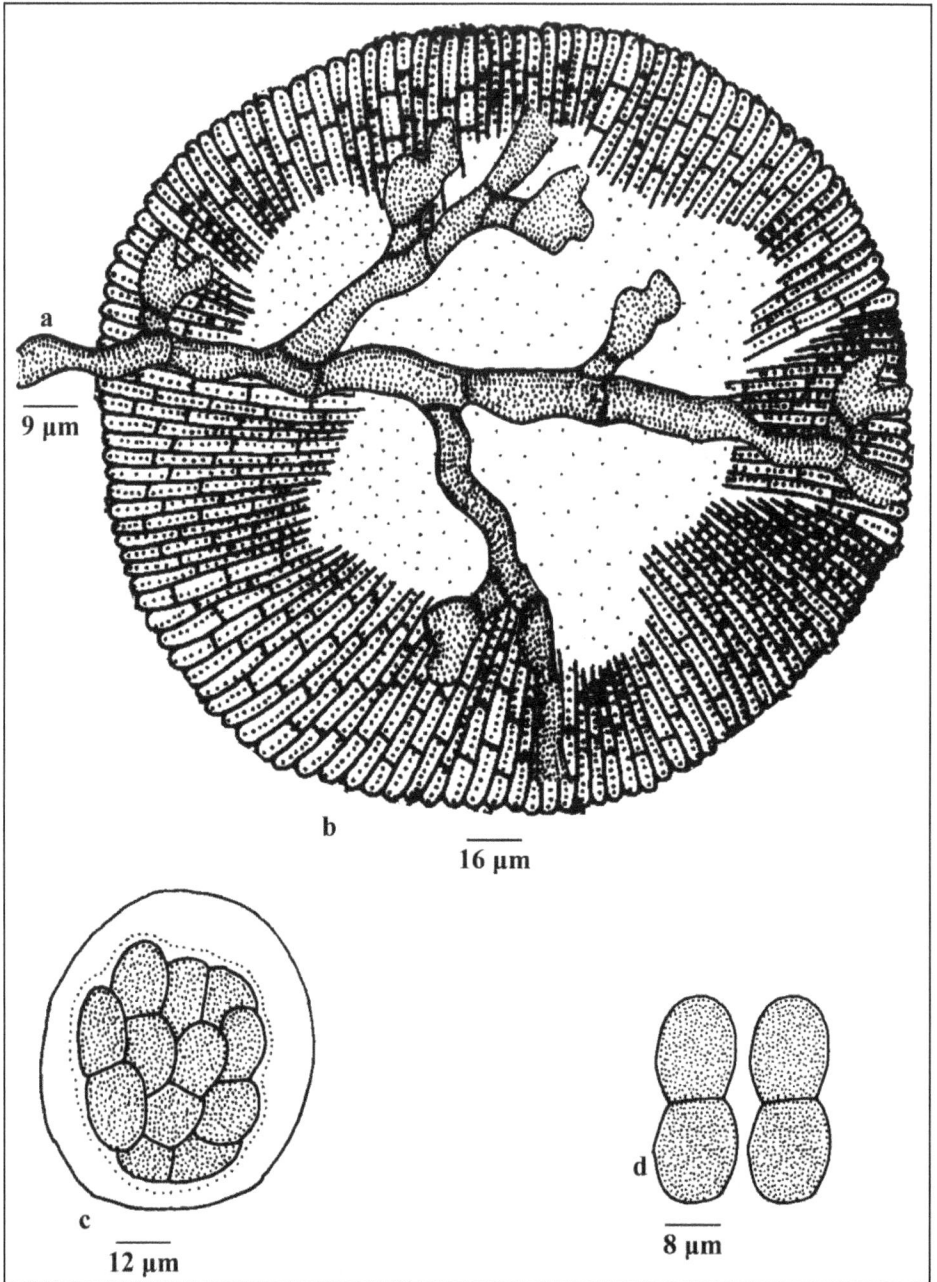

Figure 2.27: *Asterina naraveliae* **Hosag.** *et al.*

Asterina phyllanthigena Hosag., Zoos' Print J. 19: 1522, 2004 (Figure 2.28).

Colonies epiphyllous, thin to subdense, up to 1 mm in diameter, confluent. Hyphae flexuous, often form a loose net, cells 16-20 × 3-4 µm. Appressoria few, scattered, two celled, straight to curved, 9-11 µm long; stalk cells cylindrical to cuneate, 2-3 µm long; head cells ovate, cylindrical, entire, 4-6 × 4-5 µm. Thyriothecia scattered to loosely grouped, orbicular, up to 80 µm in diameter, stellately dehisced at the center, margin crenate; asci few, globose, octosporous, up to 30 µm in diameter; ascospores oblong, conglobate, brown, uniseptate, slightly constricted at the septum, 11-18 × 7-9 µm, wall smooth. Pycnothyria similar to thyriothecia, smaller; pycnothyriospores pyriform, brown, 9-15 × 7-9 µm, wall smooth.

Distribution: Observed on the leaves of *Phyllanthus* sp. (Euphorbiaceae).

Asterina pongalaparensis Hosag., C. K. Biju & Abraham, Indian Phytopathol. 54: 138, 2001 (Figure 2.29).

Colonies epiphyllous, dense, up to 2 mm in diameter, rarely confluent. Hyphae flexuous to crooked, branching irregular at acute angles, loosely to closely reticulate, cells 17-33 × 4-7 µm. Appressoria alternate, unilateral, two celled, straight to curved, gibbous at the base, 13-17 µm long; stalk cells cylindrical to cuneate 4-7 µm long; head cells clavate, ovate, hamate, straight to curved, 3-5 time sublobate to lobate, 9-13 × 4-7 µm. Thyriothecia scattered, orbicular, up to 154 µm in diameter, stellately dehisced at the center, margin crenate to fimbriate, fringed, hyphae flexuous; asci globose, octosporous, up to 30 µm in diameter; ascospores oblong, conglobate, brown, uniseptate, constricted at the septum, 15-22 × 9-11 µm, wall echinulate.

Distribution: Observed on the leaves of *Jasminum* sp. (Oleaceae).

Asterina pusilla Sydow in Sydow & Sydow, Philippine J. Sci. 8: 488, 1913 (Figure 2.30).

Colonies epiphyllous, thin, dense, crustose, up to 2 mm in diameter, confluent. Hyphae substraight to undulate, branching opposite at acute to wide angles, loosely reticulate, cells 20-26 × 4-7 µm. Appressoria unicellular, alternate scattered, antrorse, subantrorse, lobate, 9-11 × 7-9 µm. Thyriothecia scattered to connate, orbicular, up to 120 µm in diameter, stellately dehisced at the center, margin fimbriate; asci globose, up to 30 µm in diameter; ascospores brown, conglobate, uniseptate, constricted at the septum 18-20 × 9-11 µm, wall smooth. Pycnothyria numerous, mixed with thyriothecia; pycnothyriospores brown, globose, pyriform, 9-13 × 7-9 µm.

Distribution: Observed on the leaves of *Premna corymbosa* Rottl. and Willd. (Verbenaceae).

Asterina saracae Hosag., Abraham & Crane, Mycotaxon 68:19, 1998 (Figure 2.31).

Colonies epiphyllous, thin, up to 2 mm in diameter, confluent. Hyphae substraight, branching opposite to irregular at acute to wide angles, loosely reticulate, cells 13-20 × 4-7 µm. Appressoria alternate, about 20 per cent opposite, unicellular, conoid to ampulliform, broadly rounded at the apex, entire, 11-15 × 4-7 µm. Thyriothecia scattered to loosely grouped, often connate, up to 200 µm in diameter,

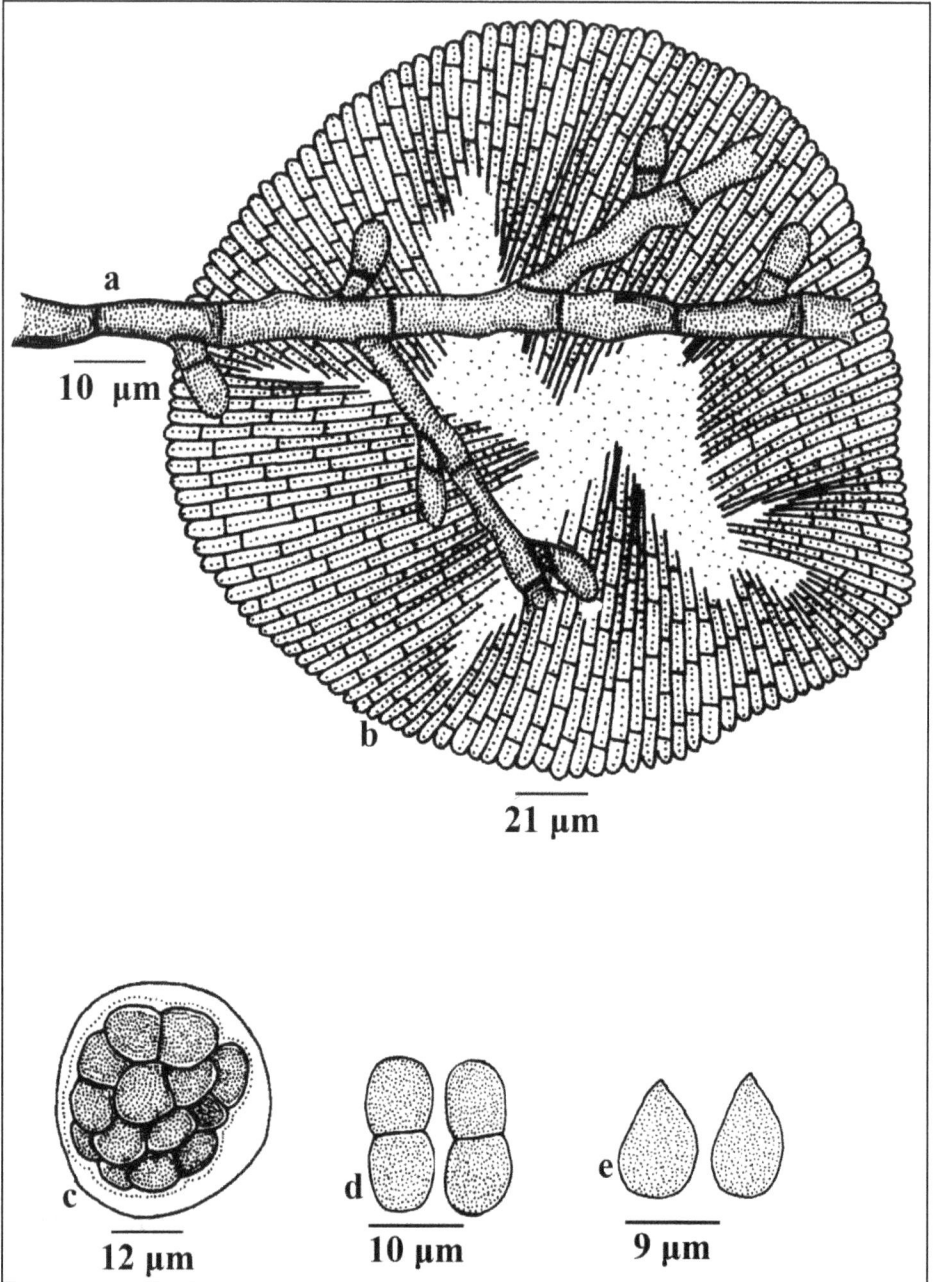

Figure 2.28: *Asterina phyllanthigena* Hosag.

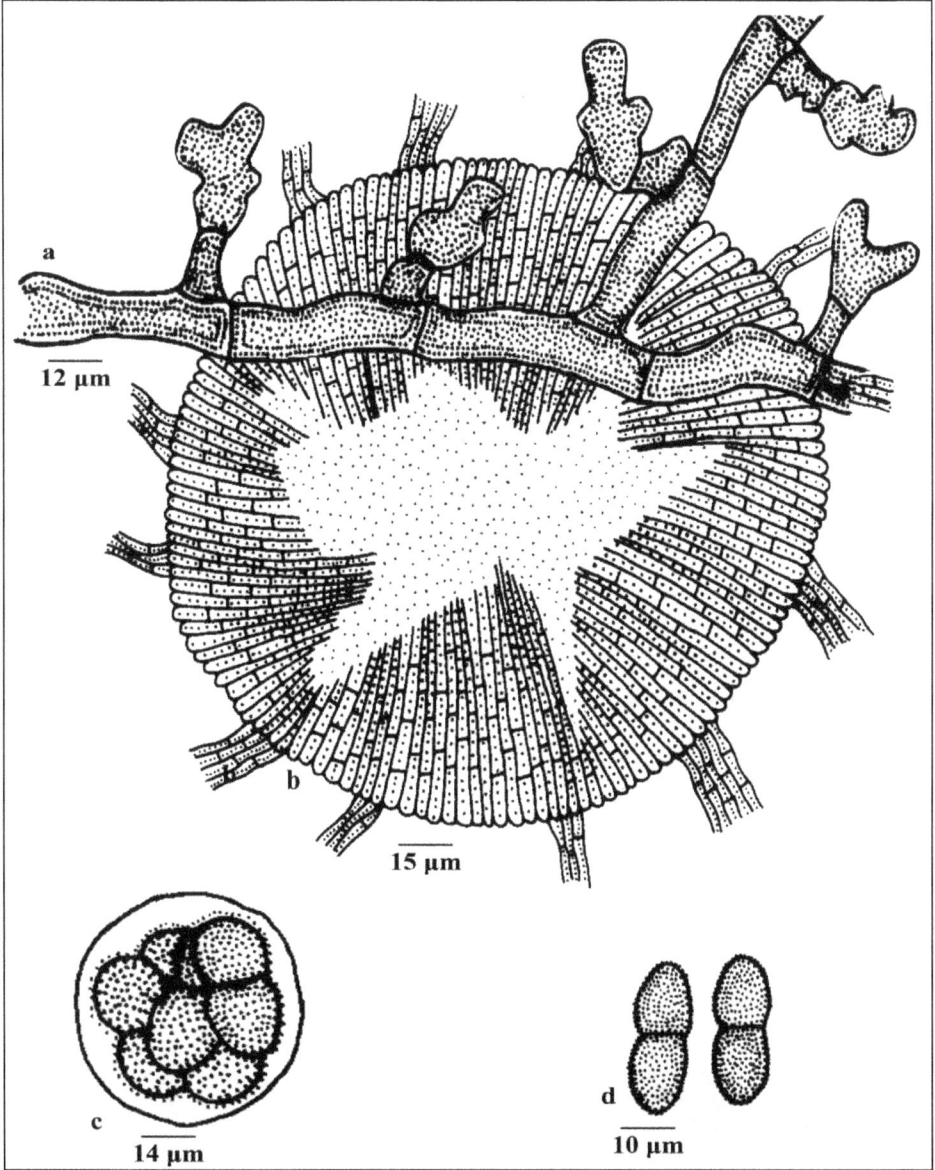

Figure 2.29: *Asterina pongalaparensis* **Hosag.** *et al.*

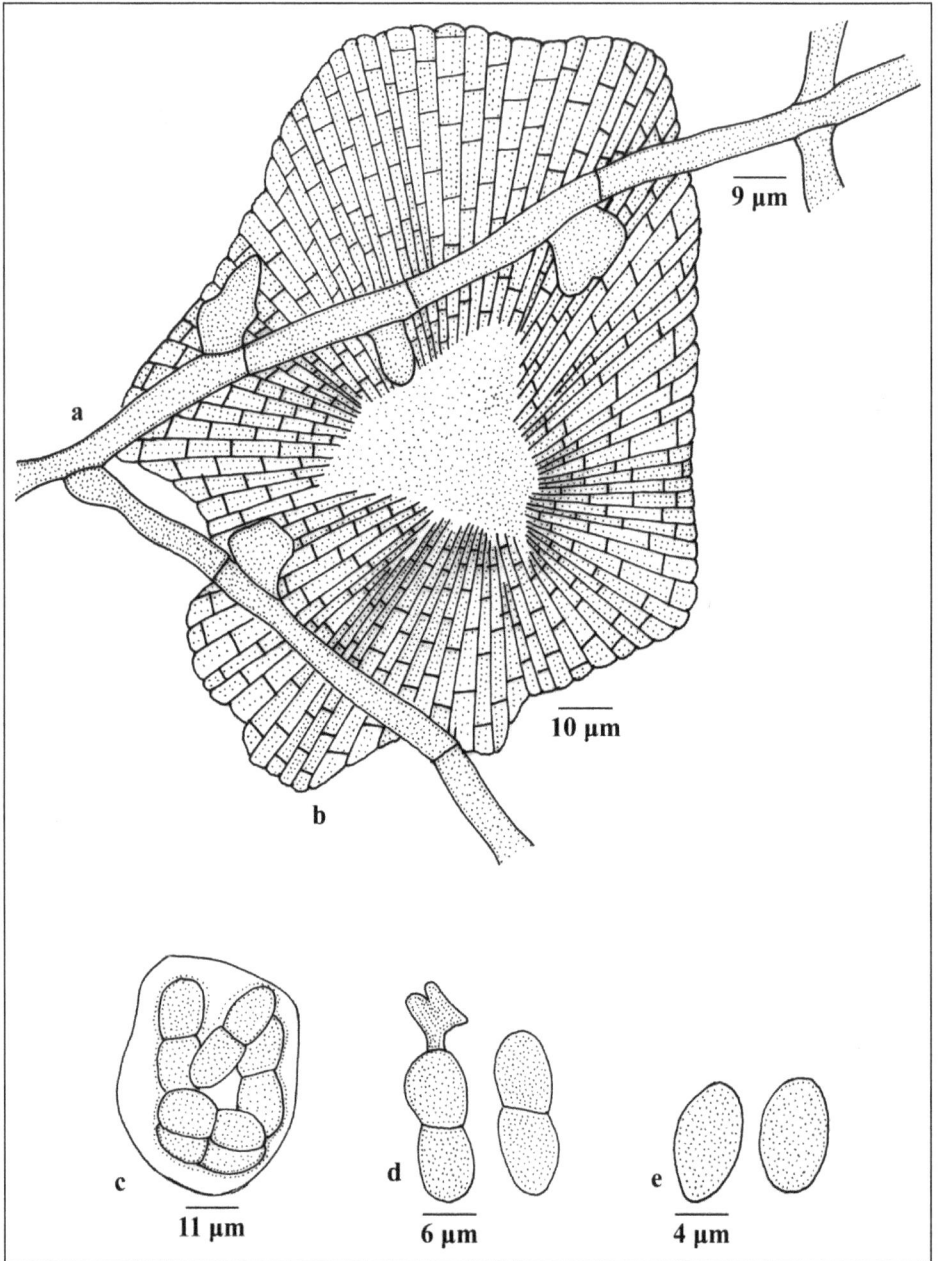

Figure 2.30: *Asterina pusilla* Sydow.

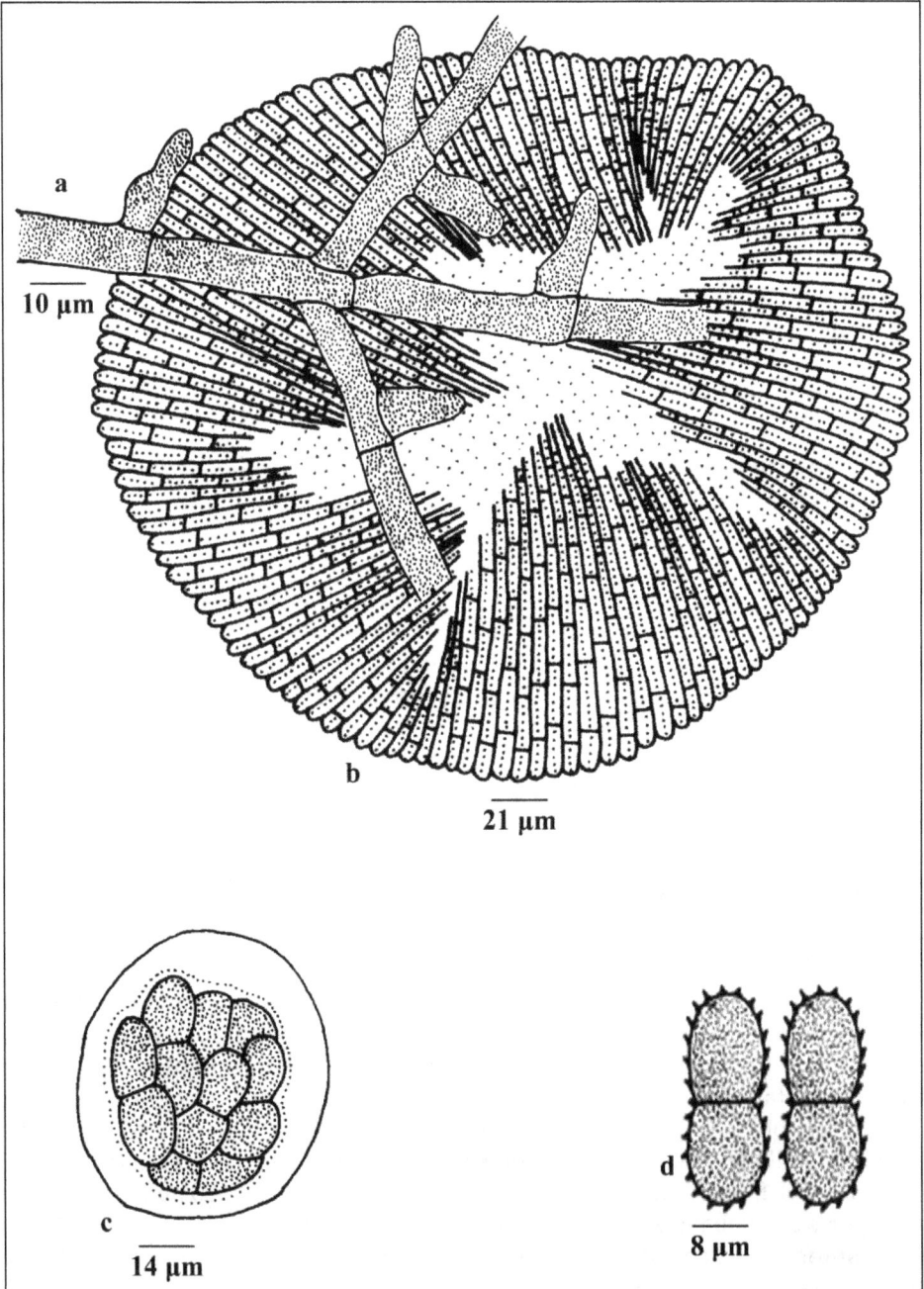

Figure 2.31: *Asterina saracae* Hosag. *et al.*

stellately dehisced at the center, margin crenate; asci globose, octosporous, up to 33 μm in diameter; ascospores conglobate, uniseptate constricted at the septum, 22-26 × 11-13 μm, wall echinulate.

Distribution: Observed on the leaves of *Saraca asoca* (Roxb.) de Willd. (Caesalpiniaceae).

Asterina tertia Racib. in Theiss., Die Gattung *Asterina*, 7:103,1913 (Figure 2.32).

Colonies amphigenous, thin, up to 3 mm in diameter, confluent. Hyphae crooked branching, irregular at acute angles, loosely to closely reticulate, cells 18-30 × 4-7μm. Appressoria unicellular, alternate, unilateral, sessile, ovate, lobate, mammiform, 6-10 ×4-7μm. Thyriothecia scattered, orbicular, 120-160 μm in diam., dehisce stellately at the center, margin crenate; asci octosporous, up to 45 μm in diameter; ascospores brown, 1-septate, constricted at the septum, 20-26 × 9-11 μm. Pycnothyria many, similar to thyriothecia; pycnothyriospore pyriform, brown, 11-22 × 7-9μm, wall smooth.

Distribution: Observed on the leaves of *Adhatoda zeylanica* Medikus (*A. vasica* Nees) (Acanthaceae).

Asterina visci Hosag., Zoos' print J. 17: 863, 2002. (Figure 2.33)

Colonies amphigenous, dense, up to 2 mm in diameter, rarely confluent. Hyphae substraight to flexuous, branching irregular at acute to wide angles, loosely reticulate, cells 13-18 × 4-7 μm. Appressoria unicellular, alternate, about 2 per cent opposite, ovate, conoid, entire, furcate, 7-11× 4-7 μm. Thyriothecia scattered to connate, orbicular, up to 150 μm in diameter; asci globose, octosporous up to 26 μm in diameter; ascospores conglobate, oblong, brown, uniseptate, constricted at the septum, 15-18 × 9-11 μm, wall minutely echinulate in the matured ascospores. Pycnothyria similar to thyriothecia, smaller; pycnothyriospores globose, brown, 11-17 × 9-11 μm, wall smooth.

Distribution: Observed on the leaves of *Viscum* sp. (Loranthaceae).

Asterina wrightiae Sydow in Sydow & Petrak, Ann. Mycol. 29: 236,1931. (Figure 2.34)

Colonies epiphyllous, dense, crustose, up to 4 mm in diameter. Hyphae substraight to flexuous, branching irregular at acute to wide angles, loosely reticulate, cells 7-15 × 2-4 μm. Appressoria 2-celled, rarely unicellular, straight to curved, 11-13μm long; stalk cells cylindrical to cuneate, 4-5μm long; head cells globose, ovate, cylindrical, lobate, 7-9 × 4-7μm. Thyriothecia scattered, orbicular, up to 105 μm in diameter, stellately dehisced at the center, margin fimbriate, rarely crenate, fringed hyphae small, flexuous to crooked; asci globose, octosporous, 20-33 μm, in diameter; ascospores conglobate, brown, 1-septate, constricted at the septum, 11-19 × 7-11 μm, wall smooth; Pycnothyria numerous, similar to thyriothecia, smaller; pycnothyriospores pyriform, 11-15 × 7-9 μm, wall smooth.

Distribution: Observed on the leaves of *Wrightia tinctoria* Roxb. R. Br. (Apocyanaceae).

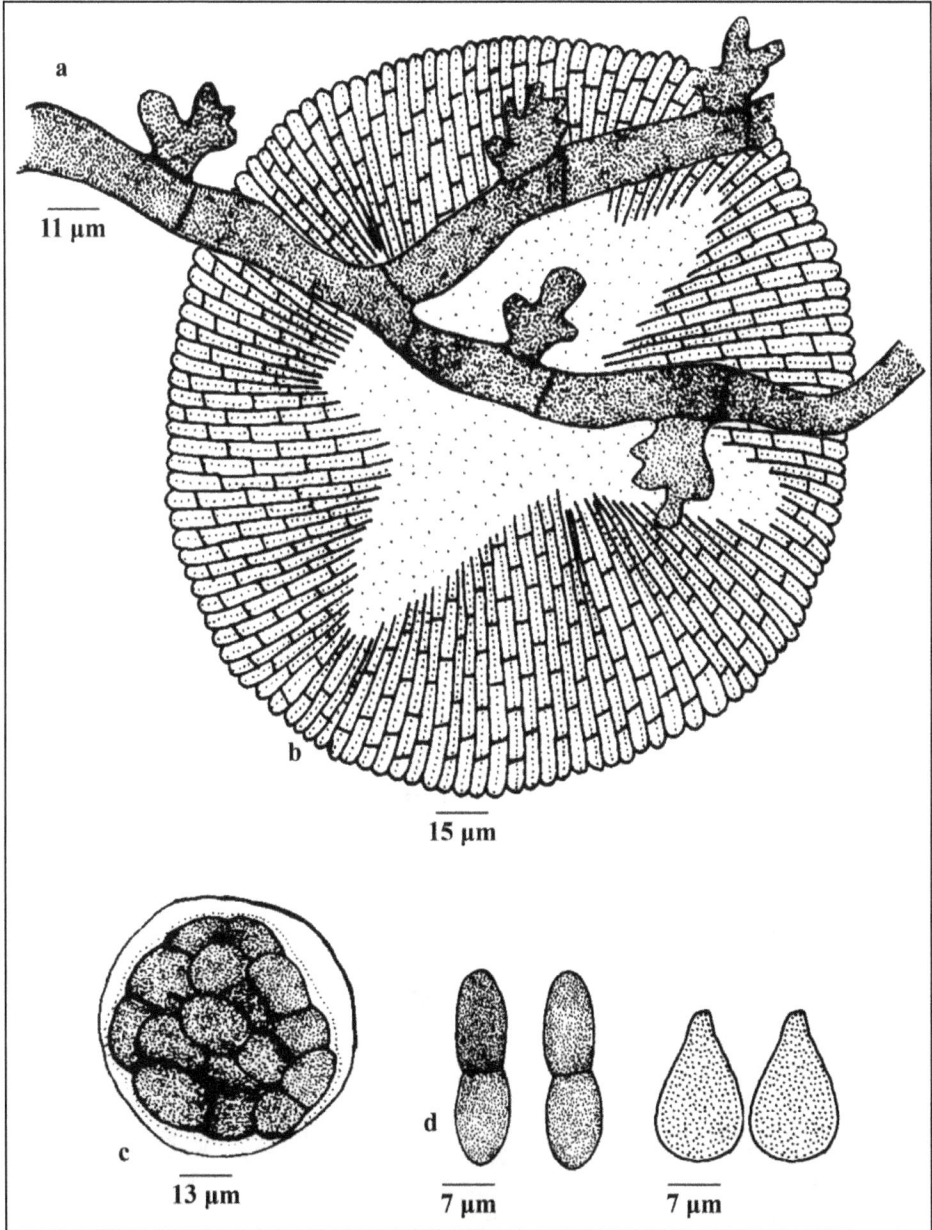

Figure 2.32: *Asterina tertia* **Racib.**

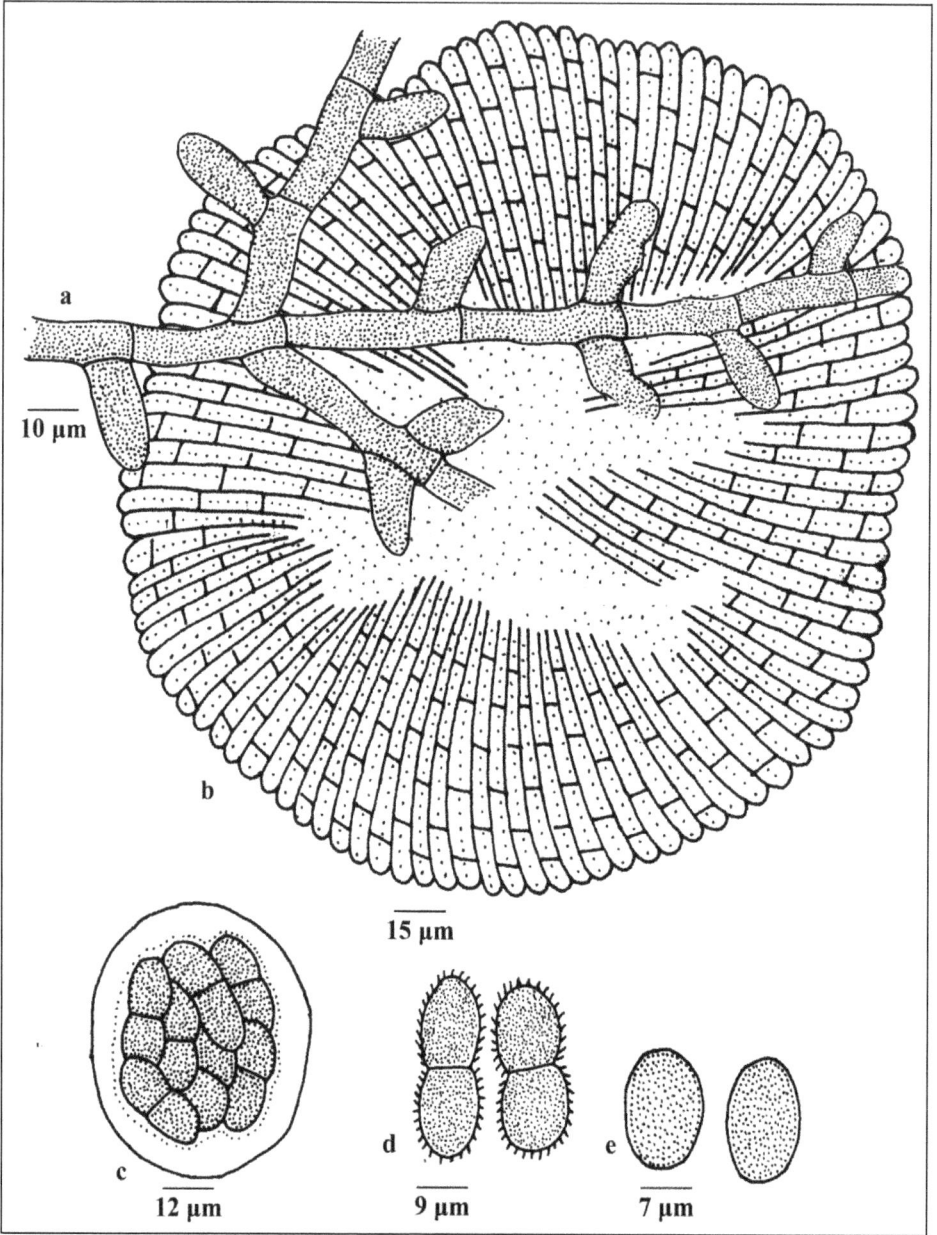

Figure 2.33: *Asterina visci* Hosag.

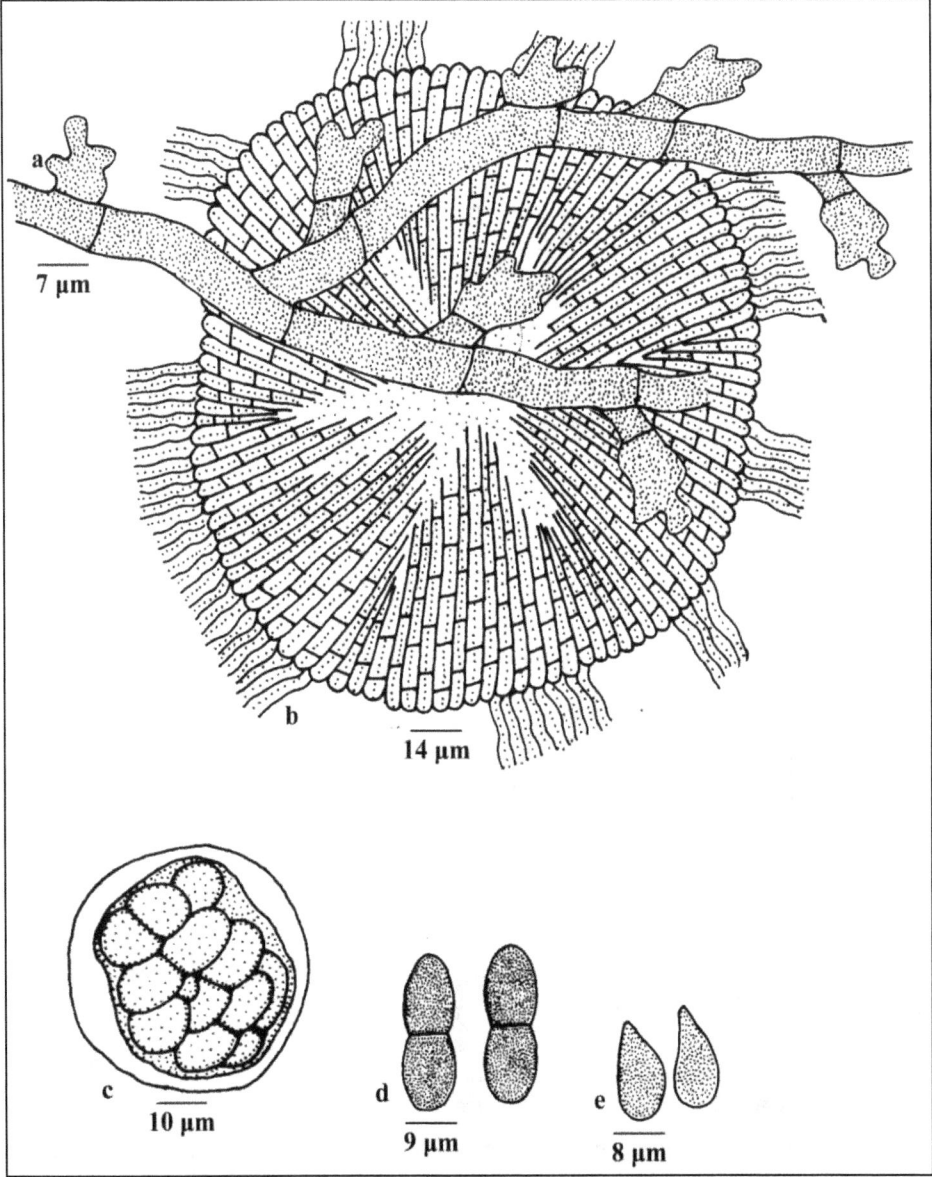

Figure 2.34: *Asterina wrightiae* **Sydow.**

THE GENUS *ASTEROLIBERTIA*

Asterolibertia Arn., Les Asterinees, 1: 161, 1918; Hansf., Mycol. Pap. 15: 189, 1946; Muller & Arx, Beitr. Krypt. Schw. 11:97, 1962; Luttrell in Ainsworth *et al.* (eds.). The Fungi. An advanced Treatise 4: 207, 1973; Arx & Muller, Stud. Mycol. 9: 43, 1975; Bilgrami, Jamaluddin & Rizwi, Fungi of India p. 54, 1991; Hosag., Abraham & C.K. Biju, J. Mycopathol. Res. 39: 61, 2001; Singh, Duke, Bhandari & Jain, J. Econ. Taxon. Bot. 30: 185, 2008.

Steyaertia Bat. and Maia, Univ. Recife, Inst. Mycol. Publ. 295:5, 1960. *Wardina* Arn., Les Asterinees 1:165, 1918.

Leaf parasites. Mycelium ectophytic, appressoria intercalary, setae absent. Thyriothecia orbicular with radiating cells, astomatous, dehisce stellately at the center; asci globose, octosporous, bitunicate; ascospores conglobate, uniseptate, brown.

Type sp.: *A. couepiae* (Henn.) Arn.

Asterolibertia mangiferae Hansf. & Thirum. Farlowia 3: 303, 1948 (Figure 2.35).

Colonies amphigenous, dense, crustose, up to 6 mm in diameter, confluent. Hyphae flexuous, branching irregular at acute to wide angles, loosely reticulate, cells 20- 33 × 4-7μm. Appressoria intercalary, scattered, ovate, globose, central pale hyaline spot visible, 7-9 × 9-11μm. Thyriothecia scattered, orbicular, up to 110 μm in diameter, stellately dehisced, margin fimbriate; asci few, globose, octosporous up to 40 μm in diameter, ascospores brown, conglobate, 1-septate, constricted at the septum, 22-26 × 11-13 μm, wall smooth.

Distribution: Observed on the leaves of *Mangifera indica* L. (Anacardiaceae).

THE GENUS *ISHWARAMYCES*

Ishwaramyces Hosag., J. Econ. Taxon. Bot. 28: 183, 2004.

Leaf parasites. Mycelium ectophytic, appressoria appears in clusters, setae absent. Thyriothecia orbicular with radiating cells, astomatous, dehisce stellately at the center; asci globose, octosporous, bitunicate; ascospores conglobate, uniseptate, brown.

Type sp. : *I. flacourtiae* Hosag. *et al.*

The genus *Ishwaramyces* differs from the genus *Asterina* in having axillary clusters of appressoria (Muller and Arx, 1962; Arx and Muller, 1975).

Ishwaramyces flacourtiae Hosag., Kamar. & Sabu in Hosag., C.K. Biju & Abraham, J. Econ. Taxon. Bot. 28: 183, 2004 (Figure 2.36).

Colonies amphigenous, mostly epiphyllous, dense, crustose, spreading up to 6 mm in diameter, confluent and cover an entire upper surface of the leaves. Hyphae straight to substraight, branching opposite at acute angles, closely reticulate, cells 11-20 × 4-7 μm. Appressoria opposite, appears in clusters, 2-celled, antrorse, 7-11 μm long; stalk cells cylindrical to cuneate, 2-4 μm long; head cells ovate, globose, entire, 4-7 × 7-9 μm. Thyriothecia scattered to grouped, orbicular, up to 300 μm in diameter, stellately dehisced at the center, margin crenate to fimbriate; asci numerous, globose,

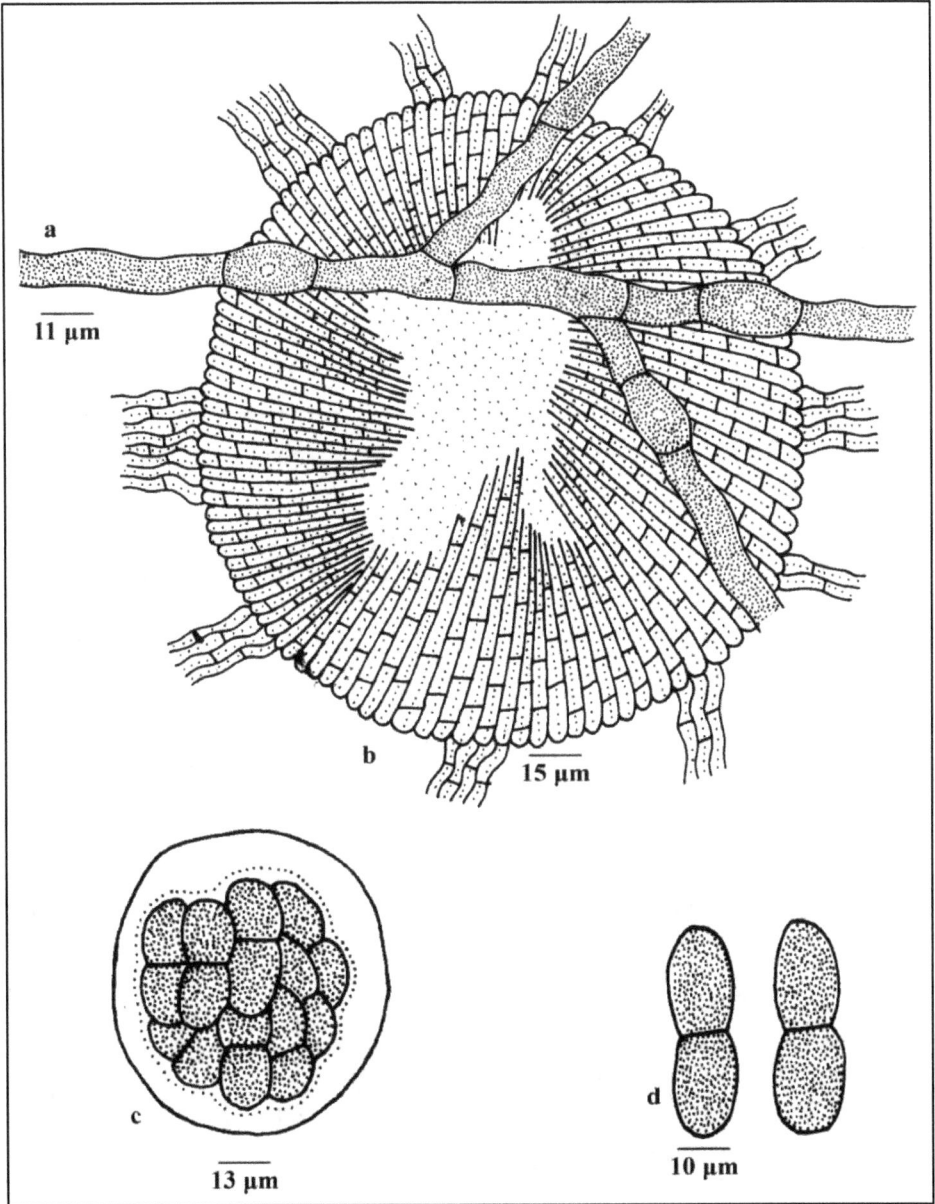

Figure 2.35: *Asterolibertia mangiferae* **Hansf. & Thirum.**

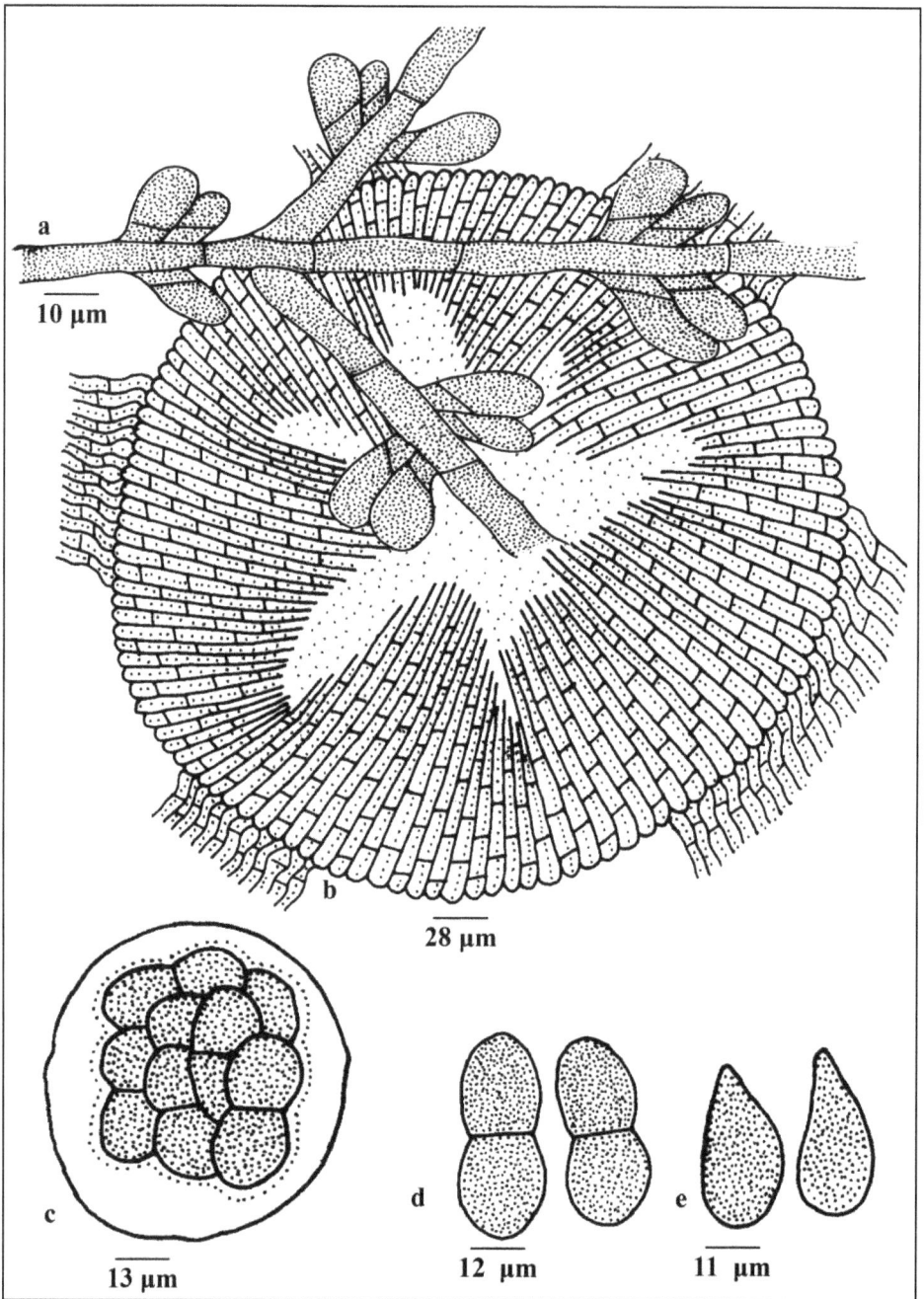

Figure 2.36: *Ishwaramyces flacourtiae* Hosag. *et al.*

octosporous, 30-40 µm in diameter; ascospores conglobate, initially hyaline, brown at maturity, uniseptate, deeply constricted at the septum, 24 -33 × 13-15 µm, wall smooth. Pycnothyria similar to thyriothecia, smaller; pycnothyriospores pyriform, brown, 13-19 × 11-13 µm, wall smooth.

Distribution: Observed on the leaves of *Flacourtia montana* Graham (Flacourtiaceae).

THE GENUS *PRILLIEUXINA*

Prillieuxina Arn., Ann. Ecol. Nat. Agric. Montpellier 16:161,1918; Hansf., Mycol. Pap. 15: 169, 1946; Muller & Arx, Beitr. Krypt. Schw. 11:132, 1962; Luttrell in Ainsworth *et al.* (eds.). The Fungi. An advanced Treatise 4: 207, 1973; Arx & Muller, Stud. Mycol. 9: 44, 1975; Bilgrami, Jamaluddin & Rizwi, Fungi of India p. 407, 1991; Hosag., Abraham & C.K. Biju, J. Mycopathol. Res. 39: 62, 2001; Singh, Duke, Bhandari & Jain, J. Econ. Taxon. Bot. 30: 191, 2008.

Leaf parasites. Mycelium ectophytic, appressoria and setae absent. Thyriothecia orbicular with radiating cells, astomatous, dehisce stellately at the center; asci globose, octosporous, bitunicate; ascospores brown, conglobate, uniseptate.

Type sp. : *P. winteriana* (Pazschke) Arn.

Prillieuxina anamirtae (Sydow & Sydow) Ryan, in Stevens & Ryan, Illinois Biol. Monographs 17 : 78, 1939 (Figure 2.37).

Asterinella anamirtae Sydow and Sydow, Ann. Mycol. 12: 558, 1914.

Colonies amphigenous, dense, up to 4 mm in diameter, confluent. Hyphae flexuous to crooked, branching irregular at acute angles, closely reticulate, cells 12-20 × 2-4 µm. Appressoria absent. Thyriothecia scattered to connate, orbicular, up to 90 µm in diameter, stellately dehisced at the center, margin crenate; asci globose, ovate, octosporous, up to 30 µm in diameter; ascospores oblong, brown, 1-septate, constricted at the septum, 22-24 × 9-11 µm, wall echinulate in matured ascospores. Pycnothyria similar to thyriothecia, scattered; pycnothyriospores pyriform, brown, 13-15 × 9-11 µm, wall smooth.

Distribution: Observed on the leaves of *Anamirta cocculus* (L.) Wight and Arn. (Menispermaceae).

Prillieuxina elaegni Hosag. & C. K. Biju in Hosag., C. K. Biju & Abraham, J. Mycopathol. Res. 40: 195, 2002 (Figure 2.38).

Colonies epiphyllous, thin, dense, up to 1 mm in diameter. Hyphae straight to flexuous, branching alternate at acute to wide angles, loosely reticulate, cells 18-22 × 3-5 µm. Appressoria absent. Thyriothecia mostly aggregated, orbicular, up to 200 µm in diameter, stellately dehisced at the center, margin crenate to fimbriate, fringed hyphae straight to substraight, run parallel; asci many, globose, octosporous, up to 30 µm in diameter; ascospores conglobate, brown, 1-septate, constricted at the septum, 18-20 × 9-11 µm, wall smooth.

Distribution: Observed on the leaves of *Elaeagnus kologa* Schlecht. (Elaeagnaceae).

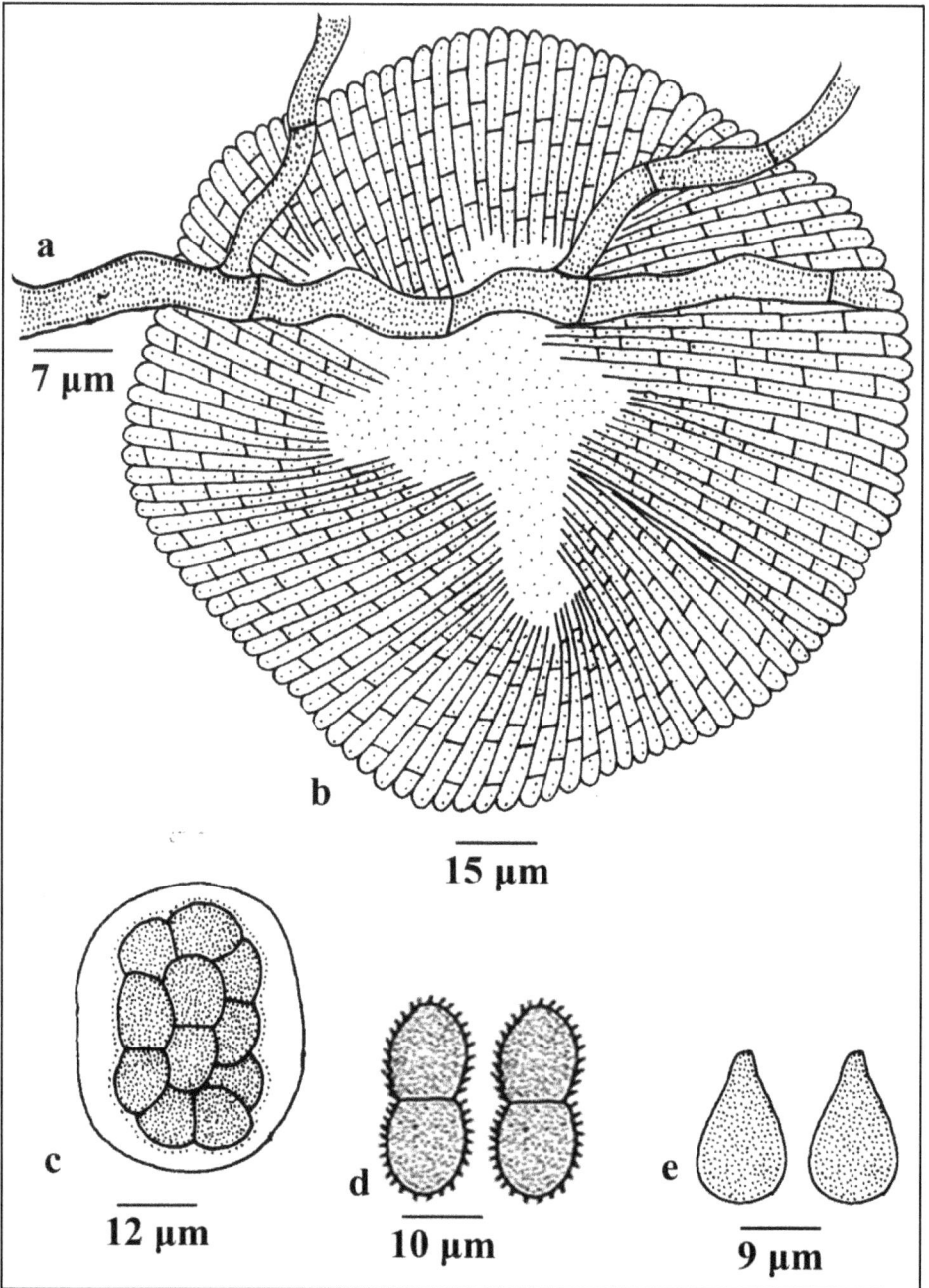

Figure 2.37: *Prillieuxina anamirtae* **(Sydow & Sydow) Ryan.**

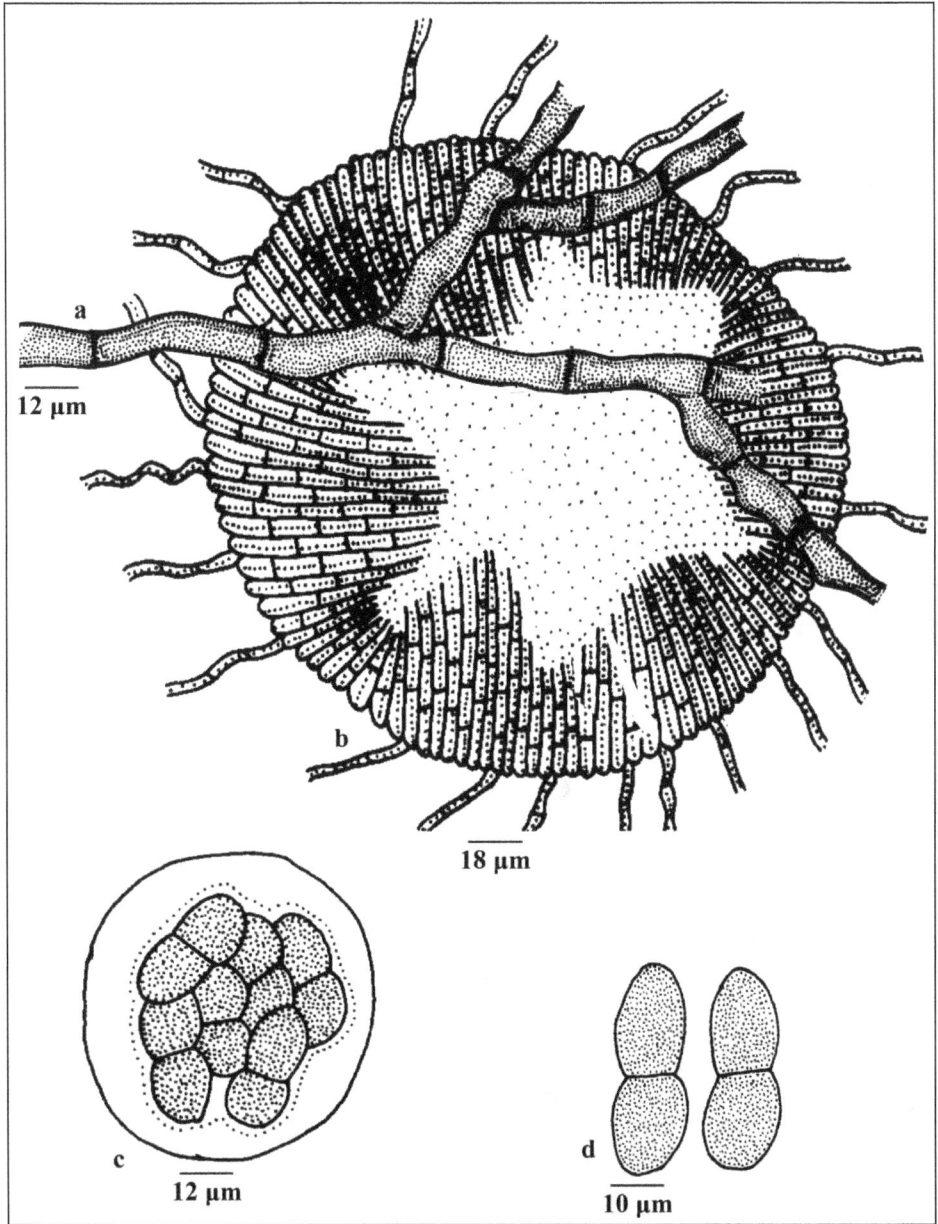

Figure 2.38: *Prillieuxina elaegni* Hosag. & C. K. Biju.

Prillieuxina ixorigena Hosag. & Chandra, Indian J. Sci. Techn. 2(6): 18, 2009 (Figure 2.39).

Colonies amphigenous, dense, up to 2 mm in diameter. Hyphae crooked, branching irregular at acute to wide angles, loosely to closely reticulate, cells 18-26 μm long and up to 4 μm broad. Appressoria absent. Thyriothecia scattered to grouped in the center of the colonies, orbicular, up to 100 μm in diameter, stellately dehisced at the center, margin crenate; asci constricted at the septum, 20-26 × 7-11 μm, wall smooth. Pycnothyria similar to thyriothecia but smaller; pycnothyriospores ovate, pyriform, 11-26 × 7-13 μm, wall smooth.

Distribution: Observed on the leaves of *Ixora coccinea* L. (Rubiaceae).

Prillieuxina polyalthiae Hosag. & Abraham, Indian phytopathol. 51: 391, 1998 (Figure 2.40).

Colonies mostly epiphyllous, rarely amphigenous, dense, crustose to velvety, up to 2 mm in diameter, confluent. Hyphae straight to substraight, branching irregular at acute to wide angles, loosely reticulate, cells 11-19 μm long and up to 2 μm broad. Appressoria absent. Thyriothecia develop all along the hyphae, orbicular, up to 310 μm in diameter, many thyriothecia join together marginally, dehisce stellately at the center, margin fimbriate; asci globose, octosporous, up to 42 μm in diameter; ascospores conglobate, brown, 1-septate, constricted at the septum, 19-22 × 7-11 μm, wall smooth.

Distribution: Observed on the leaves of *Polyalthia longifolia* (Sonn.) Thawaites. (Annonaceae).

THE FAMILY LEMBOSIACEAE

Lembosiaceae Hosag., Abraham & C.K. Biju, J. Mycopathol. Res. 39: 62, 2001.

Leaf parasites. Mycelium ectophytic, with or without appressoria, nutrient mycelium and leaf permating stroma present. Ascomata ectophytic, dimidiate, oval, ellipsoidal, "X" or "Y" shaped, elongated with radiating cells, astomatous, dehisce longitudinally at the center; asci globose, spherical, octosporous, bitunicate; ascospores two to many septate, conglobate, hyaline to brown.

Type genus: *Lembosia* Lev.

THE GENUS *LEMBOSIA*

Lembosia Lev., Ann. Sci. Nat. Bot. Ser., 3, 3: 58, 1845; Hansf., Mycol. Pap. 15: 189, 1946; Muller & Arx, Beitr. Krypt. Schw. 11:111, 1962; Luttrell in Ainsworth *et al.* (eds.). The Fungi. An advanced Treatise 4: 207, 1973; Arx & Muller, Stud. Mycol. 9: 43, 1975; Bilgrami, Jamaluddin & Rizwi, Fungi of India p. 263, 1991; Hosag., Abraham & C.K. Biju, J. Mycopathol. Res. 39: 62, 2001; Singh, Duke, Bhandari & Jain, J. Econ. Taxon. Bot. 30: 188, 2008.

Heraldoa Bat., Att. Est. Bot. Lab. Critr. Univ. Pavia 16:105, 1959.

Lembosidium Speg., Biol. Acad. Nac. Cien. Cordova. 26:342, 1923.

Lembosiellina Bat. and Maia, Atas Inst. Mycol. Recife 1:329, 1960. *Morenoella* Speg., Fungi Guar. 1: 258, 1883.

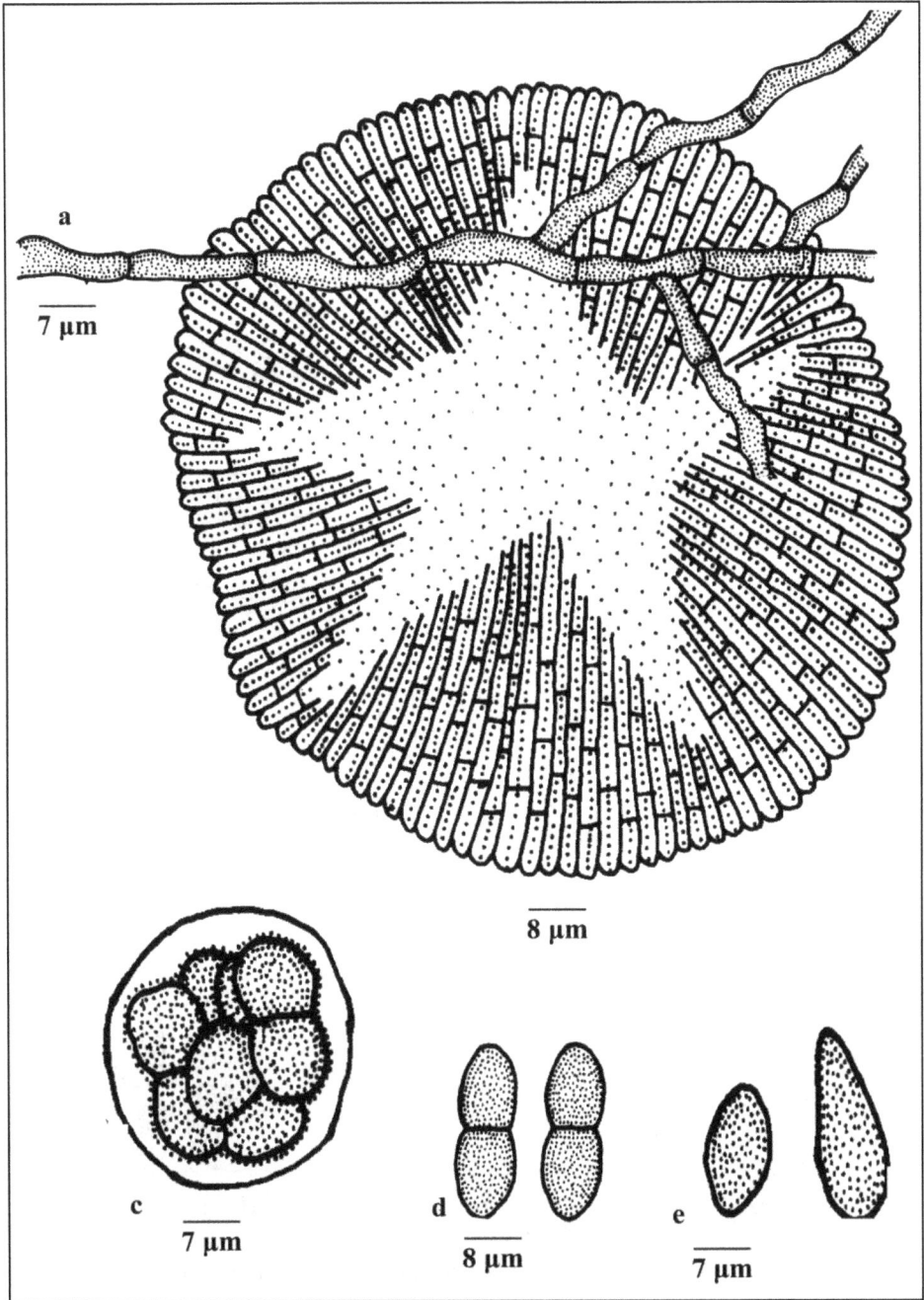

Figure 2.39: *Prillieuxina ixorigena* **Hosag. & Chandra.**

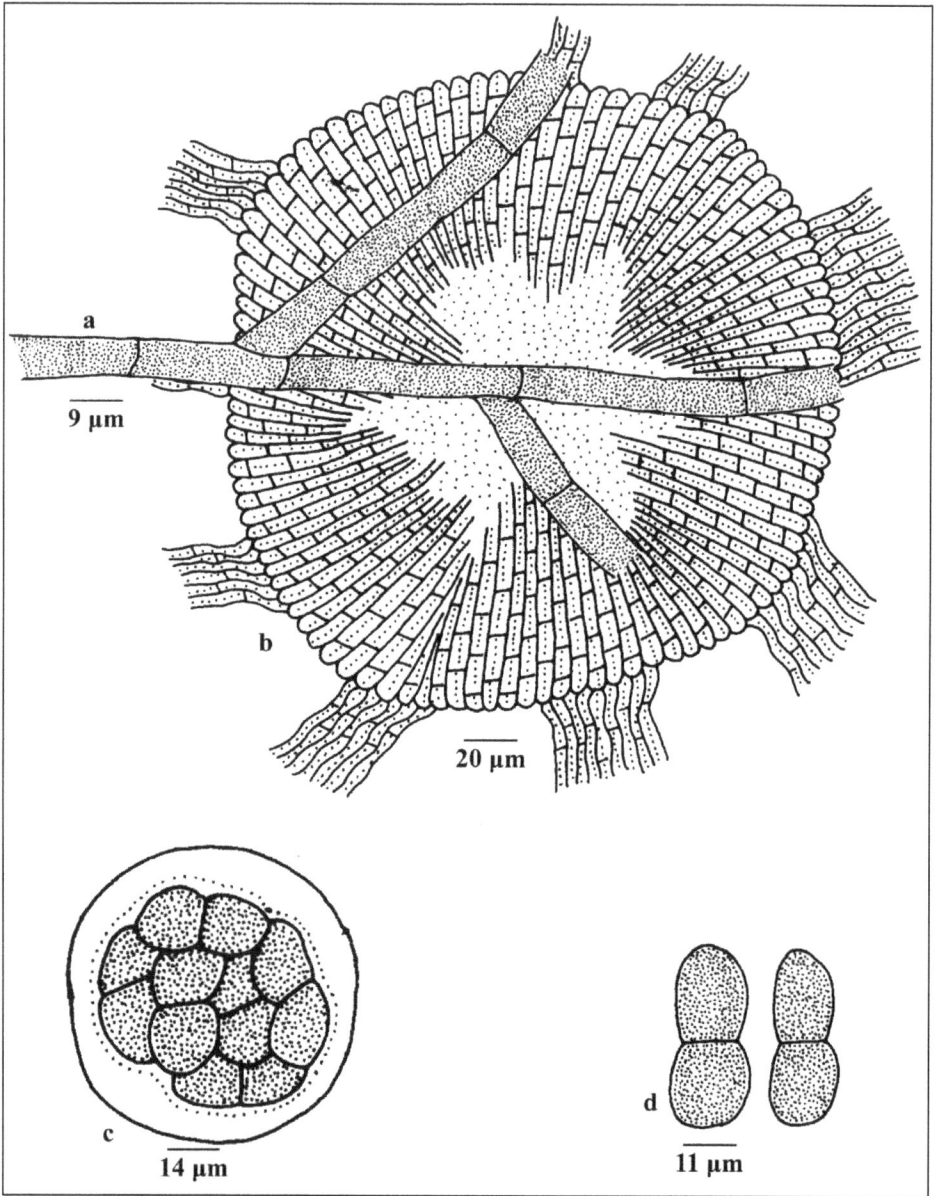

Figure 2.40: *Prillieuxina polyalthiae* **Hosag. & Abraham.**

Leaf parasites. Mycelium ectophytic, appressoria lateral. Thyriothecia oval, ellipsoidal, X or Y shaped, elongated with radiating cells, astomatous, dehisce longitudinally at the center; asci oval, octosporous, bitunicate; ascospores conglobate, uniseptate, brown.

Type sp. : *L. melastomatum* Mont.

Lembosia hosagoudarii Sivanesan & Sivas, Fungal Diversity 11: 163, 2002 (Figure 2.41).

Lembosia syzygiicola Hosag., Indian J. Forestry, 18: 276, 1995.

Colonies epiphyllous, dense, up to 2 mm in diameter. Hyphae substraight to flexuous, branching alternate at acute to wide angles, loosely reticulate, cells 22-35 × 4-7 μm. Appressoria 2-celled, scattered, alternate, straight to curved, 9-18 μm long; stalk cells cylindrical to cuneate, 4-7 μm long; head cells ovate, globose, entire, 4-11 × 4-7 μm. Thyriothecia scattered, initially circular, later elongated at maturity, 200-450 × 150-250 μm, margin fimbriate, dehiscing by a longitudinal slit at the center; asci globose, octosporous, up to 40 μm in diameter; ascospores oblong, conglobate, uniseptate, constricted at the septum, 13-30 × 4-9 μm wall smooth.

Distribution: Observed on the leaves of *Syzygium cumini* (L.) Skeels, (Myrtaceae).

Lembosia perseae Orejuela, Mycologia 36: 449, 1944 (Figure 2.42).

Colonies epiphyllous, dense, up to 2 mm in diameter, rarely confluent. Hyphae flexuous to crooked, branching irregular at acute angles, loosely reticulate, cells 15-26 × 2-4 μm. Appressoria few, scattered, alternate to unilateral, oval, globose, broad based, entire, 7-9 × 4-7 μm. Thyriothecia scattered to grouped, oval, elongated, Y shaped, 330-500 × 160-200 μm, dehisce longitudinally at the center, margin fimbriate; asci few, globose to ovate, up to 33 μm in diam.; ascospores, conglobate, brown, uniseptate, constricted at the septum, 13-20 × 7-9 μm, wall smooth.

Distribution: Observed on the leaves of *Persea macrantha* (Nees) Kosterm. (Lauraceae).

Lembosia terminaliae-chebulae Hosag., Abraham & Crane, Mycotaxon 68: 20, 1998. (Figure 2.43)

Colonies epiphyllous, dense, up to 2 mm in diameter, confluent. Hyphae substraight, branching irregular at acute to wide angles, loosely reticulate cells 14-18 × 4-7 μm. Appressoria scattered, unicellular, ovate, entire, broad based, 4-7 × 4-11 μm. Thyriothecia scattered, initially orbicular, later elongated, split longitudinally at the center, 350-500 × 165-200 μm, margin crenate to fimbriate; asci globose to ovate, octosporous, 35-46 × 15-22 μm; ascospores oblong, conglobate, 1-septate, and slightly constricted at the septum, initially hyaline, later becoming brown, 13-20 × 7-11 μm, wall tuberculate.

Distribution: Observed on the leaves of *Terminalia chebula* Retz. (Combretaceae).

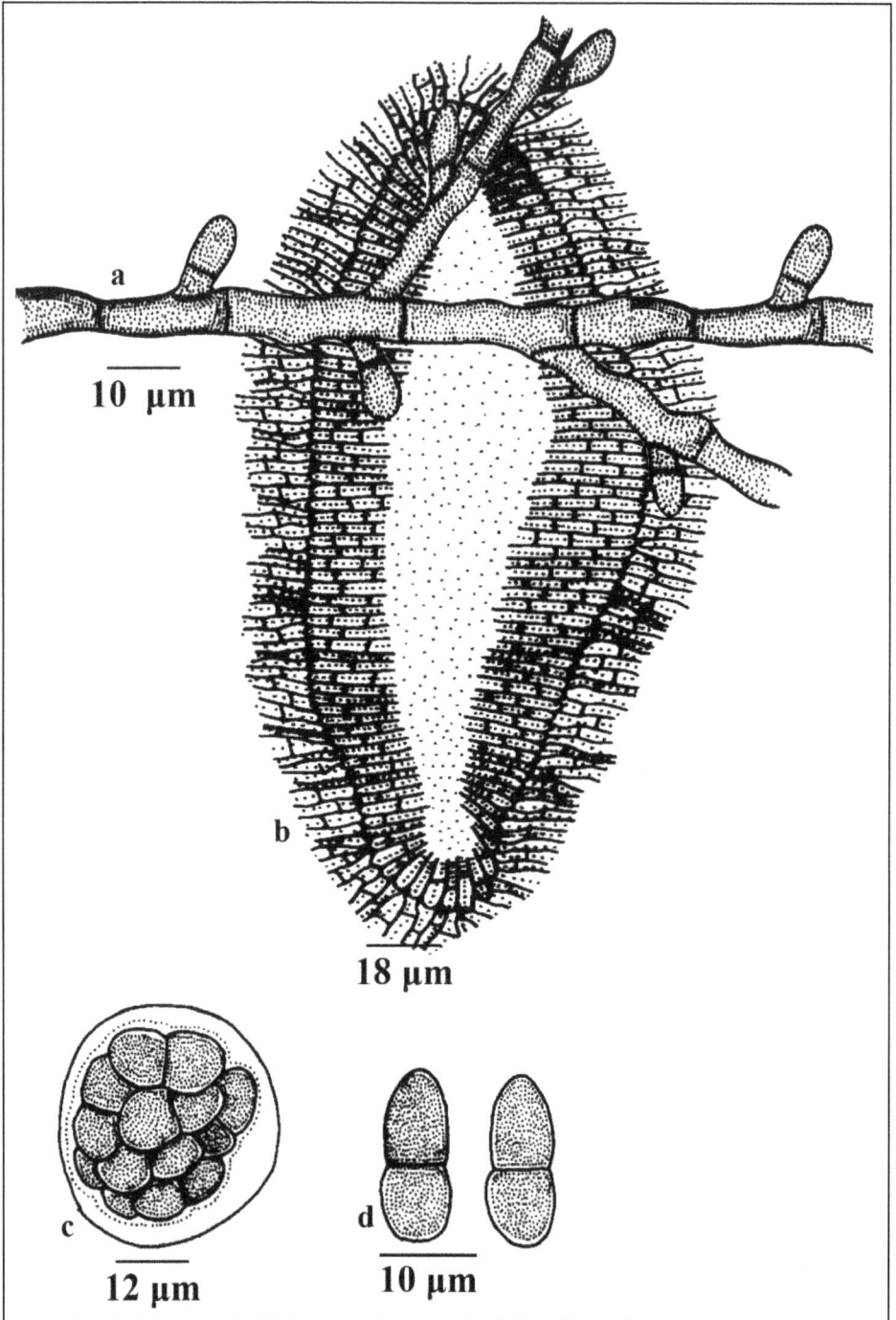

Figure 2.41: *Lembosia hosagoudarii* Sivanesan & Sivas.

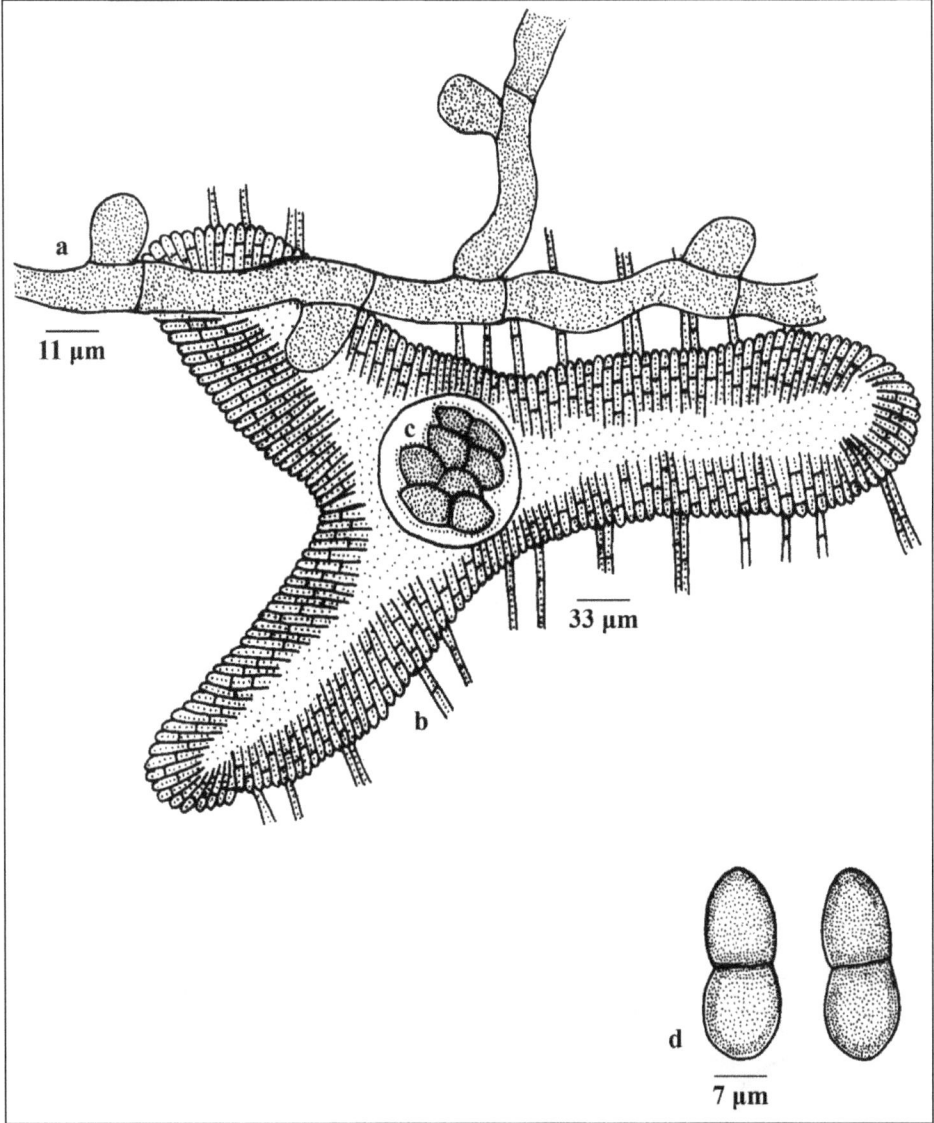

Figure 2.42: *Lembosia perseae* **Orejuela.**

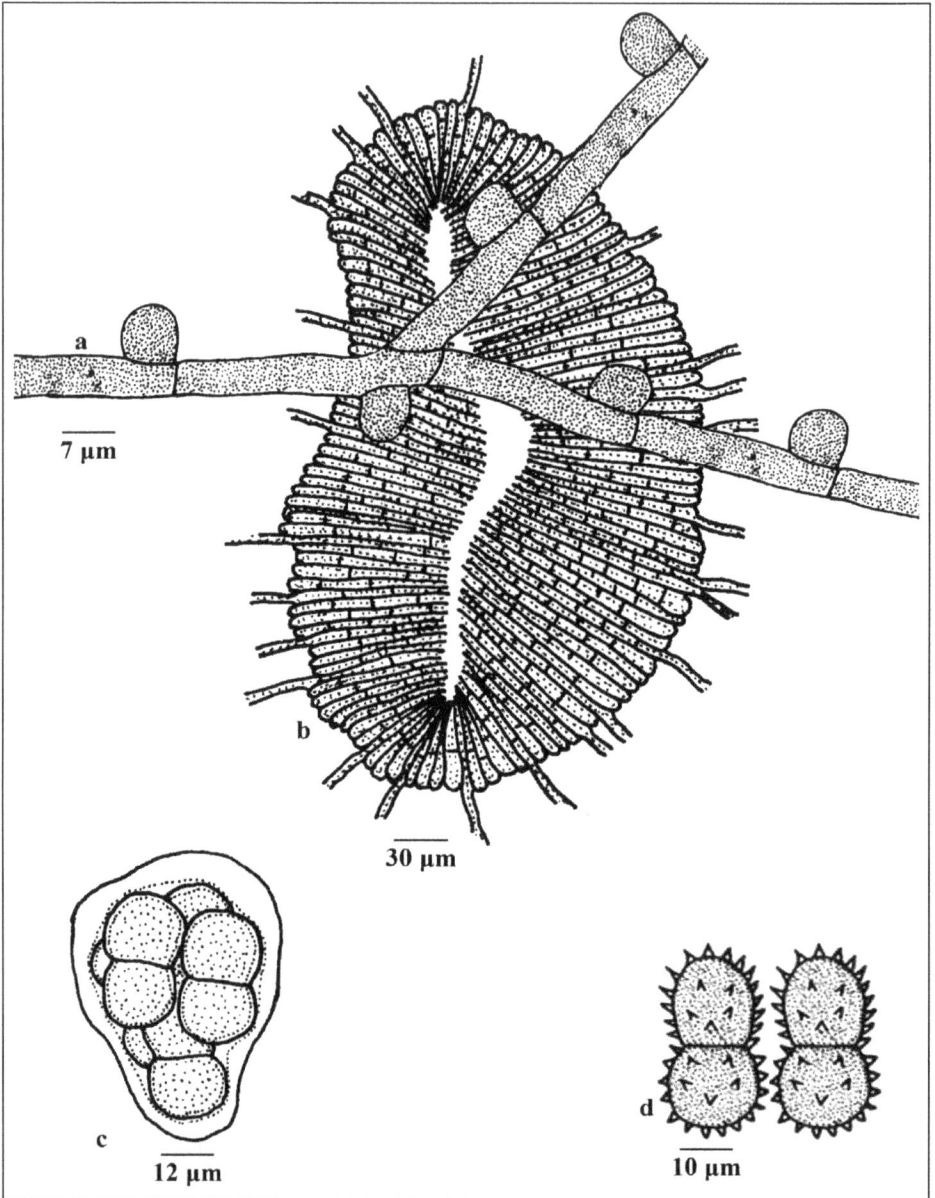

Figure 2.43: *Lembosia terminaliae-chebulae* **Hosag.** *et al.*

THE GENUS *ECHIDNODELLA*

Echidnodella Theiss. & Sydow, Ann. Mycol. 15: 422, 1917; Muller & Arx, Beitr. Krypt. Schw. 11:118, 1962; Luttrell in Ainsworth *et al.* (eds.). The Fungi. An advanced Treatise 4: 207, 1973; Arx & Muller, Stud. Mycol. 9: 46, 1975; Bilgrami, Jamaluddin & Rizwi, Fungi of India p. 185, 1991; Hosag., Abraham & C.K. Biju, J. Mycopathol. Res. 39: 62, 2001; Singh, Duke, Bhandari & Jain, J. Econ. Taxon. Bot. 30: 187, 2008.

Leaf parasites. Mycelium ectophytic, appressoria absent, hypostroma absent. Thyriothecia oval, ellipsoidal, X or Y shaped, elongated with radiating cells, astomatous, dehisce longitutionally at the center; asci oval, octosporous, bitunicate; ascospores brown, conglobate, uniseptate.

Type sp. : *E. linearis* (Sydow) Theiss. and Sydow

Echidnodella polyalthiae Hosag., J. Econ. Taxon. Bot. 28: 189, 2004 (Figure 2.44).

Colonies epiphyllous, dense, crustose, up to 4 mm in diameter, confluent. Hyphae substraight to crooked, branching irregular at acute angles, loosely to closely reticulate, cells 13-20 × 2-4 µm. Thyriothecia initially orbicular, elongated at maturity, mostly scattered, rarely connate, 370-1000 × 150-200 µm, fimbriate at the margin, fringed hyphae crooked, thyriothecia dehisce longitudinally at the center; asci ovate to globose, octosporous, 40-48 × 26-40 µm; ascospores conglobate, uniseptate, constricted at the septum, brown, 22-26 × 11-17 µm, upper cell ovate to globose, larger, lower cell smaller and measure 5-7 × 4-7 µm, wall smooth.

Distribution: Observed on the leaves of *Polyalthia* sp. (Annonaceae).

THE GENUS *ECHIDNOIDES*

Echidnoides Theiss. & Sydow, Ann. Mycol.15: 422, 1917; Hansf., Mycol. Pap. 15: 167, 1946; Arx & Muller, Stud. Mycol. 9: 46, 1975; Hosag., Abraham & C.K. Biju, J. Mycopathol. Res. 39: 62, 2001.

Lembosiodothis Hohn., Ann. Mycol.15: 369, 1917.

Maurodothella Arn., Les Asterinees 1: 124, 1918.

Leaf parasites. Mycelium ectophytic, appressoria absent, hypostroma subcuticular or intra-epidermal. Thyriothecia oval, ellipsoidal, X or Y shaped, elongated with radiating cells, astomatous, dehisce longitudinally at the center; asci oval, octosporous, bitunicate; ascospores conglobate, brown, uniseptate.

Type sp.: *E. litura* (Sydow) Theiss. and Sydow.

Echidnoides pandanicola Hosag. & Hanlin, New Botanist 22: 191, 1995 (Figure 2.45).

Colonies epiphyllous, dense, crustose, up to 5 mm in diameter often confluent. Hyphae flexuous, branching irregular at acute to wide angles, loosely reticulate, cells 20-30 × 4-7 µm. Appressoria absent. Thyriothecia scattered to grouped, often connate, initially orbicular and later elongated at maturity, 300-500 x 110-150µm, margin

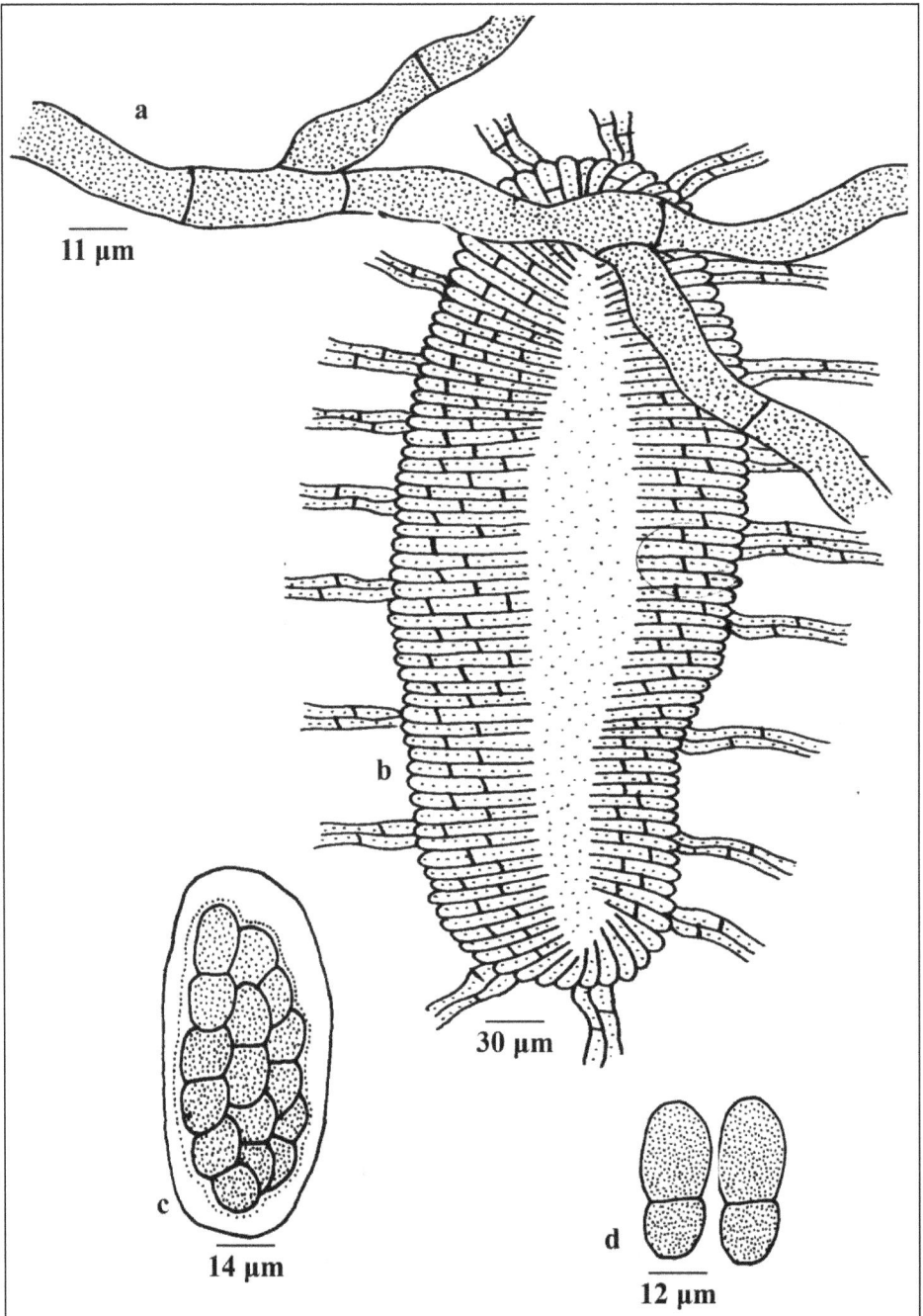

Figure 2.44: *Echidnodella polyalthiae* Hosag.

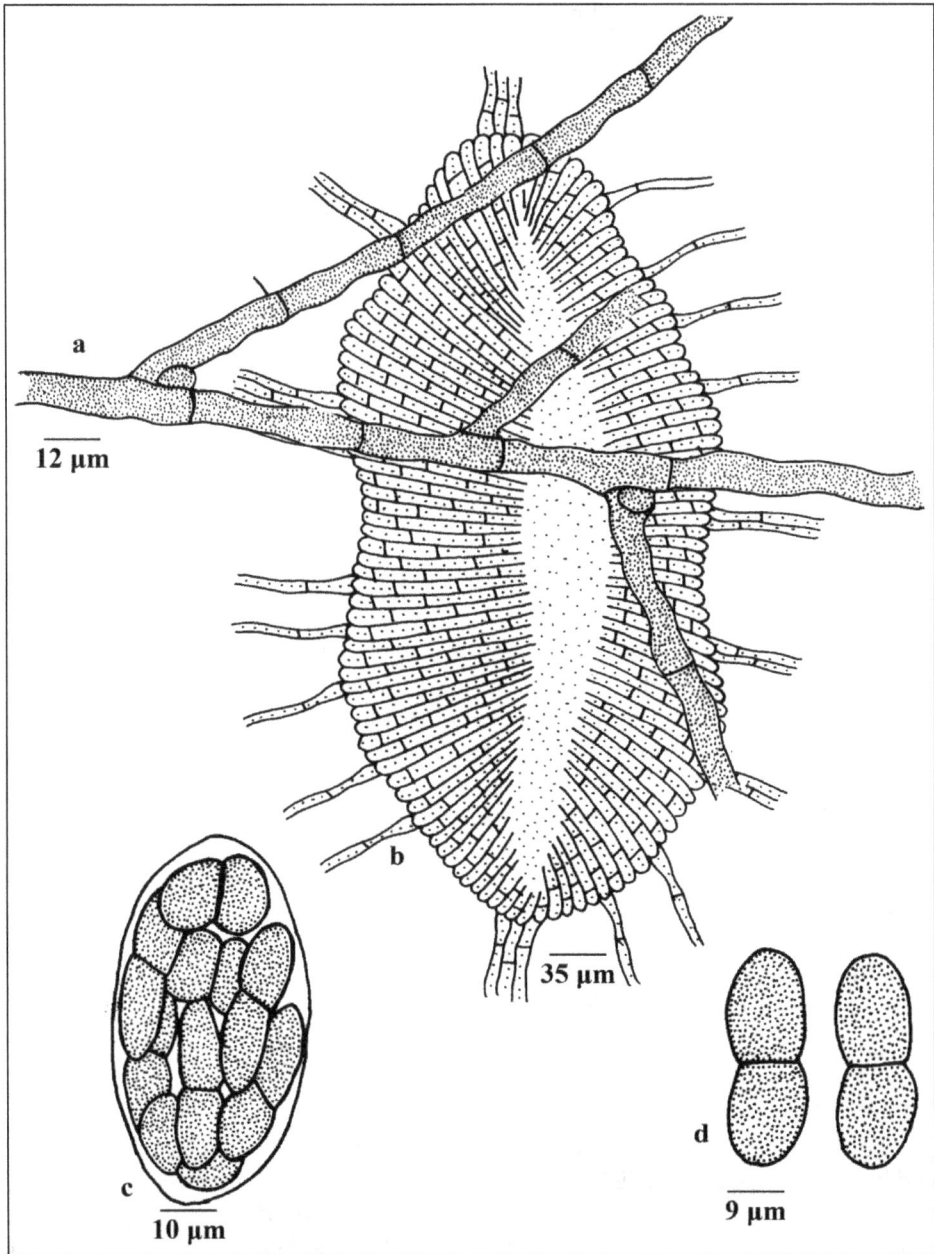

Figure 2.45: *Echidnoides pandanicola* **Hosag. & Hanlin.**

fimbiriate, fringed hyphae flexuous, longitudinally dehisced at the center; asci globose initially, slightly ovate to clavate at maturity, octosporous, 43-53 × 24-34 µm; ascospores conglobate, oblong, brown at maturity, 1-septate, constricted at the septum, 24-28 × 11-13 µm, wall smooth.

Distribution: Observed on the leaves of *Pandanus* sp. (Pandanaceae).

THE GENUS *EUPELTE*

Eupelte Sydow, Ann. Mycol. 22: 426, 1924; Hansf., Mycol. Pap. 15: 168, 1946; Muller & Arx, Beitr. Krypt. Schw. 11:137, 1962; Arx & Muller, Stud. Mycol. 9: 43, 1975; Hosag., Abraham & C.K. Biju, J. Mycopathol. Res. 39: 62, 2001.

Maurodothina Arn. ex Piroz. and Shoemaker, Can. J. Bot. 48: 1326,1970.

Anamorph: *Sporidesmium* sp.

Leaf parasites. Hyphae partly superficial and partly immersed. Conidia present, cylindrical, obclavate, broadly rounded at the apex, truncate at the base. Thyriothecia orbicular, elliptic to elongated, dehisce stellately, vertically at the center; asci clavate, spherical, octosporous, bitunicate; ascospores conglobate, uniseptate, brown.

Type sp.: *E. amicta* Syd.

Eupelte amicta Sydow, Ann. Mycol. 22; 426, 1924 (Figure 2.46).

Colonies epiphyllous, dense, crustose up to 10 mm in diameter, confluent. Hyphae partly superficial and partly immersed, superficial hyphae brown, septate, flexuous, regularly branched at acute to wide angles, loosely to closely reticulate, cells 15-31 × 3-5 µm. External mycelium enters the host through the stomata extends up to palisade tissues. Conidiophores arise from the external mycelium, brown, 0-1 septate, erect, simple, smooth, 23-28 µm long; conidiogenous cells terminal, integrated, monoblastic, determinate; conidia brown, 0-3 septate, not constricted, straight to curved, cylindrical, obclavate, broadly rounded at the apex, truncate at the base, wall smooth, 22-31 × 9-11 µm. Thyriothecia scattered to grouped, initially orbicular, later elliptic to elongated, X or Y shaped, dehisce stellately, vertically at the center, 400-700 × 100-160 µm; asci born on basal hymenium, clavate, bitunicate, becomes spherical at maturity, octosporous, 36-40 × 14-18 µm; ascospores conglobate brown, oblong, 1-septate, constricted, 17-20 × 9-11 µm,wall smooth.

Distribution: Observed on the leaves of *Olea dioica* Roxb. (Oleaceae).

Discussion

Fungi have played major role in the establishment and maintenance of ecosystem and have developed mutualistic relationship with many other organisms. In general, much work is still to be done on the diversity of fungi unlike many taxonomic group, fungi may reach their highest level of diversity. In tropics, there is not a single area for which the fungi are even relatively well known and it is impossible to prepare regional accounts for any, but a very few group on the basis of collection that are available. Even less explored are the ecological roles that fungi play and their potential to develop into pathogens or to serve as beneficial agents. It is found that total number

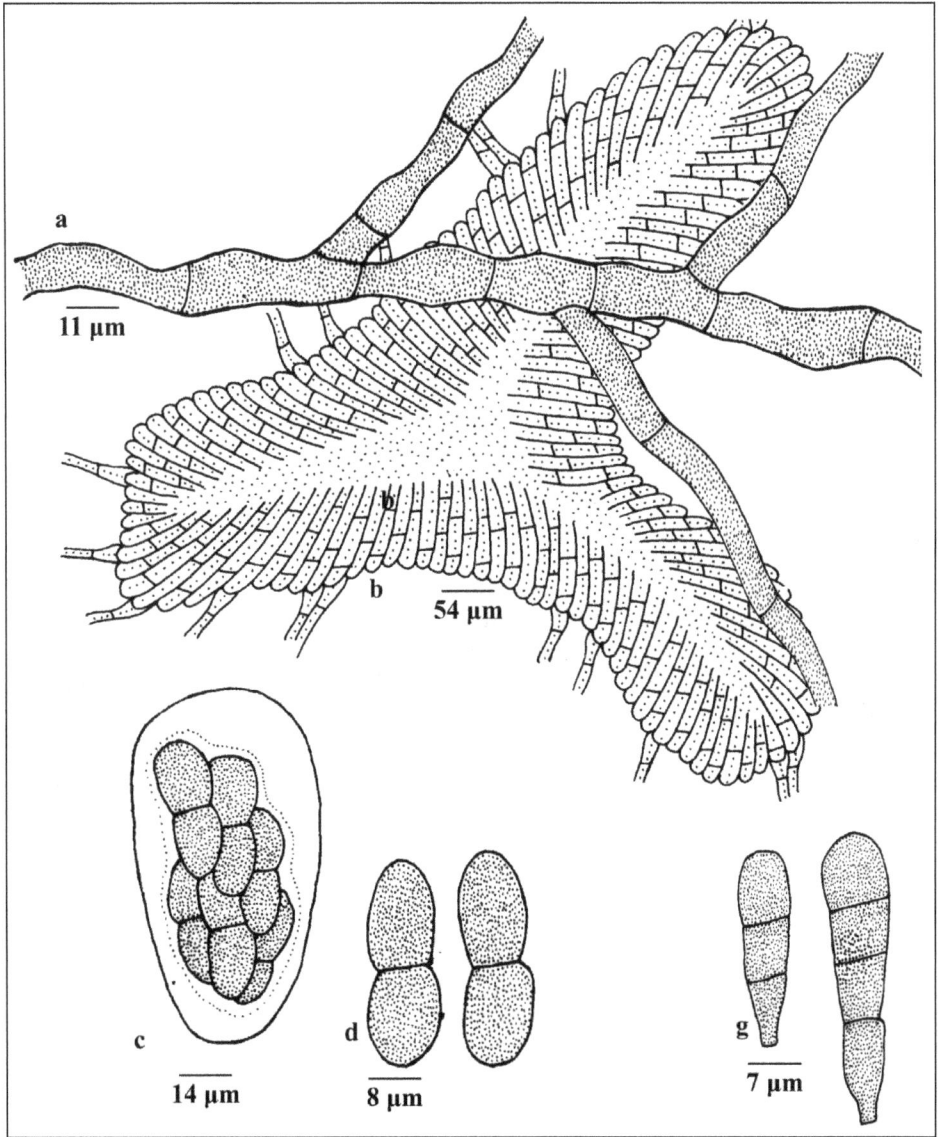

Figure 2.46: *Eupelte amicta* **Sydow.**

of fungi is also slowly decreasing (Cherfas, 1991). At many instances, it is possible that we may lose species to extinction much before they have been discovered.

Many fungi parasitize living plants, in the sense that they drive from these plants for the food materials necessary for growth but confer no benefits in returns. For some fungi the association is obligatory, they may exists outside their host in some dormant form such as spore but their growth under natural condition is confined to their appropriate host. Whatever the capabilities of the fungus outside the plants, its parasitism inevitably involves an interference with host physiology and frequent changes occur which are detrimental to the plants. The net result is a deviation from the normal functioning of these physiological processes which is called as disease. In this respect a parasite is also a pathogen.

Fungi which parasitize plants would appear to gain two main advantages by doing so they avoid competition for food and to certain extent and they are protected from those changes in environmental conditions which are adverse to their mycelial growth. Success in parasitism may be judged by the extent to which these advantages are enjoyed. Black mildews are mostly ectoparasites and do not cause destructable pathogenic effect on the host plants. However, the black colonies of these fungi increase the temperature in the infected parts and cause physiological imbalance. They decrease the photosynthetic efficiency of the plants affect the hormonal and phenolic compound level and in short affect the efficiency of the plants (Hosagoudar *et al.*, 2002).

Asterinaceous fungi are host specific because they must circumvent, tolerate and overcome the specific resistance factors of the particular host. The resistance factors may be physical barrier, fungitoxic chemicals produced by the host either before or after the response to the infection and environmental influence. The general answer for restricting the speciation to the host family level was the pathogens were adapted to a species, genus or family may be that the plants within that taxon have similar types of defence chemicals (Deacon, 1997). The species concept of Asterinaceous fungi were based on the respective host plants and also on the morphological aspects of the fungus (Diodge, 1942; Hansford, 1946a,b; Hosagoudar and Goos, 1996).

About 30 genera have been assigned to Asterinaceae (Arx and Muller, 1975). The new family Lembosiaceae in the order Asterinales has been segregated from Asterinaceae to include the genera having ellipsoidal to elongated, X or Y shaped thyriothecia split or dehisce longitudinally (Hosagoudar *et al.*, 2001a). In the present study, about 46 species of Asterinaceous fungi belonging to 8 genera were observed in Sirumalai hills (Table 2.1). *Asterina* is the dominant genus in the family Asterinaceae and it consists of 34 species. The dominant distribution of *Asterina* in the different areas of Sirumalai hills revealed its wider distribution. *Prillieuxina* was one among the genus of the family Asterinaceae represented by 4 species distributed in different forest areas of Sirumalai hills. The genus *Asterolibertia* and *Ishwaramyces* was represented by single species. *Lembosia*, the dominant genus of the family Lembosiaceae consist of 3 species. *Echidnodella*, *Echinoides* and *Eupelte* were represented by single species (Figure 2.47).

Table 2.1: List of Asterinaceous Fungi and their Respective Host Taxa of Sirumalai Hills

Sl.No.	Name of the Taxa	Name of the Host Plant	Name of the Host Family
1.	*Asterina balakrishnanii*	*Solanum torvum*	Solanaceae
2.	*Asterina chukrasiae*	*Chukrasia tabularis*	Meliaceae
3.	*Asterina cipadessae*	*Cipadessa baccifera*	Meliaceae
4.	*Asterina cissi*	*Cissus* sp.	Vitaceae
5.	*Asterina clausenicola*	*Melicope lunu-ankenda*	Rutaceae
6.	*Asterina combreti*	*Calycopteris floribunda*	Combretaceae
7.	*Asterina congesta*	*Santalum album*	Santalaceae
8.	*Asterina dallasica*	*Trema orientalis*	Ulmaceae
9.	*Aterina deightonii*	*Dendrophthoe* sp.	Loranthaceae
10.	*Asterina diospyri*	*Diospyros* sp.	Ebenaceae
11.	*Asterina diplocarpa*	*Sida cordata*	Malvaceae
12.	*Asterina elaeocarpi*	*Elaeocarpus tuberculatus*	Elaeocarpaceae
13.	*Asterina girardiniae*	*Girardinia diversifolia*	Urticaceae
14.	*Asterina helicteridis*	*Helicteres isora*	Sterculiaceae
15.	*Asterina hibisci*	*Hibiscus* sp.	Malvaceae
16.	*Asterina hydnocarpi*	*Hydnocarpus* sp.	Flacourtiaceae
17.	*Asterina indica*	*Symplocos cochinchinensis*	Symplocaceae
18.	*Asterina jambolana*	*Syzygium* sp.	Myrtaceae
19.	*Asterina lanneae*	*Lannea coromandelica*	Anacardiaceae
20.	*Asterina lawsoniae*	*Lawsonia inermis*	Lythraceae
21.	*Asterina lobulifera* var. *indica*	*Glochidion* sp.	Euphorbiaceae
22.	*Asterina lobulifera*	*Glochidion* sp.	Euphorbiaceae
23.	*Asterina malloticola*	*Mallotus philippensis*	Euphorbiaceae
24.	*Asterina melicopecola*	*Melicope lunu-ankenda*	Rutaceae
25.	*Asterina mimusopsidicola*	*Mimusops elengi*	Sapotaceae
26.	*Asterina murrayae*	*Murraya koenigii*	Rutaceae
27.	*Asterina naraveliae*	*Naravelia zeylanica*	Ranunculaceae
28.	*Asterina phyllanthigena*	*Phyllanthus* sp.	Euphorbiaceae
29.	*Asterina pongalaparensis*	*Jasminum* sp.	Oleaceae
30.	*Asterina pusilla*	*Premna corymbosa*	Verbenaceae
31.	*Asterina saracae*	*Saraca asoca*	Fabaceae
32.	*Asterina tertia*	*Adhatoda zeylanica*	Acanthaceae
33.	*Asterina visci*	*Viscum* sp.	Loranthaceae
34.	*Asterina wrightiae*	*Wrightia tinctoria*	Apocyanaceae
35.	*Asterolibertia mangiferae*	*Mangifera indica*	Anacardiaceae

Contd...

Table 2.1–*Contd...*

Sl.No.	Name of the Taxa	Name of the Host Plant	Name of the Host Family
36.	*Ishwaramyces flacourtiae*	*Flacourtia montana*	Flacourtiaceae
37.	*Prillieuxina anamirtae*	*Anamirta cocculus*	Menispermaceae
38.	*Prillieuxina elaegni*	*Elaeagnus kologa*	Elaeagnaceae
39.	*Prillieuxina ixorigena*	*Ixora coccinea*	Rubiaceae
40.	*Prillieuxina polyalthiae*	*Polyalthia longifolia*	Annonaceae
41.	*Lembosia hosagoudarii*	*Syzygium cumini*	Myrtaceae
42.	*Lembosia perseae*	*Persea macrantha*	Lauraceae
43.	*Lembosia terminaliae-chebulae*	*Terminalia chebula*	Combretaceae
44.	*Echidnodella polyalthiae*	*Polyalthia* sp.	Annonaceae
45.	*Echidnoides pandanicola*	*Pandanus* sp.	Pandanaceae
46.	*Eupelte amicta*	*Oleadioica* sp.	Oleaceae

Asterinaceous fungi have wide range of distribution. *Asterina pusilla* and *Prillieuxnia anamirtae* were reported from Philippines, the similar species were also

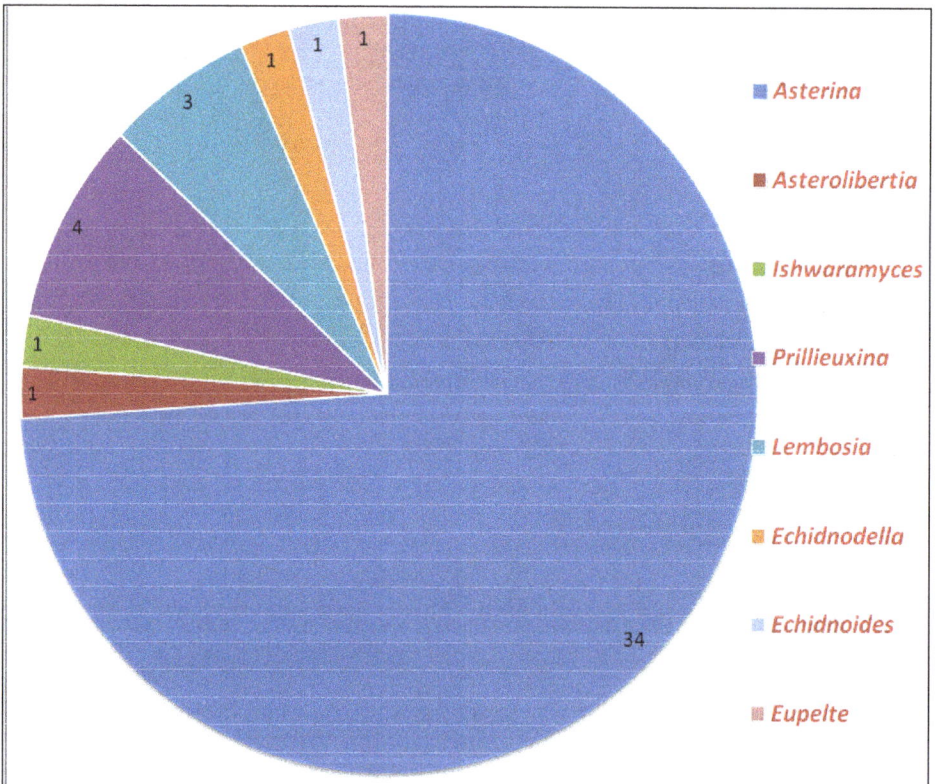

Figure 2.47: Distribution of Asterinaceous Fungi in Sirumalai Hills.

reported from Sirumalai hills. *Asterina murrayae* was reported from of Sri Lanka (Hansford, 1947) and it was also found in Sirumalai. *Asterina combreti* reported from Africa (Sydow, 1910) was also found in Sirumalai. *Asterina dallasica* was reported from Borneo Islands (Petrak, 1954) and it was also found in Sirumalai. This result revealed the affinity of Asterinaceous fungal flora to other part of the country.

In Sirumalai hills more than 1000 species of flowering plants belonging to approximately 120 families are reported. Asterinaceous fungi are known to occur on 42 species of flowering plants belonging to 31 families. Three species of Asterinaceous fungi are known to occur in the family Euphorbiaceae and Rutaceae. Anacardiaceae, Meliaceae, Malvaceae, Myrtaceae, Flacourtiaceae, Combretaceae, Loranthaceae and Oleaceae and are found to be infected with 2 species. Records indicated that, species are restricted to a narrow host range mainly to a particular host species or genus, but rarely extended to related genera within a single host family (Hansford, 1961). These fungi are host specific and differ widely in their known host range with a few exceptions of common host selections. Differences in host range indicate phylogenetic divergence. The members of Asterinaceae and Meliolaceae are found mainly on Rubiaceae, Combretaceae, Malvaceae, Myrtaceae, Oleaceae, Euphorbiaceae and Rutaceae. Overlapping of host plant families between the two parasites may indicate that competition is occurring. The structural dissimilarities among them in the ascomata characters of dry dehiscence or slimy diffluence to release and disperse ascospores may influence their host range and distribution (Thaung, 2006).

The leaves are the most important part of the body due to photosynthetic activity largely inherent to them. Plant leaves provide a very suitable habitat for the growth and development of fungal organisms and such leaf inhabiting fungi are known as foliicolous/foliar. Actually, the fungal pathogens attack the living leaves and reduce their productivity of photosynthates (foods and other valuable substances) by damaging photosynthetically active regions and also by bringing about quantitative reduction and qualitative dearrangement of living tissues of the host in multiple ways. The area of leaf invaded by foliicolous hyphomycetes usually becomes distinct due to the presence of fungus itself resulting in various kinds of local host responses ranging from discoloration to necrosis.

Majority of the fungi grow and sporulate at temperature greater than 15-20° C (Cooke and Whipps, 1993). Most of the fungi are adopted to grow in warm climates. Humidity plays a much more predominant role in influencing fungal diversity than does temperature alone. The effect of seasonality on microfungal diversity is connected intimately with the effect of water and temperature. Significant seasonal variation in either temperature or water availability will result in a fungal biota different from that developing when conditions are constant (Lodge and Cantrell, 1995). Factors such as the aspect of the site, topographic variation and soil type will affect microfungal diversity.

References

Arx, J. A. V. and Muller, E. 1975. A re-evaluation of the bitunicate ascomycetes with key to the families and genera. *Stud. Mycol.* 9: 1-159.

Blackwell, M. 2011. "The Fungi: 1, 2, 3. 5.1 million species?" _American J. Bot._ 98 (3): 426–438.

Chandra Prabha, A. 2009. Diversity and distribution of Asterinaceous fungi in Kerala state. _Ph.D. Thesis_ submitted to the University of Kerala, Thiruvananthapuram, Kerala, India.

Cherfas, J. 1991. Disappearing Mushrooms. Another mass extinction? _Science_ 254: 1458.

Cooke, M. C.1884. Some exotic fungi. _Grevillea_ 12: 85.

Cooke, R. C. and Whipps, J. M. 1993. _Ecophysiology of Fungi._ Blackwell Scientific Publishers, Oxford, England.

Crane, J. L. and Jones, A. G. 1997. An annotated catalogues of types of the University of Illionis Mycological collection (ILL). _Illinois Biol. Monograph_ 58: 1-365.

Deacon, J. W. 1997. _Modern Mycology._ Blackwell Science Ltd. Oxford, London.

Doidge, E. M. 1942. A revision of the South African Microthyriaceae. _Bothalia_ 4: 273-344.

Fries, E. M. 1849. _Summa vegetabilium sandinaviae._ Sectioposterior. _Holmiae et Lipsiae._ pp. 313.

Gamble, J.S. and Fischer, C.E.C. 1915-35. _Flora of Presidency of Madras_, Bishen Singh Mahendrapal Singh, Dehradun.

Ganeshaiah, K.N. 2012. _Plants of Western Ghats, Vol. 1 and_ 2. National Bio-resource Development Board, DBT, New Delhi.

Hansford, C. G. 1946a. Contribution towards the fungus flora of Uganda –VIII. New records. _Proc. Linn. Soc. London_ 157:132-212.

Hansford, C. G. 1946b. The foliicolous Ascomycetes, their parasites and associated fungi. _Mycol. Pap._ 15: 1-240.

Hansford, C. G. 1947. New or interesting tropical fungi-I. _Proc. Linn. Soc. London_ 158: 28-50.

Hansford, C. G. 1961. The Meliolineae. A monograph. _Sydowia_ 2: 1-806.

Hawksworth, D. L. 2006. The fungal dimension of biodiversity: magnitude, significance, and conservation. _Mycological Research_ 95 (6): 641–655.

Hawksworth, D. L. and Colwell, R. R. 1992. Microbial diversity: Biodiversity among microorganisms and its relevance. _Biodiver. Conserv._ 1: 221-226.

Hosagoudar, V. B. 2003a. Asterinaceae of India. _Zoos´ Print J._ 18: 1280-1285.

Hosagoudar, V. B. 2003b. Studies on foliicolous fungi – XII. New species, new records and hyperparasites. _Zoos´ Print J._ 18: 1037-1040.

Hosagoudar, V. B. and Abraham, T. K. 1998a. Four new foliicolous Ascomycetes from Kerala, India. _Mycol. Res._ 102: 184-186.

Hosagoudar, V. B. and Abraham, T. K. 1998b. Some interesting foliicolous thyriotheceous Ascomycetes from Kerala. _Indian Phytopathol._ 51: 389-392.

Hosagoudar, V. B. and Abraham, T. K. 2000. A list of *Asterina* Lev. species based on the literature. *J. Econ. Taxon. Bot.* 24: 557-587.

Hosagoudar, V. B., Abraham, T. K. and Pushpangadam, P. 1996a. *Fungi of the Kerala.* Tropical Botanic Garden and Research Institute, Palode, Thiruvananthapuram.

Hosagoudar, V. B., Abraham, T. K. and Biju, C. K. 1999. Notes on some foliicolous fungi from Kerala, India. *J. Mycopathol. Res.* 37: 25-28.

Hosagoudar, V. B., Abraham, T. K. and Biju, C. K. 2001a. Re-evaluation of the family Asterinaceae. *J. Mycopathol. Res.* 39: 61-63.

Hosagoudar, V. B. and Agarwal, D. K. 2003. Studies on foliicolous fungi – IX. *Indian Phytopathol.* 56: 98-101.

Hosagoudar,V. B. and Agarwal, D. K. 2008. *Taxonomic studies of Meliolales. Identification manual.* International Book Distributors, Dehradun, India.

Hosagoudar, V. B., Balakrishnan, N. P. and Goos, R. D. 1996a. Some *Asterina* species from Southern India. *Mycotaxon* 59: 167-187.

Hosagoudar, V. B., Balakrishnan, N. P. and Goos, R. D. 1996b. Some *Asterinella, Asterostomella* and *Echidnodella* species from Southern India. *Mycotaxon* 58: 489-498.

Hosagoudar, V. B., Biju, C. K. and Abraham, T. K. 2002. Diversity of foliicolous micromycobionts in munnar and wyanad forest regions of Kerala. *J. Mycopathol. Res.* 40: 191-196.

Hosagoudar, V. B., Biju, C. K. and Abraham, T. K. 2004. Studies on foliicolous fungi-II. *J. Econ. Taxon. Bot.* 28: 183-186.

Hosagoudar, V. B. and Chandra Prabha, A. 2008a. *Bramhamyces,* a new anamorphic genus India. *Indian J. Sci and Technol.,* 2(6): 17.

Hosagoudar, V. B. and Chandra Prabha, A. 2008b. New Asterinaceae members from Kerala, India. *Indian J. Sci and Technol.,* 2(6): 15-16.

Hosagoudar, V. B. and Chandra Prabha, A. 2008c. Two new *Asterina* species from Kerala, India. *J. Adv. Pollen Spore Res. Indian J. Sci and Technol.,* 2(6): 18.

Hosagoudar, V. B. and Chandra Prabha, A. 2008d. Two new *Prillieuxina* species from Kerala, India (in press).

Hosagoudar, V. B. and Goos, R. D. 1994. Some *Asterina, Asterostomella* and *Lembosia* species from Southern India. *Mycotaxon* 52: 467-473.

Hosagoudar, V. B. and Kapoor, J. N. 1984. New Technique of mounting Meliolaceous fungi. *Indian Phytopathol.* 38: 548-549.

Joseph, J. 1999. Forest profile changes of Sirumalai hills by GIS approach. *Ph.D. Thesis* submitted to Gandhigram Rural Institute (Deemed University), Gandhigram, India.

Karuppusamy, S., 2007. Medicinal plants used by paliyan tribes of Sirumalai hills of southern India. *Nat. Prod. Radi.* 6 (5): 436 - 442.

Karuppusamy, S., Rajasekaran, K. M. and Kumuthakalavalli, R. 1999. Orchids of Sirumalai hills. *J. Swamy. Bot. Cl.*, 16: 73-74.

Karuppusamy, S., Rajasekaran, K. M. and Karmegam, N. 2001a. Endemic flora of Sirumalai hills (Eatern Ghats), South India. *J. Econ. Taxon. Bot.* 25 (2): 367-373.

Karuppusamy, S., Rajasekaran, K. M. and Karmegam, N. 2001b. Enumeration, ecology and ethnobotany of ferns of Sirumalai hills, South India. *J. Econ. Taxon. Bot.* 25, 631-634.

Karuppusamy, S., Rajasekaran, K. M. and Karmegam, N. 2002. Evaluation of Phytomedicines from street herbal vendors in Tamil Nadu, South India. *Ind. J. Trad. Knowl.* 1: 26 - 31.

Kirk, P. M., Cannon, P. F., David, J. C. and Stalpers, J. A. 2001. *Ainsworth and Bisby's Dictionary of Fungi- 9th edition*. CAB international, Oxon, UK.

Kottaimuthu, R., Ganesan, R., Natarajan, K., Brabhu and Vimala, M. 2008. Addition to the flora of Eastern Ghats, Tamil Nadu, India. *Ethnobotanical Leaflets* 12: 299 - 304.

Krishnankutty, N., Sujatha, T. R. and Jeyakumar, G. 2003. Leaf litter retention, transport and decomposition in tropical forest stream of Sirumalai hill, Eastern Ghats, South India. *Trop. Ecol.*, 44: 169 - 174.

Leveille, J. H. 1845. Champignons exotiques. *Ann. Sci. Nat. Ser.* 3: 38-71.

Lodge, D. J. and Cantrell, S. 1995. Fungal communities in wet tropical forest; variation in time and space. *Can. J. Bot.* 73: 1391-1398.

Maruthupandian, A., Mohan, V. R., Kottaimuthu, R. 2011. Ethno-medicinal plants used for the treatment of diabetes and jaundies by Palliyar tribals in Sirumalai hills, Western Ghats, Tamil Nadu, India. *Ind. J. Nat. Pro. and Resources.*, 2(4): 493 - 497.

Mibey, R. K. and Hawksworth, D. L. 1997. Meliolaceae and Asterinaceae of Shimba Hills, Kenya. *Mycol. Pap.* 174: 1-108.

Muller, E. and Arx, J. A. V. 1962. Die Gattungen der didymosporen Kryptogamenfl. Schweiz 11:1-922 *Pyrenomyceten*. Beitr.

Mueller, G. M. and Schmit, J. P. 2006. "Fungal biodiversity: what do we know? What can we predict?" *Biodiversity and Conservation* 16: 1–5.

Nayar, T. S., Rasiya Beegam, A. Mohanan, N. and Rajkumar, G. 2006. *Flowering plants of Kerala: A Handbook*. Tropical Botanic Garden and Research Institute, Thiruvananthapuram, Kerala.

Ouyang, Y., Song, B. and Hu, Y. 1995. Studies on the taxonomy of *Asterina* in China – I. *Acta Mycol. Sinica* 14: 241-247.

Ouyang, Y., Song, B. and Hu, Y. 1996. Studies on the taxonomy of *Asterina* in China – I. *Acta Mycol. Sinica* 15: 86-92

Pallithanam, J. M. 2001. *A pocket flora of Sirumalai hills, South India*. The Rapinat Herbarium, St. Joseph's college, Tiruchirapalli, pp 410.

Petrak, F. 1954. Beitrage Zur Pilzflora Von Britisch Nord, Borneo. *Sydowia* 8:12-26.

Rajasekaran, K. M. 2004. Ecological study of the flora of Sirumalai hill. *Ph.D. Thesis* submitted to Gandhigram Rural Institute (Deemed University), Gandhigram, India.

Rangaswamy, G. 1975. *Diseases of Crop Plants in India*. Prentice-Hall of India, Pvt. Ltd, New Delhi.

Saccardo, P. A. 1883. *Sylloge Fungorum* 2: 813.

Saccardo, P. A. 1924. *Sylloge Fungorum* 24: 450-451.

Sharma, B.D. and Sanjappa, M. 1993. *Flora of India Vol. 4*, Botanical Survey of India, Calcutta.

Silva, M. D. and Pereira, O.L. 2008. Black mildews disease of the Neotropical Orchid, *Epidendrum secundum* caused by *Lembosia epidendri* from Minas Gerais, *Brazil. Mycotaxon* 104: 385-390.

Sivanesan, A. and Shivas, R. G. 2002. New species of *Lembosia* and *Lembosina* from Australia. *Fungal Diversity* 11:159- 168.

Song, B. and Hosagoudar, V. B. 2003. A list of *Lembosia* species based on the literature. *Guizhou Science* 21: 93-101.

Stevens, F. L. and Ryan, M. H. 1939. The Microthyriaceae. *Illinois Biol. Monograph*. 17: 1-138.

Sydow, H. 1910. *Engl. Bot. Jahrb*. 45: 264.

Sydow, H.1938. Novae Fungorum species- XXVI. *Ann. Mycol*. 36: 156-253.

Thaung, M. M. 2006. Biodiversity of Phylloplane Ascomycetes in Burma. *Austral. Mycol*. 25: 5-23.

Theissen, F. 1913a. Hemisphariales. *Ann. Mycol*. 11: 468-469.

Theissen, F.1913b. *Lembosia-* Studien. *Ann. Mycol*. 11: 425-467.

Yamamoto, W. 1956. The Formosan species of Microthyriaceae -I. Sci. Rep. *Hyogo Univ. Agric. Ser. Agric. Biol*. 2: 33-36.

Chapter 3
Floristic Diversity of Makangiri District of Odisha, India

Deepak Kumar Sahu, Sanjit Biswas and Nabin Kumar Dhal

CSIR-Institute of Minerals and Materials Technology,
Bhubaneswar – 13, Odisha
E-mail: nkdhal@immt.res.in, nkd.radha@gmail.com

ABSTRACT

Extensive and intensive field surveys have been conducted during the year 2008 to 2011 in different forest pockets of the under-explored Malkangiri district with a view to assess the floristic wealth of the district. The survey includes a total of 1296 plant specimens of which 763 different species belonging to 505 genera and 147 families were identified. Out of these, 745 species are angiosperms including 596 dicots and 149 monocots, 16 fern species (belonging to 13 genera and 13 families) and 2 gymnosperms. The collection embraces as many as 437 herbs, 87 shrubs, 160 trees and 79 climber species. Out of total 147 families, the 10 dominant families in descending order are Fabaceae (62 sp.), Poaceae (57 sp.), Asteraceae (37 sp.), Euphorbiaceae (33 sp.), Acanthaceae (32 sp.), Rubiaceae (30 sp.), Cyperaceae (29 sp.), Caesalpiniaceae and Lamiaceae (20 sp.), Malvaceae (19 sp.) and Scrophulariaceae (18 sp.). Furthermore, some threatened taxa like *Blepharispermum subsessile, Stemona tuberosa, Litsea glutinosa*, etc. were also collected having immense medicinal value. Such diversified species indicate the floral diversity richness of the district that should be conserved as these invaluable wild resources are vanishing from the wild at an alarming rate.

Keywords: Eastern Ghats, Malkangiri, Flora, Odisha.

Introduction

Malkangiri, the southernmost district of Odisha was carved out as a separate district on 2nd October from the erstwhile Koraput district in 1992. It lies between 81° 10′ to 82° 00′ E longitudes and 17° 45′ to18° 40′ N latitudes (Figure 3.1) extending over an area of 5791 sq. km. It is surrounded on east by Koraput district of Odisha and Visakhapatnam and East Godavari districts of Andhra Pradesh, on west by Bastar district of Chhattisgarh, on north by Koraput district of Odisha and on south by East

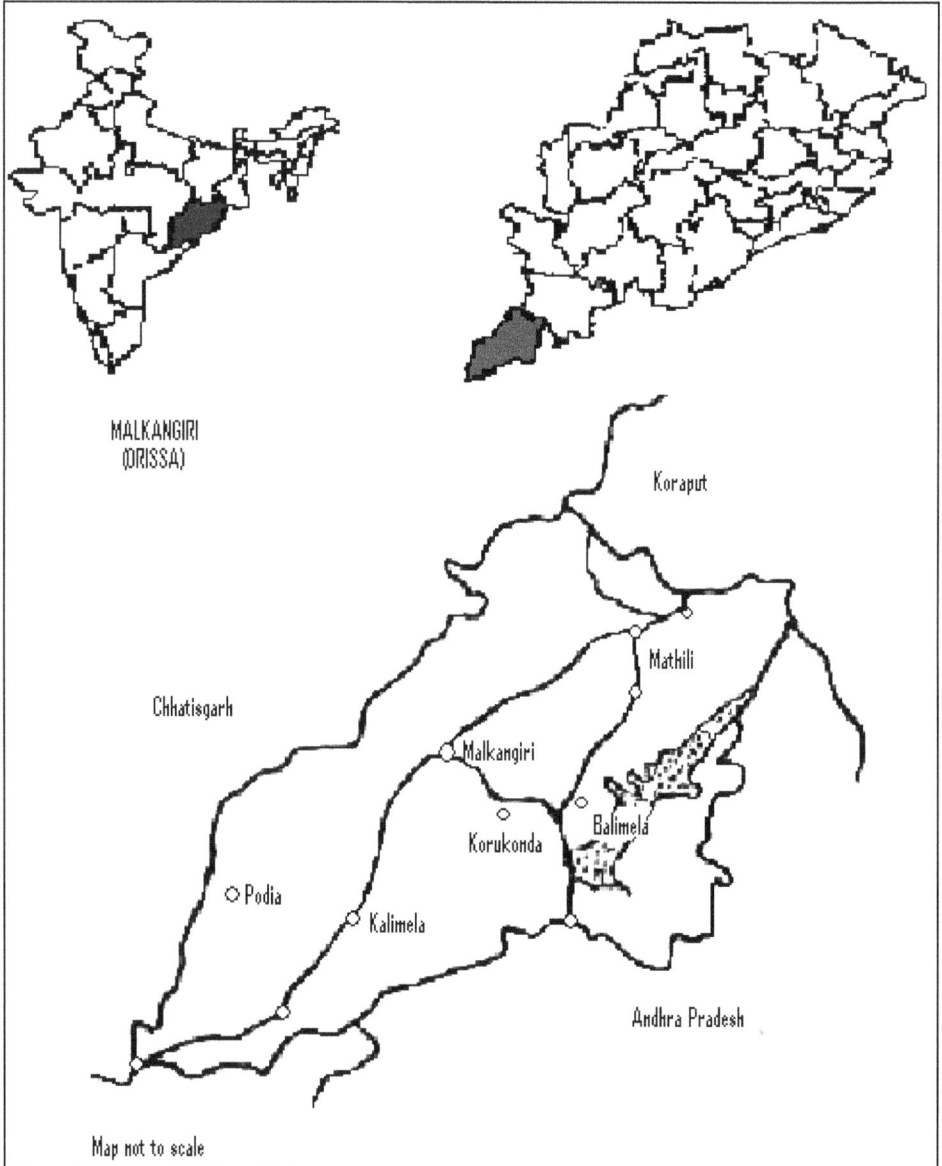

Figure 3.1: Showing Study Area.

Godavari and Khammam districts of Andhra Pradesh. Being a part of Eatern Ghats of India the district possesses 38.19 per cent (2,212 sq. km.) of forest cover with undulating topography and a large number of seasonal and perennial hill streams. The entire rainwater is being drained through a number of small rivulets that ultimately drain to major rivers such as Sabari, Sileru and Potteru etc.

The luxuriant forest pockets have been used as hiding dwellings by the Naxalites, Maoists and other criminal elements of Odisha, Andhra Pradesh and Chhattisgarh states. Their presence coupled with improper tracks and hilly terrain has in fact obstructed the plant explorers from carrying out their field works for which the district remained botanically under-explored.

Previous Botanical Exploration

As Malkangiri was part of the erstwhile Koraput, the Flora of the Presidency of Madras (Gamble and Fischer 1915-36) should have contained enough references to the floristic elements but the truth is unfortunately contrary. The Botany of Bihar and Orissa by Haines (1921-25) did not contain any reference to the floristic elements of this area because it was included in Madras Presidency when the Botany was written. Mooney (1950) while bringing out a supplement to the Botany of Bihar and Orissa, did not explore erstwhile Ganjam and Koraput districts, because the supplement had the same geographical jurisdiction as that of Haines. Though, extensive field works were carried out in Ganjam and Koraput districts (Saxena and Brahmam, 1994-96), Malkangiri subdivision somehow could not be botanised perfectly owing to the presence of Naxalites. A few attempts were made previously (Aminuddin and Girach, 1991; Hemadri and Rao, 1989-91; Mishra and Das, 1988; Prusty and Behera, 2007; Prusty, 2007; Pattanaik *et al.*, 2006, 2007, 2008), but their works were mostly confined to ethnobotanical notings only. Seeing the potentialities of non-timber forest products in changing the rural economy, a few workers have studied Malkangiri along with other districts of erstwhile Koraput for evaluating marketing strategies (Teki Surayya *et al.*, 2003).

Vegetation

Malkangiri falls under Deccan region of Hooker's sketch of Flora of British India, but Gamble (1915-36) placed it in the sal region of Madras presidency. The existing vegetation has surveyed now is found to fall broadly under following three types as per the classification of revised forest types of India (Champion and Seth, 1968). These are

    I. Tropical semi-evergreen forest

    II. Tropical dry deciduous forest

    III. Tropical moist deciduous forest

Some of the important plant species of tropical semi-evergreen forest are *Diospyros malabarica, Ficus benghalensis, Syzygium cumini, Mangifera indica, Schleichera oleosa Bridelia retusa, Ardisia solanacea, Smilax zeylanica* etc.

Some of the important plant species of tropical dry deciduous forest are *Terminalia alata, Terminalia bellirica, Bombax ceiba, Cochlospermum religiosum, Sterculia urens,*

Sterculia villosa, Oroxylum indicum, Firmiana colorata, Desmodium oojeinensis, Semecarpus anacardium, Boswellia serrata, Nyctanthes arbor-tristis, Lannea coromandelica etc.

Some of the important plant species of tropical moist deciduous forest are *Pterocarpus marsupium, Xylia xylocarpa, Haldinia cordifolia, Emblica officinalis, Madhuca indica, Lagerstroemia parviflora, Schleichera oleosa, Dillenia pentagyna, Careya arborea,Terminalia alata, Helicteres isora, Dalbergia latifolia* etc.

Besides the above, several subtypes and seral types like bamboo brakes, riparian fringing forest, scrubs, grass lands etc. are noticed mostly either developed or resulted due to biotic influence.

Mathili-Malkangiri forest range finds unique place in the vegetation map of Orissa as it represents a transition zone between Sal (*Shorea robusta*) and Teak (*Tectona grandis*).

Methodology

Systematic field surveys were undertaken in different forest ranges of Malkangiri during 2008-11. Intensive and extensive explorations along with critical studies of specimens both in the field and in the Herbarium (IMMT, Herbarium) were carried out. The field trips were organized in such a way so as to cover all the areas of the district at regular intervals in different seasons. As a result it became possible to record the seasonal variations in the vegetation including distributional patterns and collect most of the plants in different developmental stages of their life cycle. After critical study, the specimens were identified following The Flora of British India (Hooker, 1872-97), Flora of the Presidency of Madras (Gamble and Fischer, 1915-36), The Botany of Bihar and Orissa (Haines, 1921-25), Flora of Orissa (Saxena and Brahmam, 1994-96) and a host of recent monographs and reviews. The identity of the specimens was confirmed with the help of the authentic specimens of the IMMT Herbarium before they are incorporated. For a very few critically difficult specimens, Central National Herbarium (CAL), Kolkata was visited and identity was confirmed. Specimens collected by the earlier workers from this area and deposited in different herbaria were also consulted before preparing this account. All the specimens are housed and preserved in the Herbarium of the IMMT (RRL-B), Bhubaneswar, Odisha, India.

Results and Discussion

During the 3 years (2008 to 2011) of floristic exploration of Malkangiri district of Orissa (India) a total of 1296 plant specimens were collected of which 763 species belonging to 505 genera and 147 families were identified. Out of these, 745 species are angiosperms including 596 dicots and 149 monocots, 16 fern species (belongs to 13 genera and families) and 2 gymnosperms (Table 3.1). The collection embraces as many as 437 herbs, 87 shrubs, 160 trees and 79 climber species.

Dominant Families

Out of total 147 families, the 10 dominant families in descending order are Fabaceae (62 sp.), Poaceae (57 sp.), Asteraceae (37 sp.), Euphorbiaceae (33 sp.), Acanthaceae (32 sp.), Rubiaceae (30 sp.), Cyperaceae (29 sp.), Caesalpiniaceae and

Lamiaceae (20 sp.), Malvaceae (19 sp.) and Scrophulariaceae (18 sp.). The Poaceae are one of the largest families of flowering plants, comprising some 10,000 species under approximately 896 genera (Tzvelen, 1989). In respect of number of species, they rank fifth and are only exceeded by the Rubiaceae, Asteraceae, Leguminosae (Fabaceae) and the Orchidaceae (Good, 1956). In accordance with the above observations, here the family Fabaceae comes first followed by Poaceae. But, Asteraceae, Rubiaceae and Orchidaceae could not exceed Poaceae in terms of the number of species.

Table 3.1: Statistical Analysis of Floristic Data

	Class	Families	Genera	Wild Species	Planted Species	Total Species
Angiosperm	**Dicotyledoneae**	109	388	524	72	596
	Monocotyledoneae	23	102	141	8	149
Gymnosperm		2	2	–	2	2
Pteridophytes		13	13	16	–	16
	Total	147	505	681	82	763

Threatened Plants of Malkangiri

Saxena and Brahmam (1996) enlisted some of the red listed plants of Orissa. Among the plants 3 species namely *Acampe rigida*, *Leucas clarkei* and *Stemona tuberosa* found in Malkangiri can also be kept under threatened category as they are found occasionally or rarely. Ved *et al.* (2007) identified 41 red listed medicinal plant species of Orissa. Out of that 12 medicinal species such as *Blepharispermum subsessile*, *Celastrus paniculata*, *Gloriosa superba*, *Litsea glutinosa*, *Oroxylum indicum*, *Piper longum*, *Pterocarpus marsupium*, *Rauvolfia serpentina*, *Saraca asoca*, *Schrebera swietenioides*, *Stemona tuberosa*

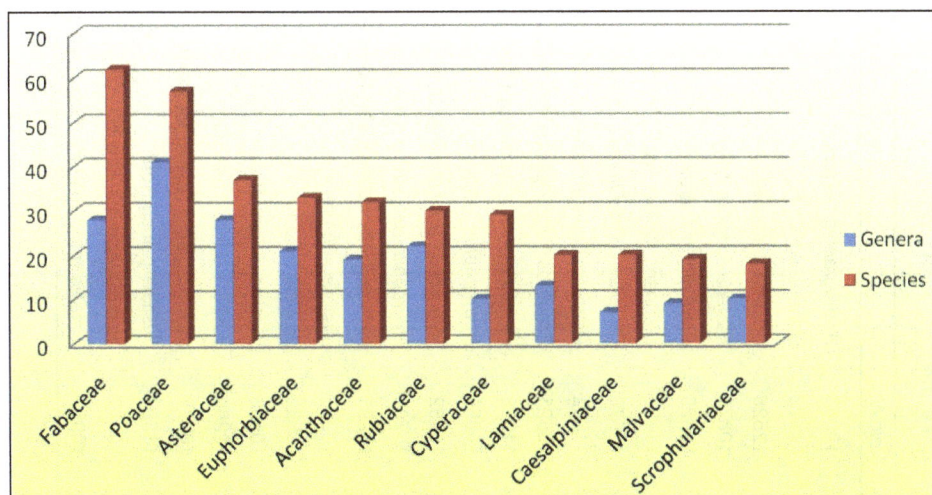

Figure 3.2: Ten Dominant Families of Phanerogams in the District.

Table 3.2: Check List of Plants

Sl.No.	Field No.	Botanical Name	Family	Habit	Locality
1.	10266	*Andrographis paniculata* (Burm.f.) Wall. ex Nees	Acanthaceae	Herb	Biralaxmanpur
2.	11122	*Barleria cristata* L.	Acanthaceae	Herb	Chitapari
3.	12525	*Barleria prionitis* L.	Acanthaceae	Herb	Way to Jayapuriaguda
4.	10375	*Barleria strigosa* Willd.	Acanthaceae	Herb	Govindapally
5.	10625	*Blepharis maderaspatensis* (L.) Roth	Acanthaceae	Herb	Undragonda
6.	11493	*Blepharis repens* (Vahl) Roth	Acanthaceae	Herb	Motu
7.	10578	*Dicliptera bupleuroides* Nees	Acanthaceae	Herb	Nalagunthi
8.	11892	*Dipteracanthus prostratus* (Poir.) Nees	Acanthaceae	Herb	Chitapari
9.	10525	*Eranthemum purpurascens* Nees	Acanthaceae	Herb	Bandhaguda
10.	11068	*Hemiadelphis polysperma* (Roxb.) Nees	Acanthaceae	Herb	Satiguda
11.	10686	*Hemigraphis latebrosa* (Heyne ex Roth) Nees	Acanthaceae	Herb	Chitrakonda
12.	10452	*Hygrophila auriculata* (Schum.) Heyne	Acanthaceae	Herb	Mathili
13.	10708	*Hygrophila balsamica* (L.f.) Raf.	Acanthaceae	Herb	Balimela
14.	10295	*Indoneesiella echioides* (L.) Sreemadh.	Acanthaceae	Herb	Dasmantpur
15.	10986	*Justicia adhatoda* L.	Acanthaceae	Shrub	Chitrakonda
16.	10592	*Justicia betonica* L.	Acanthaceae	Herb	Nalagunthi
17.	10566	*Justicia diffusa* Willd.	Acanthaceae	Herb	Nalagunthi
18.	12290	*Justicia gendarussa* Burm. f.	Acanthaceae	Herb	Orkel
19.	10577	*Justicia glabra* Koenig ex Roxb.	Acanthaceae	Shrub	Nalagunthi
20.	11369	*Justicia glauca* Rottl.	Acanthaceae	Herb	Goiparbat
21.	10377	*Justicia japonica* Thunb.	Acanthaceae	Herb	Govindapally
22.	10957	*Justicia quinqueangularis* Koenig ex Roxb.	Acanthaceae	Herb	Chitapari

Contd...

Table 3.2–*Contd...*

Sl.No.	Field No.	Botanical Name	Family	Habit	Locality
23.	10689	*Lepidagathis fasciculata* (Retz.) Nees	Acanthaceae	Herb	Chitrakonda
24.	11497	*Lepidagathis hamiltoniana* Nees	Acanthaceae	Herb	Motu
25.	10942	*Lepidagathis incurva* Buch.-Ham. ex D. Don	Acanthaceae	Herb	Balimela
26.	10326	*Peristrophe paniculata* (Forssk.) Brummitt	Acanthaceae	Herb	Dasmantpur
27.	10590	*Petalidium barlerioides* (Roth) Nees	Acanthaceae	Shrub	Nalagunthi
28.	10351	*Phaulopsis imbricata* (Forssk.) Sw.	Acanthaceae	Herb	Govindapally
29.	11877	*Rhinacanthus nasutus* (L.) Kurz.	Acanthaceae	Herb	Satiguda
30.	12531	*Ruellia tuberosa* L.	Acanthaceae	Herb	Malkangiri
31.	10397	*Rungia pectinata* (L.) Nees	Acanthaceae	Herb	Kamalapadar
32.	12293	*Thunbergia grandiflora* (Roxb. ex Rottl.) Roxb.	Acanthaceae	Climber	Orkel
33.	11037	*Adiantum incisum* Forssk.	Adiantaceae	Herb	Chitapari
34.	11217	*Adiantum philippense* L.	Adiantaceae	Herb	Tamasapalli
35.	12160	*Trianthema portulacastrum* L.	Aizoaceae	Herb	Tamasapalli
36.	10487	*Sagittaria guayanensis* H.B.K. ssp. *Lappula* (D. Don) Bogin	Alismataceae	Herb	Chalanguda
37.	10345	*Achyranthes aspera* L.	Amaranthaceae	Herb	Govindapally
38.	10661	*Aerva lanata* (L.) Juss. ex Schultes	Amaranthaceae	Herb	Balimela
39.	10804	*Aerva sanguinolenta* (L.) Bl.	Amaranthaceae	Herb	Govindapally
40.	10432	*Alternanthera sessilis* (L.) R.Br. ex DC.	Amaranthaceae	Herb	Sitakund
41.	10411	*Amaranthus caudatus* L.	Amaranthaceae	Herb	Mudulipada
42.	10985	*Amaranthus spinosus* L.	Amaranthaceae	Herb	Chitrakonda
43.	12170	*Amaranthus viridis* L.	Amaranthaceae	Herb	Tamasapalli
44.	10283	*Celosia argentea* L.	Amaranthaceae	Herb	Balimela

Contd...

Table 3.2–*Contd...*

Sl.No.	Field No.	Botanical Name	Family	Habit	Locality
45.	11732	*Celosia argentea* L. var. *cristata* (L.) Kuntze	Amaranthaceae	Herb	Bhubanpalli
46.	10433	*Cyathula prostrata* (L.) Bl.	Amaranthaceae	Herb	Sitakund
47.	10402	*Gomphrena celosioides* Mart.	Amaranthaceae	Herb	Mudulipara
48.	10706	*Anacardium occidentale* L.	Anacardiaceae	Tree	Balimela
49.	10588	*Buchanania lanzan* Spreng.	Anacardiaceae	Tree	Nalagunthi
50.	11083	*Lannea coromandelica* (Houtt.) Merr.	Anacardiaceae	Tree	Satiguda
51.	10642	*Mangifera indica* L.	Anacardiaceae	Tree	Undragonda
52.	10318	*Semecarpus anacardium* L.f.	Anacardiaceae	Tree	Dasmantpur
53.	12187	*Annona reticulata* L.	Annonaceae	Shrub	Malkangiri
54.	11099	*Annona squamosa* L.	Annonaceae	Shrub	Tamasapalli
55.	12189	*Polyalthia longifolia* (Sonn.) Thw.	Annonaceae	Tree	Tamasapalli
56.	12354	*Centella asiatica* (L.) Urban	Apiaceae	Herb	Tamasapalli
57.	11073	*Foeniculum vulgare* Mill.	Apiaceae	Herb	Satiguda
58.	12365	*Pimpinella heyneana* (Wall. ex DC.) Kurz.	Apiaceae	Herb	Goiparbat
59.	11364	*Aganosma caryophyllata* (Roxb. ex Sims.) G. Don	Apocynaceae	Climber	Goiparbat
60.	12505	*Alstonia scholaris* (L.) R. Br.	Apocynaceae	Tree	Satiguda
61.	10405	*Alstonia venenata* R. Br.	Apocynaceae	Tree	Mudulipara
62.	10554	*Cascabela thevetia* (L.) Lipp.	Apocynaceae	Shrub	Kalimela
63.	11236	*Catharanthus pusillus* (Murr.) G. Don	Apocynaceae	Herb	Tamasapalli
64.	12528	*Catharanthus roseus* (L.) G. Don	Apocynaceae	Herb	MV-42
65.	12532	*Ervatamia divaricata* (L.) Burkill	Apocynaceae	Shrub	MV-42
66.	11085	*Holarrhena pubescens* (Buch.-Ham.) Wall.ex G.Don	Apocynaceae	Tree	Satiguda

Contd...

Table 3.2–*Contd...*

Sl.No.	Field No.	Botanical Name	Family	Habit	Locality
67.	11889	*Ichnocarpus frutescens* (L.) R.Br.	Apocynaceae	Climber	Chitapari
68.	11151	*Plumeria rubra* L.	Apocynaceae	Shrub	Balimela
69.	10697	*Rauvolfia serpentina* (L.) Benth.ex Kurz.	Apocynaceae	Herb	Chitrakonda
70.	10951	*Wrightia arborea* (Dennst.) Mabb.	Apocynaceae	Tree	Chitapari
71.	10488	*Aponogeton natans* (L.) Engl. & Krause	Aponogetonaceae	Herb	Chalanguda
72.	10312	*Amorphophalus bulbifer* (Roxb.) Bl.	Araceae	Herb	Dasmantpur
73.	11262	*Colocasia esculanta* (L.) Schott	Araceae	Herb	Dasmantpur
74.	12386	*Lasia spinosa* (L.) Thw.	Araceae	Herb	Dasmantpur
75.	11760	*Pistia stratiotes* L.	Araceae	Herb	Balisagar
76.	11417	*Plesmonium margaritiferum* (Roxb.) Schott	Araceae	Herb	Koilipari
77.	10438	*Remusatia vivipara* (Roxb.) Schott	Araceae	Herb	Mudulipada
78.	11219	*Theriophonum minutum* (Willd.) Baill.	Araceae	Herb	Tamasapalli
79.	12536	*Araucaria columnaris* (Forst. f.) Hook.	Araucariaceae	Tree	MV-42
80.	12166	*Borassus flabellifer* L.	Arecaceae	Tree	Tamasapalli
81.	11897	*Caryota urens* L.	Arecaceae	Tree	Nuaguda
82.	12177	*Cocos nucifera* L.	Arecaceae	Tree	Tamasapalli
83.	10479	*Phoenix acaulis* Buch- Ham. ex Roxb.	Arecaceae	Shrub	Kamalapadar
84.	12167	*Phoenix sylvestris* (L.) Roxb.	Arecaceae	Tree	Tamasapalli
85.	10414	*Calotropis gigantea* R. Br.	Asclepiadaceae	Shrub	Mudulipara
86.	11088	*Leptadenia reticulata* (Retz.) Wt. & Arn.	Asclepiadaceae	Climber	Satiguda
87.	12182	*Oxystelma esculenta* (L.f.) R. Br. ex Schult.	Asclepiadaceae	Climber	Tamasapalli
88.	10313	*Pergularia daemia* (Forsk.) Chiov.	Asclepiadaceae	Climber	Dasmantpur

Contd...

Contd...

Table 3.2–*Contd...*

Sl.No.	Field No.	Botanical Name	Family	Habit	Locality
89.	12107	*Tectaria cicutaria* (L.) Copel.	Aspidiaceae	Herb	Chitrakonda
90.	10293	*Acanthospermum hispidum* DC.	Asteraceae	Herb	Dasmantpur
91.	10451	*Adenostemma lavenia* (L.) Kuntze	Asteraceae	Herb	Mathili
92.	11066	*Ageratum conyzoides* L.	Asteraceae	Herb	Satiguda
93.	11372	*Bidens biternata* (Lour.) Merr. & Sherff.	Asteraceae	Herb	Goiparbat
94.	12114	*Bidens pilosa* L.	Asteraceae	Herb	Chitrakonda
95.	11477	*Blainvillea acmella* (L.) Philipson	Asteraceae	Herb	Motu
96.	10526	*Blepharispermum subsessile* DC.	Asteraceae	Herb	Bandhaguda
97.	11880	*Blumea aurita* (L.f.) DC.	Asteraceae	Herb	Satiguda
98.	10571	*Blumea fistulosa* (Roxb.) Kurz.	Asteraceae	Herb	Nalagunthi
99.	10678	*Blumea hieracifolia* (D. Don) DC. var. *hamiltonii* (DC.) C. B. Cl.	Asteraceae	Herb	Chitrakonda
100.	10644	*Blumea lacera* (Burm.f.) DC.	Asteraceae	Herb	Kalimela
101.	11032	*Blumea laciniata* (Roxb.) DC.	Asteraceae	Herb	Chitapari
102.	12527	*Blumea oxyodonta* DC.	Asteraceae	Herb	MV-42
103.	11900	*Blumea virens* Wall. ex DC.	Asteraceae	Herb	Nuaguda
104.	10379	*Caesulia axillaris* Roxb.	Asteraceae	Herb	Kamalapadar
105.	10362	*Chromolaena odorata* (L.) R. King & H. Robins	Asteraceae	Shrub	Govindapally
106.	10968	*Cyathocline purpurea* (Buch.-Ham. ex D.Don) Kuntze	Asteraceae	Herb	Chitrakonda
107.	10649	*Eclipta prostrata* (L.) L.	Asteraceae	Herb	Kalimela
108.	11444	*Elephantopus scaber* L.	Asteraceae	Herb	Pulkelkonda
109.	11346	*Emilia sonchifolia* (L.) DC.	Asteraceae	Herb	Goiparbat
110.	11725	*Epaltes divaricata* (L.) Cass.	Asteraceae	Herb	Bhubanpalii

Table 3.2–*Contd...*

Sl.No.	Field No.	Botanical Name	Family	Habit	Locality
111.	10614	*Gnaphalium polycaulon* Pers.	Asteraceae	Herb	Undragonda
112.	10314	*Guizotia abyssinica* (L.f.) Cass.	Asteraceae	Herb	Dasmantpur
113.	11077	*Gynura aurantiaca* (Bl.) DC.	Asteraceae	Herb	Satiguda
114.	10465	*Parthenium hysterophorus* L.	Asteraceae	Herb	Mudulipara
115.	12119	*Siegesbeckia orientalis* L.	Asteraceae	Herb	Chitrakonda
116.	10380	*Sphaeranthus indicus* L.	Asteraceae	Herb	Kamalapadar
117.	11476	*Spilanthes paniculata* Wall. ex DC.	Asteraceae	Herb	Motu
118.	12352	*Synedrella nodiflora* (L.) Gaertn.	Asteraceae	Herb	Tamasapalli
119.	12175	*Tagetes erecta* L.	Asteraceae	Herb	Tamasapalli
120.	10646	*Tridax procumbens* L.	Asteraceae	Herb	Kalimela
121.	10473	*Vernonia cinerea* (L.) Less	Asteraceae	Herb	Kamalapadar
122.	11886	*Vernonia divergens* (Roxb.) Edgew.	Asteraceae	Herb	Chitapari
123.	11489	*Vicoa indica* (L.) DC.	Asteraceae	Herb	Motu
124.	10417	*Wedelia chinensis* (Osbeck) Merr.	Asteraceae	Herb	Mudulipara
125.	12360	*Wedelia utricifolia* DC.	Asteraceae	Herb	Goiparbat
126.	10330	*Xanthium indicum* Koenig	Asteraceae	Herb	Govindapally
127.	10332	*Diplazium polypodioides* Bl.	Athyriaceae	Herb	Govindapally
128.	11758	*Azolla pinnata* R.Br.	Azollaceae	Herb	Balisagar
129.	11401	*Impatiens balsamina* L.	Balsaminaceae	Herb	Goiparbat
130.	10334	*Barringtonia acutangula* (L.) Gaertn.	Barringtoniaceae	Tree	Govindapally
131.	10704	*Careya arborea* Roxb.	Barringtoniaceae	Tree	Balimela
132.	10523	*Basella alba* L.	Basellaceae	Climber	Bandhaguda

Contd...

Table 3.2–*Contd...*

Sl.No.	Field No.	Botanical Name	Family	Habit	Locality
133.	10303	*Begonia picta* Sm.	Begoniaceae	Herb	Dasmantpur
134.	11873	*Heterophragma quadriloculare* (Roxb.) K. Schum.	Bignoniaceae	Tree	Satiguda
135.	12267	*Millingtonia hortensis* L.f.	Bignoniaceae	Tree	Chitapari
136.	11778	*Oroxylum indicum* (L.) Vent.	Bignoniaceae	Tree	Malaguda
137.	10682	*Stereospermum chelonoides* (L. f.) DC.	Bignoniaceae	Tree	Chitrakonda
138.	10289	*Stereospermum colais* (Buch.-Ham. ex Dillw.) Mabb. var. *colais*	Bignoniaceae	Tree	Balimela
139.	10364	*Tecoma stans* (L.) Kunth	Bignoniaceae	Shrub	Govindapally
140.	11175	*Bixa orellana* L.	Bixaceae	Tree	Disariguda
141.	10952	*Bombax ceiba* L.	Bombacaceae	Tree	Chitapari
142.	10602	*Ceiba pentandra* (L.) Gaertn. var. *pentandra*	Bombacaceae	Tree	Undragonda
143.	10959	*Coldenia procumbens* L.	Boraginaceae	Herb	Chitapari
144.	10434	*Cynoglossum zeylanicum* (Vahl. ex Hornem.) Thunb. ex Lehm.	Boraginaceae	Herb	Sitakund
145.	11054	*Heliotropium indicum* L.	Boraginaceae	Herb	Satiguda
146.	11216	*Heliotropium marifolium* Retz.	Boraginaceae	Herb	Tamasapalli
147.	11494	*Heliotropium strigosum* Willd. subsp. *brevifolium* (Wall.) Kazmi	Boraginaceae	Herb	Motu
148.	10624	*Trichodesma indicum* (L.) R. Br.	Boraginaceae	Herb	Undragonda
149.	10992	*Trichodesma zeylanicum* (Burm. f.) R. Br.	Boraginaceae	Herb	Chitrakonda
150.	11076	*Brassica napus* L. var. *napus*	Brassicaceae	Herb	Satiguda
151.	11733	*Buddleja asiatica* Lour.	Buddlejaceae	Tree	Khirkoliguda
152.	12181	*Boswellia serrata* Roxb. ex Colebr.	Burseraceae	Tree	Tamasapalli
153.	11142	*Protium serratum* (Wall. ex Colebr.) Engl.	Burseraceae	Tree	Chitapari
154.	11158	*Tenagocharis latifolia* (D.Don) Buchen.	Butomaceae	Herb	Disariguda

Contd...

Table 3.2–*Contd...*

Sl.No.	Field No.	Botanical Name	Family	Habit	Locality
155.	11036	*Sarcococca saligna* (D.Don) Muell.-Arg.	Buxaceae	Shrub	Chitapari
156.	11100	*Cereus pterogonus* Lem.	Cactaceae	Shrub	Tamasapalli
157.	12539	*Opuntia stricta* (Haw.) Haw. var. *dilleni* (Ker-Gawl.) Benson	Cactaceae	Herb	Malkangiri
158.	11101	*Bauhinia acuminata* L.	Caesalpiniaceae	Shrub	Tamasapalli
159.	10502	*Bauhinia malabarica* Roxb.	Caesalpiniaceae	Tree	Chalanguda
160.	10284	*Bauhinia purpurea* L.	Caesalpiniaceae	Tree	Balimela
161.	10503	*Bauhinia racemosa* Lam.	Caesalpiniaceae	Tree	Chalanguda
162.	10560	*Bauhinia vahlii* Wt. & Arn.	Caesalpiniaceae	Climber	Nalagunthi
163.	10315	*Caesalpinia decapetala* (Roth) Alston	Caesalpiniaceae	Shrub	Dasmantpur
164.	12179	*Caesalpinia pulcherrima* (L.) Sw.	Caesalpiniaceae	Shrub	Tamasapalli
165.	12380	*Cassia absus* L.	Caesalpiniaceae	Herb	Dasmantpur
166.	12513	*Cassia alata* L.	Caesalpiniaceae	Shrub	Satiguda
167.	10532	*Cassia fistula* L.	Caesalpiniaceae	Tree	Balimela
168.	10324	*Cassia hirsuta* L.	Caesalpiniaceae	Shrub	Govindapally
169.	10296	*Cassia mimosoides* L.	Caesalpiniaceae	Herb	Dasmantpur
170.	12351	*Cassia occidentalis* L.	Caesalpiniaceae	Herb	Tamasapalli
171.	10407	*Cassia siamea* Lam.	Caesalpiniaceae	Tree	Mudulipara
172.	10536	*Cassia sophera* L.	Caesalpiniaceae	Herb	Korukonda
173.	10509	*Cassia tora* L.	Caesalpiniaceae	Herb	Chalanguda
174.	10599	*Delonix regia* (Boj. ex Hook.) Raf.	Caesalpiniaceae	Tree	Kalimela
175.	12535	*Peltophorum pterocarpum* (DC.) Baker ex Heyne	Caesalpiniaceae	Tree	MV-42
176.	12185	*Saraca asoca* (Roxb.) de Wilde	Caesalpiniaceae	Tree	Malkangiri

Contd...

Table 3.2–*Contd...*

Sl.No.	Field No.	Botanical Name	Family	Habit	Locality
177.	10528	*Tamarindus indica* L.	Caesalpiniaceae	Tree	Govindapally
178.	11211	*Sphenoclea zeylanica* Gaertn.	Campanulaceae	Herb	Tamasapalli
179.	12245	*Wahlenbergia marginata* (Thunb.) A.DC.	Campanulaceae	Herb	Chitapari
180.	12259	*Cannabis sativa* L.	Cannabinaceae	Herb	Chitapari
181.	12526	*Canna indica* L.	Cannaceae	Herb	MV-42
182.	11415	*Capparis zeylanica* L.	Capparaceae	Climber	Kolipari
183.	11004	*Cleome monophylla* Linn.	Capparaceae	Herb	Chitrakonda
184.	11060	*Cleome viscosa* L.	Capparaceae	Herb	Satiguda
185.	12164	*Carica papaya* L.	Caricaceae	Shrub	Tamasapalli
186.	10436	*Drymaria cordata* (L.) Willd. ex Roem. & Schultes	Caryophyllaceae	Herb	Sitakund
187.	10522	*Casuarina equisetifolia* L.	Casuarinaceae	Tree	Bandhaguda
188.	12264	*Cassine glauca* (Rottb.) Kuntze	Celastraceae	Tree	Chitapari
189.	11119	*Celastrus paniculata* Willd.	Celastraceae	Shrub	Chitapari
190.	10605	*Maytenus emarginatus* (Willd.) Ding Hou	Celastraceae	Shrub	Undragonda
191.	12515	*Ceratophyllum demersum* L.	Ceratophyllaceae	Herb	Way 2 Jayapuriaguda
192.	11215	*Cheilanthes tenuifolia* (Burm.f.) Sw.	Cheilanthaceae	Herb	Tamasapalli
193.	10443	*Chenopodium ambrosioides* L.	Chenopodiaceae	Herb	Mudulipara
194.	10955	*Cochlospermum religiosum* (L.) Alston.	Cochlospermaceae	Tree	Chitapari
195.	11166	*Anogeissus acuminata* (Roxb.ex DC.) Guill. & Perr.	Combretaceae	Tree	Disariguda
196.	10271	*Anogeissus latifolia* (Roxb.ex DC.) Wall. ex Guill.	Combretaceae	Tree	Biralaxmanpur
197.	10947	*Calycopteris floribunda* Lam.	Combretaceae	Climber	Balimela
198.	11018	*Combretum albidum* G. Don	Combretaceae	Climber	Chitrakonda

Contd...

Table 3.2–*Contd...*

Sl.No.	Field No.	Botanical Name	Family	Habit	Locality
199.	10576	*Combretum roxburghii* Spreng.	Combretaceae	Shrub	Nalagunthi
200.	11014	*Quisqualis indica* L.	Combretaceae	Climber	Chitrakonda
201.	10664	*Terminalia alata* Heyne ex Roth	Combretaceae	Tree	Chitrakonda
202.	10288	*Terminalia arjuna* (Roxb. ex DC.) Wt. & Arn.	Combretaceae	Tree	Balimela
203.	10670	*Terminalia bellirica* (Gaertn.) Roxb.	Combretaceae	Tree	Chitrakonda
204.	12357	*Terminalia catappa* L.	Combretaceae	Tree	Panaguda
205.	10668	*Terminalia chebula* Retz.	Combretaceae	Tree	Chitrakonda
206.	12397	*Aneilema ovalifolium* (Wight) Hook. f. ex C. B. Cl.	Commelinaceae	Herb	Dasmantpur
207.	11775	*Commelina appendiculata* C.B.Cl.	Commelinaceae	Herb	Saptadhara
208.	11383	*Commelina benghalensis* L.	Commelinaceae	Herb	Goiparbat
209.	11402	*Commelina erecta* L.	Commelinaceae	Herb	Goiparbat
210.	11270	*Cyanotis cristata* (L.) D. Don	Commelinaceae	Herb	Dasmantpur
211.	11773	*Floscopa scandens* Lour.	Commelinaceae	Herb	Saptadhara
212.	11363	*Murdannia nudiflora* (L.) Brenan	Commelinaceae	Herb	Goiparbat
213.	11395	*Murdannia spirata* (L.) Brueck.	Commelinaceae	Herb	Goiparbat
214.	10495	*Tonningia axillaris* (L.) Kuntze	Commelinaceae	Herb	Chalanguda
215.	10280	*Argyreia daltonii* C.B. Cl.	Convolvulaceae	Climber	Balimela
216.	12247	*Erycibe paniculata* Roxb.	Convolvulaceae	Climber	Chitapari
217.	10961	*Evolvulus nummularius* (L.) L.	Convolvulaceae	Herb	Chitapari
218.	10333	*Hewittia sublobata* (L.f.) Kuntze	Convolvulaceae	Climber	Govindpalli
219.	10511	*Ipomoea aquatica* Forssk.	Convolvulaceae	Herb	Pangam
220.	10714	*Ipomoea carnea* Jacq.	Convolvulaceae	Shrub	Dasmantpur

Contd...

Table 3.2–*Contd...*

Sl.No.	Field No.	Botanical Name	Family	Habit	Locality
221.	11468	*Ipomoea eriocarpa* R.Br.	Convolvulaceae	Climber	Pulkelkonda
222.	11001	*Ipomoea hederifolia* Linn.	Convolvulaceae	Climber	Chitrakonda
223.	11716	*Ipomoea nil* (L.) Roth.	Convolvulaceae	Climber	Motu
224.	10418	*Ipomoea pestigridis* L.	Convolvulaceae	Climber	Mudulipara
225.	12299	*Ipomoea violacea* L.	Convolvulaceae	Climber	Orkel
226.	12102	*Merremia emarginata* (Burm.f.) Hall.f.	Convolvulaceae	Herb	Nuaguda
227.	11450	*Merremia hederacea* (Burm.f.) Hall.f.	Convolvulaceae	Climber	Pulkelkonda
228.	12305	*Merremia tridentata* (L.) Hall. f. ssp. *tridentata*	Convolvulaceae	Herb	Balimela
229.	12263	*Merremia umbellata* (L.) Hall.f.	Convolvulaceae	Climber	Chitapari
230.	12113	*Merremia vitifolia* (Burm.f.) Hall.f.	Convolvulaceae	Climber	Chitrakonda
231.	11096	*Kalanchoe pinnata* (Lam.) Pers.	Crassulaceae	Herb	Satiguda
232.	12353	*Benincasa hispida* (Thunb.) Cogn.	Cucurbitaceae	Climber	Tamasapalli
233.	12503	*Citrullus lanatus* (Thunb.) Matsum & Nakai	Cucurbitaceae	Climber	Satiguda
234.	10328	*Cucumis sativus* L.	Cucurbitaceae	Climber	Govindapally
235.	12163	*Cucurbita maxima* Duch. ex Lam.	Cucurbitaceae	Climber	Tamasapalli
236.	10282	*Diplocyclos palmatus* (L.) Jeff.	Cucurbitaceae	Climber	Balimela
237.	12174	*Lagenaria siceraria* (Molina) Standley	Cucurbitaceae	Climber	Tamasapalli
238.	12165	*Luffa aegyptiaca* Mill.	Cucurbitaceae	Climber	Tamasapalli
239.	10553	*Momordica charantia* L.	Cucurbitaceae	Climber	Kalimela
240.	11361	*Mukia maderaspatana* (L.) Roem.	Cucurbitaceae	Climber	Goiparbat
241.	11377	*Trichosanthes cucumerina* L.	Cucurbitaceae	Climber	Goiparbat
242.	12504	*Thuja orientalis* L.	Cuppresaceae	Shrub	Satiguda

Contd...

Table 3.2–*Contd...*

Sl.No.	Field No.	Botanical Name	Family	Habit	Locality
243.	11731	*Cuscuta reflexa* Roxb.	Cuscutaceae	Climber	Bhubanpalli
244.	11232	*Bulbostylis barbata* (Rottb.) C.B. Cl.	Cyperaceae	Herb	Tamasapalli
245.	11265	*Carex filicina* Nees	Cyperaceae	Herb	Dasmantpur
246.	11230	*Cyperus brevifolius* (Rottb.) Hassk.	Cyperaceae	Herb	Tamasapalli
247.	11227	*Cyperus compressus* L.	Cyperaceae	Herb	Tamasapalli
248.	12183	*Cyperus cyperoides* (L.) Kuntze	Cyperaceae	Herb	Tamasapalli
249.	12252	*Cyperus difformis* L.	Cyperaceae	Herb	Chitapari
250.	10368	*Cyperus diffusus* Vahl var. *diffusus*	Cyperaceae	Herb	Govindapally
251.	12295	*Cyperus flavidus* Retz.	Cyperaceae	Herb	Tunibeda
252.	11384	*Cyperus iria* L.	Cyperaceae	Herb	Goiparbat
253.	10423	*Cyperus nutans* Vahl	Cyperaceae	Herb	Mudulipara
254.	10950	*Cyperus rotundus* L. subsp. *rotundus*	Cyperaceae	Herb	Chitapari
255.	11224	*Cyperus squarosus* L.	Cyperaceae	Herb	Tamasapalli
256.	10467	*Cyperus tenuispica* Steud.	Cyperaceae	Herb	Mudulipara
257.	11228	*Cyperus triceps* Endl.	Cyperaceae	Herb	Tamasapalli
258.	12298	*Eleocharis atropurpurea* (Retz.) Presl.	Cyperaceae	Herb	Tunibeda
259.	11213	*Eleocharis dulcis* (Burm.f.) Hensch.	Cyperaceae	Herb	Tamasapalli
260.	11005	*Eleocharis geniculata* (L.) Roem. & Schult.	Cyperaceae	Herb	Chitrakonda
261.	10616	*Fimbristylis aestivalis* (Retz.) Vahl	Cyperaceae	Herb	Undragonda
262.	11134	*Fimbristylis dichotoma* (L.) Vahl	Cyperaceae	Herb	Chitapari
263.	11006	*Fimbristylis ferruginea* (L.) Vahl	Cyperaceae	Herb	Chitrakonda
264.	10611	*Fimbristylis littoralis* Gaud.	Cyperaceae	Herb	Undragonda

Contd...

Table 3.2–*Contd...*

Sl.No.	Field No.	Botanical Name	Family	Habit	Locality
265.	11428	*Fimbristylis schoenoides* (Retz.) Vahl	Cyperaceae	Herb	Koilipari
266.	11425	*Fuirena ciliaris* (L.) Roxb.	Cyperaceae	Herb	Koilipari
267.	10492	*Indocourtoisia cyperoides* (Roxb.) Bennet & Raizada	Cyperaceae	Herb	Chalanguda
268.	11440	*Lipocarpha sphaecelata* (Vahl) Kunth	Cyperaceae	Herb	Pulkelkonda
269.	11064	*Scirpus articulatus* L.	Cyperaceae	Herb	Satiguda
270.	10491	*Scirpus grossus* L. f.	Cyperaceae	Herb	Chalanguda
271.	12521	*Scirpus juncoides* Roxb.	Cyperaceae	Herb	Bhalukhupli
272.	11426	*Scleria biflora* Roxb.	Cyperaceae	Herb	Koilipari
273.	12374	*Dillenia aurea* Sm.	Dilleniaceae	Tree	Chalanguda
274.	12190	*Dillenia indica* L.	Dilleniaceae	Tree	Tamasapalli
275.	10938	*Dillenia pentagyna* Roxb.	Dilleniaceae	Tree	Balimela
276.	10270	*Dioscorea bulbifera* L.	Dioscoreaceae	Climber	Biralaxmanpur
277.	10299	*Dioscorea glabra* Roxb.	Dioscoreaceae	Climber	Dasmantpur
278.	11463	*Dioscorea oppositifolia* L.	Dioscoreaceae	Climber	Pulkelkonda
279.	11373	*Dioscorea pentaphylla* L.	Dioscoreaceae	Climber	Goiparbat
280.	11456	*Dioscorea puber* Bl.	Dioscoreaceae	Climber	Pulkelkonda
281.	11167	*Shorea robusta* Gaertn.f.	Dipterocarpaceae	Tree	Disariguda
282.	10692	*Diospyros malabarica* (Desr.) Kostel.	Ebenaceae	Tree	Chitrakonda
283.	10501	*Diospyros melanoxylon* Roxb.	Ebenaceae	Tree	Chalanguda
284.	10940	*Diospyros montana* Roxb.	Ebenaceae	Tree	Balimela
285.	10272	*Diospyros sylvatica* Roxb.	Ebenaceae	Tree	Biralaxmanpur
286.	11770	*Rotula aquatica* Lour.	Ehretiaceae	Herb	Saptadhara

Contd...

Table 3.2–*Contd...*

Sl.No.	Field No.	Botanical Name	Family	Habit	Locality
287.	11065	*Bergia ammannioides* Roxb.	Elatinaceae	Herb	Satiguda
288.	10656	*Eriocaulon quinquangulare* L.	Eriocaulaceae	Herb	Balimela
289.	10349	*Eriocaulon sollyanum* Royle	Eriocaulaceae	Herb	Govindapally
290.	11107	*Acalypha indica* L.	Euphorbiaceae	Tree	Tamasapalli
291.	11767	*Antidesma acidum* Retz.	Euphorbiaceae	Tree	Malaguda
292.	12104	*Baliospermum montanum* (Willd.) Muell.-Arg.	Euphorbiaceae	Shrub	Chitrakonda
293.	11247	*Breynia retusa* (Dennst.) Alston	Euphorbiaceae	Shrub	Kamalapadar
294.	11362	*Bridelia montana* (Roxb.) Willd.	Euphorbiaceae	Tree	Goiparbat
295.	10378	*Bridelia retusa* (L.) Spreng.	Euphorbiaceae	Tree	Udilibed Way
296.	11017	*Chrozophora rottleri* (Geisel.) Juss.	Euphorbiaceae	Herb	Chitrakonda
297.	11132	*Cleistanthus collinus* (Roxb.) Benth. ex Hook. f.	Euphorbiaceae	Tree	Chitapari
298.	10653	*Croton bonplandianus* Baill.	Euphorbiaceae	Herb	Kalimela
299.	12237	*Croton roxburghii* Balak.	Euphorbiaceae	Tree	Balimela
300.	10267	*Euphorbia hirta* L.	Euphorbiaceae	Herb	Biralaxmanpur
301.	12538	*Euphorbia milii* Des Moul.	Euphorbiaceae	Herb	MV-42
302.	11009	*Euphorbia nivulia* Buch.-Ham.	Euphorbiaceae	Tree	Chitrakonda
303.	11102	*Euphorbia thymifolia* L.	Euphorbiaceae	Herb	Tamasapalli
304.	10519	*Euphorbia tirucalli* Linn.	Euphorbiaceae	Shrub	Bandhaguda
305.	11111	*Homonoia riparia* Lour.	Euphorbiaceae	Shrub	Chitapari
306.	11020	*Jatropha gossypifolia* L.	Euphorbiaceae	Shrub	Chitrakonda
307.	12117	*Macaranga peltata* (Roxb.) Muell.-Arg.	Euphorbiaceae	Tree	Chitrakonda
308.	11011	*Mallotus philippensis* (Lam.) Muell. Arg.	Euphorbiaceae	Tree	Chitrakonda

Contd...

Table 3.2–*Contd...*

Sl.No.	Field No.	Botanical Name	Family	Habit	Locality
309.	11008	*Pedilanthus tithymaloides* (L.) Poit.	Euphorbiaceae	Herb	Chitrakonda
310.	10518	*Phyllanthus acidus* (L.) Skeels	Euphorbiaceae	Tree	Bandhaguda
311.	10278	*Phyllanthus airy-shawii* Brunel & Roux	Euphorbiaceae	Herb	Balimela
312.	11094	*Phyllanthus amarus* Schum. & Thonn.	Euphorbiaceae	Herb	Satiguda
313.	10257	*Phyllanthus emblica* L.	Euphorbiaceae	Tree	Biralaxmanpur
314.	11254	*Phyllanthus reticulatus* Poir.	Euphorbiaceae	Herb	Kamalapadar
315.	10622	*Phyllanthus urinaria* L.	Euphorbiaceae	Herb	Undragonda
316.	11144	*Phyllanthus virgatus* Forst.f.	Euphorbiaceae	Herb	Chitapari
317.	10803	*Ricinus communis* L.	Euphorbiaceae	Shrub	Govindapally
318.	11097	*Sauropus quadrangularis* (Willd.) Muell.-Arg. var. *quadrangularis*	Euphorbiaceae	Herb	Satiguda
319.	11420	*Sebastiania chamaelea* (L.) Muell.- Arg.	Euphorbiaceae	Herb	Koiipari
320.	11059	*Securinega virosa* (Roxb. ex Willd.) Baill.	Euphorbiaceae	Tree	Satiguda
321.	10447	*Tragia involucrata* L.	Euphorbiaceae	Climber	Mudulipada
322.	11177	*Trewia nudiflora* L.	Euphorbiaceae	Tree	Disariguda
323.	10390	*Abrus precatorius* L.	Fabaceae	Climber	Kamalapadar
324.	10337	*Aeschynomene americana* L.	Fabaceae	Herb	Govindapally
325.	11755	*Aeschynomene aspera* L.	Fabaceae	Herb	Balisagar
326.	11022	*Aeschynomene indica* L.	Fabaceae	Herb	Chitrakonda
327.	11378	*Alysicarpus vaginalis* (L.) DC. var. *vaginalis*	Fabaceae	Herb	Goiparbat
328.	10652	*Alysicarpus vaginalis* (L.) DC. var.*numularifolius* Miq.	Fabaceae	Herb	Kalimela
329.	10481	*Atylosia scarabaeoides* (L.) Benth.	Fabaceae	Climber	Goiparbat

Contd...

Table 3.2–*Contd...*

Sl.No.	Field No.	Botanical Name	Family	Habit	Locality
330.	11024	*Atylosia volubilis* (Blanco) Gamble	Fabaceae	Climber	Chitrakonda
331.	12356	*Butea monosperma* (Lam.) Taub.	Fabaceae	Tree	Panaguda
332.	10958	*Butea superba* Roxb.	Fabaceae	Climber	Chitapari
333.	11447	*Cajanus cajan* (L.) Huth	Fabaceae	Shrub	Pulkelkonda
334.	12197	*Clitoria ternatea* L.	Fabaceae	Climber	Tamasapalli
335.	11728	*Crotalaria acicularis* Buch.-Ham. ex Benth.	Fabaceae	Herb	Bhubanpalli
336.	10316	*Crotalaria albida* Heyne. ex Roth	Fabaceae	Herb	Saptadhara
337.	10359	*Crotalaria pallida* Ait.	Fabaceae	Shrub	Govindapally
338.	10477	*Crotalaria prostrata* Rottl. ex Willd.	Fabaceae	Herb	Kamalapadar
339.	12371	*Crotalaria quinquefolia* L.	Fabaceae	Herb	Chalanguda
340.	11457	*Crotalaria ramosissima* Roxb.	Fabaceae	Herb	Pulkelkonda
341.	10679	*Crotalaria spectabilis* Roth	Fabaceae	Herb	Chitrakonda
342.	12266	*Dalbergia lanceolaria* L.f.	Fabaceae	Tree	Chitapari
343.	11051	*Dalbergia latifolia* Roxb.	Fabaceae	Tree	Satiguda
344.	11751	*Dalbergia sissoo* Roxb.	Fabaceae	Tree	Malkangiri
345.	10628	*Dalbergia volubilis* Roxb.	Fabaceae	Shrub	Undragonda
346.	11081	*Desmodium gangeticum* (L.) DC.	Fabaceae	Herb	Satiguda
347.	10294	*Desmodium heterocarpon* (L.) DC. var. *heterocarpon*	Fabaceae	Herb	Dasmantpur
348.	10367	*Desmodium motorium* (Houtt.) Merr.	Fabaceae	Herb	Govindapally
349.	10997	*Desmodium oojeinensis* (Roxb.) Ohashi	Fabaceae	Tree	Chitrakonda
350.	11894	*Desmodium pulchellum* (L.) Benth.	Fabaceae	Herb	Nuaguda
351.	11351	*Desmodium triflorum* (L.) DC.	Fabaceae	Herb	Goiparbat

Contd...

Table 3.2–*Contd...*

Sl.No.	Field No.	Botanical Name	Family	Habit	Locality
352.	10991	*Desmodium triquetrum* (L.) DC.	Fabaceae	Herb	Chitrakonda
353.	10256	*Desmodium velutinum* (Willd.) DC.	Fabaceae	Shrub	Biralaxmanpur
354.	10964	*Erythrina variegata* L.	Fabaceae	Tree	Chitapari
355.	10369	*Flemingia bracteata* (Roxb.) Wight	Fabaceae	Herb	Govindapally
356.	12400	*Galactia longifolia* Benth.	Fabaceae	Climber	Satiguda
357.	10297	*Indigofera astragalina* DC.	Fabaceae	Herb	Dasmantpur
358.	10972	*Indigofera cassioides* Rottl. ex DC.	Fabaceae	Shrub	Chitrakonda
359.	11474	*Indigofera glabra* L.	Fabaceae	Herb	Motu
360.	10490	*Indigofera linifolia* (L.f.) Retz.	Fabaceae	Herb	Chalanguda
361.	11475	*Indigofera linnaei* Ali	Fabaceae	Herb	Motu
362.	10382	*Indigofera nummularifolia* (L.) Liv. ex Alston.	Fabaceae	Herb	Kamalapadar
363.	11388	*Indigofera prostrata* Willd.	Fabaceae	Herb	Goiparbat
364.	10336	*Millettia extensa* (Benth.) Baker	Fabaceae	Climber	Govindapally
365.	10425	*Millettia racemosa* (Roxb.) Benth.	Fabaceae	Climber	Sitakund
366.	11090	*Mucuna nigricans* (Lour.) Steud.	Fabaceae	Climber	Satiguda
367.	10975	*Mucuna pruriens* (L.) DC.	Fabaceae	Climber	Chitrakonda
368.	11135	*Pongamia pinnata* (L.) Pierre	Fabaceae	Tree	Chitapari
369.	12373	*Pseudarthria viscida* (L.) Wt. & Arn.	Fabaceae	Herb	Chalanguda
370.	10500	*Pterocarpus marsupium* Roxb.	Fabaceae	Tree	Chalanguda
371.	12196	*Pueraria tuberosa* (Willd.) DC.	Fabaceae	Climber	Tamasapalli
372.	11496	*Rothia indica* (L.) Druce	Fabaceae	Herb	Motu
373.	11479	*Sesbania bispinosa* (Jacq.) W.F.Wight	Fabaceae	Shrub	Motu

Contd...

Table 3.2–*Contd...*

Sl.No.	Field No.	Botanical Name	Family	Habit	Locality
374.	11498	*Sesbania grandiflora* (L.) Poir.	Fabaceae	Tree	Motu
375.	12372	*Smithia conferta* J. E. Sm.	Fabaceae	Herb	Chalanguda
376.	10381	*Tephrosia purpurea* (L.) Pers.	Fabaceae	Herb	Kamalapadar
377.	11717	*Tephrosia villosa* (L.) Pers.	Fabaceae	Herb	Motu
378.	11484	*Teramnus labialis* (L.f.) Spreng. var. *labialis*	Fabaceae	Climber	Motu
379.	12234	*Teramnus mollis* Benth.	Fabaceae	Climber	Sadasivpur
380.	11419	*Uraria lagopodioides* (L.) Desv. ex DC.	Fabaceae	Herb	Koilipari
381.	11743	*Uraria rufescens* (DC.) Schindl.	Fabaceae	Herb	Khirkoliguda
382.	10410	*Vigna radiata* (L.) Wilczek	Fabaceae	Climber	Mudulipara
383.	11055	*Vigna umbellata* (Thunb.) Ohwi & Ohashi	Fabaceae	Climber	Satiguda
384.	10420	*Zornia gibbosa* Spanoghe	Fabaceae	Herb	Mudulipara
385.	10591	*Casearia elliptica* Willd.	Flacourtiaceae	Tree	Nalagunthi
386.	11178	*Casearia graveolens* Dalz.	Flacourtiaceae	Tree	Disariguda
387.	11033	*Canscora decussata* (Roxb.) Schult & Schult.f.	Gentianaceae	Herb	Chitapari
388.	12376	*Canscora diffusa* (Vahl.) R.Br.	Gentianaceae	Herb	Chalanguda
389.	12378	*Exacum petiolare* Griseb.	Gentianaceae	Herb	Dasmantpur
390.	10650	*Hoppea fastigiata* (Griseb.) C.B. Cl.	Gentianaceae	Herb	Kalimela
391.	11350	*Chirita hamosa* R.Br.	Gesneriaceae	Herb	Goiparbat
392.	11053	*Hydrilla verticillata* (L.f.) Royle	Hydrocharitaceae	Herb	Satiguda
393.	11057	*Ottelia alismoides* (L.) Pers.	Hydrocharitaceae	Herb	Satiguda
394.	11747	*Vallisneria natans* (Lour.) Hara	Hydrocharitaceae	Herb	Khirkoliguda
395.	10456	*Hydrolea zeylanica* (L.) Vahl	Hydrophyllaceae	Herb	Mathili

Contd...

Table 3.2–*Contd...*

Sl.No.	Field No.	Botanical Name	Family	Habit	Locality
396.	11242	*Curculigo orchioides* Gaertn.	Hypoxidaceae	Herb	Kamalapadar
397.	11347	*Acrocephalus hispidus* (L.) Nicolson & Sivadasan	Lamiaceae	Herb	Goiparbat
398.	10483	*Anisochilus carnosus* (L. f.) Wall.	Lamiaceae	Herb	Goiparbat
399.	10370	*Anisomeles indica* (L.) Kuntze.	Lamiaceae	Herb	Dasmantpur
400.	10396	*Anisomeles indica* (L.) Kuntze. var. *mollissima* Benth.	Lamiaceae	Herb	Kamalapadar
401.	12110	*Colebrookia oppositifolia* Sm.	Lamiaceae	Shrub	Chitrakonda
402.	11410	*Eusteralis stellata* (Lour.) Panig. & Raizada var. *roxburghiana* (Keng) Bennet	Lamiaceae	Herb	Kolipari
403.	10255	*Hyptis suaveolens* (L.) Poit.	Lamiaceae	Herb	Biralaxmanpur
404.	10463	*Leonotis nepetifolia* (L.) R. Br.	Lamiaceae	Herb	Mudulipara
405.	11487	*Leucas aspera* (Willd.) Link	Lamiaceae	Herb	Motu
406.	10404	*Leucas cephalotes* (Roth) Spreng.	Lamiaceae	Herb	Mathili
407.	11488	*Leucas clarkei* Hook.f.	Lamiaceae	Herb	Motu
408.	10341	*Leucas indica* (L.) R. Br. ex Vatke	Lamiaceae	Herb	Govindapally
409.	11448	*Leucas mollissima* Wall. ex Benth. var. *scaberula* Hook. f.	Lamiaceae	Herb	Pulkelkonda
410.	10268	*Ocimum canum* Sims	Lamiaceae	Herb	Biralaxmanpur
411.	12533	*Ocimum gratissimum* L.	Lamiaceae	Herb	MV-42
412.	12355	*Ocimum sanctum* L.	Lamiaceae	Herb	Tamasapalli
413.	11390	*Orthosiphon rubicundus* (D.Don) Benth.	Lamiaceae	Herb	Goiparbat
414.	11749	*Perilla frutescens* (L.) Britton	Lamiaceae	Herb	Malkangiri
415.	10409	*Plectranthus mollis* (Ait.) Spreng.	Lamiaceae	Herb	Mudulipara
416.	10462	*Pogostemon benghalensis* (Burm.f.) Kuntze	Lamiaceae	Herb	Mudulipada
417.	12396	*Litsea glutinosa* (Lour.) Robins.	Lauraceae	Tree	Dasmantpur

Contd...

Table 3.2–*Contd...*

Sl.No.	Field No.	Botanical Name	Family	Habit	Locality
418.	10513	*Utricularia aurea* Lour.	Lentibulariaceae	Herb	Pangam
419.	10291	*Asparagus racemosus* Willd.	Liliaceae	Climber	Dasmantpur
420.	11371	*Chlorophytum arundinaceum* Baker	Liliaceae	Herb	Goiparbat
421.	10274	*Gloriosa superba* L.	Liliaceae	Climber	Biralaxmanpur
422.	10275	*Urginea indica* (Roxb.) Kunth.	Liliaceae	Herb	Biralaxmanpur
423.	11437	*Hugonia mystax* L.	Linaceae	Shrub	Pulkelkonda
424.	10648	*Lobelia alsinoides* Lam.	Lobeliaceae	Herb	Kalimela
425.	10601	*Dendrophthoe falcata* (L. f.) Etting.	Loranthaceae	Shrub	Undragonda
426.	10983	*Scurrula parasitica* L.	Loranthaceae	Shrub	Chitrakonda
427.	10489	*Viscum articulatum* Burm.f.	Loranthaceae	Herb	Chalanguda
428.	11240	*Lygodium flexuosum* (L.) Sw.	Lygodiaceae	Climber	Kamalapadar
429.	12364	*Lygodium microphyllum* R. Br.	Lygodiaceae	Climber	Goiparbat
430.	11078	*Ammannia baccifera* L.	Lythraceae	Herb	Satiguda
431.	10276	*Lagerostroemia parviflora* Roxb.	Lythraceae	Tree	Balimela
432.	12192	*Lagerstroemia reginae* Roxb.	Lythraceae	Tree	Malkangiri
433.	12296	*Lawsonia inermis* L.	Lythraceae	Shrub	Orkel
434.	10647	*Nesaea brevipes* Koehne	Lythraceae	Herb	Kalimela
435.	10385	*Rotala densiflora* (Roth ex Roem. & Sch.) Koehne	Lythraceae	Herb	Kamalapadar
436.	10386	*Rotala indica* (Willd.) Koehne	Lythraceae	Herb	Kamalapadar
437.	10681	*Woodfordia fruticosa* (L.) Kurz.	Lythraceae	Shrub	Chitrakonda
438.	12186	*Michelia champaca* L.	Magnoliaceae	Tree	Malkangiri
439.	12278	*Aspidopterys indica* (Roxb.) Hochr.	Malpighiaceae	Climber	Chitrakonda

Contd...

Table 3.2–*Contd...*

Sl.No.	Field No.	Botanical Name	Family	Habit	Locality
440.	11460	*Abelmoschus moschatus* Medic.	Malvaceae	Herb	Pulkelkonda
441.	10406	*Abutilon indicum* (L.) Sweet ssp. *indicum*	Malvaceae	Herb	Mudulipara
442.	10424	*Abutilon persicum* (Burm. f.) Merr.	Malvaceae	Herb	Mudulipara
443.	11052	*Gossypium arboreum* L.	Malvaceae	Shrub	Satiguda
444.	10515	*Hibiscus aculeatus* Roxb.	Malvaceae	Herb	Bandhaguda
445.	11445	*Hibiscus cannabinus* L.	Malvaceae	Herb	Pulkelkonda
446.	10292	*Hibiscus lobatus* (Murray) Kuntze	Malvaceae	Herb	Dasmantpur
447.	10707	*Hibiscus rosa-sinensis* L.	Malvaceae	Shrub	Balimela
448.	10499	*Hibiscus sabdariffa* Linn.	Malvaceae	Herb	Chalanguda
449.	10971	*Kydia calycina* Roxb.	Malvaceae	Tree	Chitrakonda
450.	11480	*Pavonia odorata* Willd.	Malvaceae	Herb	Motu
451.	10348	*Sida acuta* Burm.f.	Malvaceae	Herb	Govindapally
452.	10371	*Sida cordata* (Burm. f.) Borssum	Malvaceae	Herb	Govindapally
453.	10623	*Sida cordifolia* L.	Malvaceae	Herb	Undragonda
454.	12393	*Sida rhombifolia* L. ssp. *retusa* (L.) Borssum	Malvaceae	Herb	Dasmantpur
455.	10365	*Sida rhombifolia* L. ssp. *rhombifolia*	Malvaceae	Herb	Govindapally
456.	10304	*Thespesia lampas* (Cav.) Dalz. & Gibs.	Malvaceae	Shrub	Dasmantpur
457.	10400	*Urena lobata* L. ssp. *lobata*	Malvaceae	Shrub	Kamalapadar
458.	10327	*Urena lobata* L. ssp. *sinuata* (L.) Borssum	Malvaceae	Shrub	Dasmantpur
459.	11069	*Marsilea minuta* L.	Marsileaceae	Herb	Satiguda
460.	10448	*Martynia annua* L.	Martyniaceae	Herb	Mudulipara
461.	11154	*Melastoma malabathricum* L.	Melastomataceae	Shrub	Disariguda

Contd...

Table 3.2–*Contd...*

Sl.No.	Field No.	Botanical Name	Family	Habit	Locality
462.	12250	*Memecylon umbellatum* Burm. f.	Melastomataceae	Tree	Chitapari
463.	11441	*Osbeckia chinensis* L.	Melastomataceae	Herb	Pulkelkonda
464.	10392	*Osbeckia muralis* Naud.	Melastomataceae	Herb	Kamalapadar
465.	12383	*Sonerila tenera* Royle	Melastomataceae	Herb	Dasmantpur
466.	10979	*Azadirachta indica* A. Juss.	Meliaceae	Tree	Chitrakonda
467.	10419	*Cipadessa baccifera* (Roth) Miq.	Meliaceae	Tree	Mudulipara
468.	12184	*Melia azedarach* L.	Meliaceae	Tree	Tamasapalli
469.	12231	*Soymida febrifuga* (Roxb.) A. Juss	Meliaceae	Tree	Sadasivpur
470.	11029	*Cissampelos pareira* L. var. *hirsuta* (Buch.-Ham. ex DC.) Forman	Menispermaceae	Climber	Chitapari
471.	11478	*Cocculus hirsutus* (L.) Diels	Menispermaceae	Climber	Motu
472.	11356	*Stephania japonica* (Thunb.) Miers. var. *timoriensis* (DC.) Forman	Menispermaceae	Climber	Goiparbat
473.	12291	*Tinospora cordifolia* (Willd.) Hook.f. & Thoms.	Menispermaceae	Climber	Orkel
474.	11438	*Tinospora sinensis* (Lour.) Merr.	Menispermaceae	Climber	Pulkelkonda
475.	10455	*Nymphoides hydrophylla* (Lour.) Kuntze	Menyanthaceae	Herb	Mathili
476.	11753	*Nymphoides indica* (L.) Kuntze	Menyanthaceae	Herb	Balisagar
477.	10415	*Acacia auriculiformis* A. Cunn. ex Benth.	Mimosaceae	Tree	Mudulipara
478.	11439	*Acacia catechu* (L. f.) Willd.	Mimosaceae	Tree	Pulkelkonda
479.	10273	*Acacia farnesiana* (L.) Willd.	Mimosaceae	Tree	Biralaxmanpur
480.	11117	*Acacia lenticularis* Buch.-Ham. ex Benth.	Mimosaceae	Tree	Chitapari
481.	12235	*Acacia nilotica* (L.) Willd ex Delile ssp. *indica* (Benth.) Brenan	Mimosaceae	Tree	Balimela
482.	10530	*Acacia pennata* (L.) Willd.	Mimosaceae	Climber	Balimela

Contd...

Table 3.2—Contd...

Sl.No.	Field No.	Botanical Name	Family	Habit	Locality
483.	12275	Acacia sinuata (Lour.) Merr.	Mimosaceae	Shrub	Chitrakonda
484.	12176	Adenanthera pavonina L.	Mimosaceae	Tree	Tamasapalli
485.	11016	Albizia procera (Roxb.) Benth.	Mimosaceae	Tree	Nuaguda
486.	10427	Albizzia amara (Roxb.) Boiv.	Mimosaceae	Tree	Sitakund
487.	12239	Albizzia lebbeck (L.) Benth.	Mimosaceae	Tree	Chitapari
488.	11499	Dicrostachys cinerea (L.) Wright & Arn.	Mimosaceae	Tree	Motu
489.	12169	Leucaena leucocephala (Lam.) de Wit.	Mimosaceae	Tree	Tamasapalli
490.	10607	Mimosa pudica L.	Mimosaceae	Herb	Undragonda
491.	12289	Pithecellobium dulce (Roxb.) Benth.	Mimosaceae	Tree	Orkel
492.	10555	Samanea saman (Jacq.) Merr.	Mimosaceae	Tree	Kalimela
493.	11056	Xylia xylocarpa (Roxb.) Taub.	Mimosaceae	Tree	Satiguda
494.	11071	Glinus lotoides L.	Molluginaceae	Herb	Satiguda
495.	10974	Glinus oppositifolius (L.) A. DC.	Molluginaceae	Herb	Chitrakonda
496.	10688	Mollugo pentaphylla L.	Molluginaceae	Herb	Chitrakonda
497.	10978	Artocarpus heterophyllus Lam.	Moraceae	Tree	Chitrakonda
498.	10675	Ficus amplissima Sm.	Moraceae	Tree	Chitrakonda
499.	10976	Ficus benghalensis L. var. benghalensis	Moraceae	Tree	Chitrakonda
500.	12508	Ficus elastica Roxb. ex Hornem.	Moraceae	Tree	Satiguda
501.	11174	Ficus hispida L. f.	Moraceae	Tree	Disariguda
502.	11138	Ficus microcarpa L.f.	Moraceae	Tree	Chitapari
503.	11374	Ficus mollis Vahl	Moraceae	Tree	Gojparbat
504.	10673	Ficus racemosa L.	Moraceae	Tree	Chitrakonda

Contd...

Table 3.2–*Contd...*

Sl.No.	Field No.	Botanical Name	Family	Habit	Locality
505.	10677	*Ficus religiosa* L.	Moraceae	Tree	Chitrakonda
506.	10538	*Ficus semicordata* Buch.- Ham. ex J. E. Sm.	Moraceae	Tree	Balimela
507.	10319	*Ficus tinctoria* Forst. f. subsp. *parasitica* (Willd.) Corner	Moraceae	Tree	Dasmantpur
508.	11881	*Ficus virens* Ait. var. *virens*	Moraceae	Tree	Satiguda
509.	11155	*Morus australis* Poir.	Moraceae	Tree	Disariguda
510.	11162	*Streblus asper* Lour.	Moraceae	Tree	Disariguda
511.	10626	*Moringa oleifera* Lam.	Moringaceae	Tree	Undragonda
512.	12168	*Musa paradisiaca* L.	Musaceae	Herb	Tamasapalli
513.	10446	*Ardisia solanacea* Roxb.	Myrsinaceae	Shrub	Mudulipara
514.	11357	*Eucalyptus tereticornis* Sm.	Myrtaceae	Tree	Goiparbat
515.	10937	*Eucalyptus torelliana* F. v. Muell	Myrtaceae	Tree	Balimela
516.	12162	*Psidium guajava* L.	Myrtaceae	Tree	Tamasapalli
517.	11137	*Syzygium cumini* (L.) Skeels	Myrtaceae	Tree	Chitapari
518.	11139	*Syzygium roxburghianum* Raizada	Myrtaceae	Tree	Chitapari
519.	10690	*Nephrolepis delicatula* (Decne.) Pichi-Sermoli	Nephrolepidaceae	Herb	Chitrakonda
520.	10620	*Boerhavia diffusa* L.	Nyctaginaceae	Herb	Undragonda
521.	10552	*Bougainvillea spectabilis* Willd.	Nyctaginaceae	Shrub	Kalimela
522.	10524	*Mirabilis jalapa* Linn.	Nyctaginaceae	Herb	Bandhaguda
523.	12194	*Nelumbo nucifera* Gaertn.	Nymphaeaceae	Herb	Malkangiri
524.	10512	*Nymphaea pubescens* Willd.	Nymphaeaceae	Herb	Pangam
525.	11129	*Olax scandens* Roxb.	Olacaceae	shrub	Chitapari
526.	10970	*Chionanthus intermedius* (Wight) Bedd.	Oleaceae	Tree	Chitrakonda

Contd...

Table 3.2–*Contd...*

Sl.No.	Field No.	Botanical Name	Family	Habit	Locality
527.	12292	*Jasminum flexile* Vahl	Oleaceae	Climber	Orkel
528.	10264	*Nyctanthes arbor-tristis* L.	Oleaceae	Tree	Biralaxmanpur
529.	10967	*Schrebera swietenioides* Roxb.	Oleaceae	Tree	Chitrakonda
530.	11757	*Ludwigia adscendens* (L.) Hara	Onagraceae	Herb	Balisagar
531.	10338	*Ludwigia hyssopifolia* (G. Don) Exell	Onagraceae	Herb	Dasmantpur
532.	10474	*Ludwigia octovalvis* (Jacq.) Raven	Onagraceae	Herb	Kamalapadar
533.	11404	*Ludwigia perennis* L.	Onagraceae	Herb	Goiparbat
534.	11130	*Opilia amantacea* Roxb.	Opiliaceae	Shrub	Chitapari
535.	11181	*Acampe praemorsa* (Roxb.) Blatter. & Mc Cann.	Orchidaceae	Herb	Disariguda
536.	11249	*Acampe rigida* (Buch.-Ham. ex Sm.) Hunt	Orchidaceae	Herb	Kamalapadar
537.	10583	*Aerides odoratum* Lour.	Orchidaceae	Herb	Nalagunthi
538.	10527	*Dendrobium crepidatum* Lindl.	Orchidaceae	Herb	Mudulipara
539.	11481	*Geodorum densiflorum* (Lam.) Schltr.	Orchidaceae	Herb	Motu
540.	10718	*Habenaria commelinifolia* Wall. ex Lindl.	Orchidaceae	Herb	Dasmantpur
541.	10301	*Habenaria plantaginea* Lindl.	Orchidaceae	Herb	Dasmantpur
542.	11768	*Luisia filiformis* Hook.f.	Orchidaceae	Herb	Malaguda
543.	11392	*Nervilia aragoana* Gaud.	Orchidaceae	Herb	Goiparbat
544.	10311	*Vanda tessellata* (Roxb.) Hook. ex G. Don	Orchidaceae	Herb	Dasmantpur
545.	12103	*Zeuxine longilabris* (Lindl.) Benth. ex Hook. f.	Orchidaceae	Herb	Chitrakonda
546.	11387	*Aeginetia indica* L.	Orobanchaceae	Herb	Goiparbat
547.	10279	*Biophytum reinwardtii* (Zucc.) Klotz.	Oxalidaceae	Herb	Balimela
548.	12388	*Pandanus fascicularis* Lam.	Pandanaceae	Shrub	Dasmantpur

Contd...

Table 3.2–*Contd...*

Contd...

Sl.No.	Field No.	Botanical Name	Family	Habit	Locality
549.	10556	*Argemone mexicana* L.	Papaveraceae	Herb	Kalimela
550.	10470	*Ceratopteris thalictroides* (L.) Brongn.	Parkeriaceae	Herb	Kamalapadar
551.	11063	*Passiflora foetida* L.	Passifloraceae	Climber	Satiguda
552.	10353	*Sesamum orientale* L.	Pedaliaceae	Herb	Govindapally
553.	10604	*Hemidesmus indicus* (L.) R.Br.	Periplocaceae	Climber	Undragonda
554.	10421	*Piper longum* L.	Piperaceae	Climber	Mudulipara
555.	10520	*Plumbago indica* Linn.	Plumbaginaceae	Herb	Bandhaguda
556.	10258	*Plumbago zeylanica* Linn.	Plumbaginaceae	Herb	Biralaxmanpur
557.	11231	*Alloteropsis cimicina* (L.) Stapf	Poaceae	Herb	Tamasapalli
558.	10534	*Apluda mutica* L.	Poaceae	Herb	Balimela
559.	11104	*Aristida setacea* Retz.	Poaceae	Herb	Tamasapalli
560.	12398	*Arthraxon lancifolius* (Trin.) Hochst.	Poaceae	Herb	Dasmantpur
561.	11422	*Avena sativa* L.	Poaceae	Herb	Koilipari
562.	10521	*Bambusa arundinacea* (Retz.) Willd.	Poaceae	Shrub	Bandhaguda
563.	10531	*Bothriochloa bladhii* (Retz.) S. T. Blake	Poaceae	Herb	Balimela
564.	10458	*Chrysopogon aciculatus* (Retz.) Trin.	Poaceae	Herb	Mathili
565.	12382	*Chrysopogon fulvus* (Spreng.) Chiov.	Poaceae	Herb	Dasmantpur
566.	11466	*Chrysopogon verticillatus* (Roxb.) Trin. ex Steud.	Poaceae	Herb	Pulkelkonda
567.	11183	*Coelachne simpliciuscula* (Wight & Arn. ex Steud.) Munro ex Benth.	Poaceae	Herb	Disariguda
568.	10548	*Coix gigantea* Koenig ex Roxb.	Poaceae	Herb	Korukonda
569.	10973	*Coix lacryma-jobi* Linn.	Poaceae	Herb	Chitrakonda

Table 3.2–*Contd...*

Sl.No.	Field No.	Botanical Name	Family	Habit	Locality
570.	11087	*Cynodon dactylon* (L.) Pers.	Poaceae	Herb	Satiguda
571.	11220	*Dactyloctenium aegyptium* (L.) P. Beauv.	Poaceae	Herb	Tamasapalli
572.	10572	*Dendrocalamus strictus* (Roxb.) Nees	Poaceae	Shrub	Nalagunthi
573.	11235	*Digitaria ciliaris* (Retz.) Koeler	Poaceae	Herb	Tamasapalli
574.	12377	*Echinochloa stagnina* (Retz.) P. Beauv.	Poaceae	Herb	Chalanguda
575.	10354	*Eleusine coracana* (L.) Gaertn.	Poaceae	Herb	Govindapally
576.	10988	*Elytrophorus spicatus* (Willd.) A. Camus	Poaceae	Herb	Chitrakonda
577.	11423	*Eragrostis aspera* (Jacq.) Nees	Poaceae	Herb	Kolipari
578.	11092	*Eragrostis coarctata* Stapf	Poaceae	Herb	Satiguda
579.	10645	*Eragrostis gangetica* (Roxb.) Steud.	Poaceae	Herb	Kalimela
580.	10643	*Eragrostis japonica* (Thunb.) Trin.	Poaceae	Herb	Undragonda
581.	10702	*Eragrostis tenella* (L.) P. Beauv. ex Roem. & Schult.	Poaceae	Herb	Chitrakonda
582.	10471	*Eragrostis unioloides* (Retz.) Nees ex Steud.	Poaceae	Herb	Kamalapadar
583.	10639	*Eriochloa procera* (Retz.) Hubbard	Poaceae	Herb	Undragonda
584.	11086	*Eulalia leschenaultiana* (Decne) Ohwi	Poaceae	Herb	Satiguda
585.	12363	*Hackelochloa granularis* (L.) Kuntze	Poaceae	Herb	Goiparbat
586.	10468	*Heteropogon contortus* (L.) P. Beauv. ex Roem. & Schult.	Poaceae	Herb	Kamalapadar
587.	10398	*Isachne globosa* (Thunb.) Kuntze	Poaceae	Herb	Kamalapadar
588.	10469	*Ischaemum indicum* (Houtt.) Merr.	Poaceae	Herb	Kamalapadar
589.	10494	*Ischaemum rugosum* Salisb.	Poaceae	Herb	Chalanguda
590.	10428	*Oplismenus burmannii* (Retz.) P. Beauv.	Poaceae	Herb	Sitakund
591.	10373	*Oplismenus compositus* (L.) P. Beauv.	Poaceae	Herb	Govindapally

Contd...

Table 3.2–*Contd...*

Sl.No.	Field No.	Botanical Name	Family	Habit	Locality
592.	10453	*Oryza rufipogon* Griff.	Poaceae	Herb	Mathili
593.	12276	*Panicum brevifolium* L.	Poaceae	Herb	Chitrakonda
594.	10317	*Panicum notatum* Retz.	Poaceae	Herb	Saptadhara
595.	11739	*Panicum sumatrense* Roth ex Roem. & Schult.	Poaceae	Herb	Khirkoliguda
596.	11133	*Paspalidium flavidum* (Retz.) A. Camus	Poaceae	Herb	Chitapari
597.	10654	*Paspalum scrobiculatum* L.	Poaceae	Herb	Balimela
598.	10674	*Pennisetum pedicellatum* Trin.	Poaceae	Herb	Chitrakonda
599.	10461	*Pennisetum polystachyon* (L.) Schult.	Poaceae	Herb	Mudulipara
600.	10615	*Perotis indica* (L.) Kuntze	Poaceae	Herb	Undragonda
601.	11267	*Pogonatherum paniceum* (Lam.) Hack.	Poaceae	Herb	Dasmantpur
602.	11738	*Rottboellia cochinchinensis* (Lour.) Clayton	Poaceae	Herb	Khirkoliguda
603.	12246	*Saccharum spontaneum* L.	Poaceae	Herb	Chitapari
604.	10399	*Sacciolepis indica* (L.) Chase	Poaceae	Herb	Kamalapadar
605.	11762	*Sacciolepis interrupta* (Willd.) Stapf	Poaceae	Herb	Balisagar
606.	10472	*Sacciolepis myosuroides* (R. Br.) A. Camus	Poaceae	Herb	Kamalapadar
607.	10987	*Setaria intermedia* Roem. & Schult.	Poaceae	Herb	Chitrakonda
608.	11500	*Setaria verticillata* (L.) P. Beauv.	Poaceae	Herb	Motu
609.	11062	*Sorghum vulgare* Pers.	Poaceae	Herb	Satiguda
610.	10582	*Sporobolus indicus* (L.) R. Br. var. *fertilis* (Steud.) Jovet & Guedes	Poaceae	Herb	Nalagunthi
611.	10480	*Themeda triandra* Forssk.	Poaceae	Herb	Chalanguda
612.	10990	*Thysanolaena maxima* (Roxb.) Kuntze.	Poaceae	Herb	Chitrakonda

Contd...

Table 3.2–*Contd...*

Sl.No.	Field No.	Botanical Name	Family	Habit	Locality
613.	10609	*Vetiveria zizanioides* (L.) Nash	Poaceae	Herb	Undragonda
614.	11222	*Polygala arvensis* Willd.	Polygalaceae	Herb	Tamasapalli
615.	10634	*Polygonum barbatum* L. var. *barbatum*	Polygonaceae	Herb	Undragonda
616.	12523	*Polygonum barbatum* L. var. *stagninum* (Buch.-Ham. ex Meissn.) Steward	Polygonaceae	Herb	Jayapuriaguda
617.	10412	*Polygonum chinense* L.	Polygonaceae	Herb	Mudulipara
618.	10329	*Polygonum glabrum* Willd.	Polygonaceae	Herb	Govindapally
619.	11771	*Polygonum hydropiper* L. var. *flaccidum* Steward	Polygonaceae	Herb	Malaguda
620.	11759	*Polygonum pulchrum* Bl.	Polygonaceae	Herb	Balisagar
621.	12284	*Paraleptochilus decurrens* (Bl.) Copel.	Polypodiaceae	Herb	Chitrakonda
622.	11748	*Eichhornia crassipes* (Mart.) Solms-Laub.	Pontederiaceae	Herb	Khirkoliguda
623.	11411	*Monochoria hastata* Solms- Laub.	Pontederiaceae	Herb	Kolipari
624.	11082	*Portulaca oleracea* L.	Portulacaceae	Herb	Satiguda
625.	12529	*Grevillea pteridifolia* Knight	Proteaceae	Tree	MV-42
626.	12277	*Pteris pellucida* Presl.	Pteridaceae	Herb	Chitrakonda
627.	10440	*Pteris quadriaurita* Retz.	Pteridaceae	Herb	Mudulipara
628.	12178	*Punica granatum* L.	Punicaceae	Shrub	Tamasapalli
629.	12236	*Ventilago denticulata* Willd.	Rhamnaceae	Climber	Balimela
630.	10627	*Ziziphus mauritiana* Lam.	Rhamnaceae	Tree	Undragonda
631.	10325	*Ziziphus oenoplia* (L.) Mill.	Rhamnaceae	Shrub	Dasmantpur
632.	10999	*Ziziphus rugosa* Lam.	Rhamnaceae	Tree	Chitrakonda

Table 3.2–*Contd...*

Sl.No.	Field No.	Botanical Name	Family	Habit	Locality
633.	10307	*Ziziphus xylopyrus* (Retz.) Willd.	Rhamnaceae	Tree	Dasmantpur
634.	10585	*Stemona tuberosa* Lour.	Roxburghiaceae	Climber	Nalagunthi
635.	12534	*Anthocephalus chinensis* (Lam.) A. Rich ex Walp.	Rubiaceae	Tree	MV-42
636.	11172	*Canthium dicoccum* (Gaertn.) Teijs. & Binn.	Rubiaceae	Shrub	Disariguda
637.	11171	*Catunaregam spinosa* (Thunb.) Tirveng.	Rubiaceae	Shrub	Disariguda
638.	11067	*Dentella repens* (L.) J. R. & G. Forst var. *serpyllifolia* (Wall. ex Craib.) Verdc.	Rubiaceae	Herb	Satiguda
639.	12241	*Dentella repens* (L.) J. R. & G. Forst. var *repens*	Rubiaceae	Herb	Chitapari
640.	11147	*Fagerlindia fasciculata* (Roxb.) Tirveng.	Rubiaceae	Shrub	Nakamamudi
641.	10684	*Gardenia latifolia* Ait.	Rubiaceae	Tree	Chitrakonda
642.	12370	*Gardenia resinifera* Roth	Rubiaceae	Tree	Chalanguda
643.	10277	*Haldinia cordifolia* (Roxb.) Ridsd.	Rubiaceae	Tree	Balimela
644.	11348	*Hedyotis affinis* Roem. & Schult.	Rubiaceae	Herb	Goiparbat
645.	10383	*Hedyotis corymbosa* (L.) Lam.	Rubiaceae	Herb	Kamalapadar
646.	11212	*Hedyotis diffusa* Willd.	Rubiaceae	Herb	Tamasapalli
647.	10331	*Hedyotis herbacea* L.	Rubiaceae	Herb	Dasmantpur
648.	11349	*Hedyotis ovatifolia* Cav.	Rubiaceae	Herb	Goiparbat
649.	12368	*Hedyotis pinifolia* Wall. ex G. Don	Rubiaceae	Herb	Goiparbat
650.	11123	*Hedyotis vestita* R.Br. ex G.Don	Rubiaceae	Herb	Chitapari
651.	11370	*Hymenodictyon orixense* (Roxb.) Mabb.	Rubiaceae	Tree	Goiparbat
652.	10635	*Ixora pavetta* Andr.	Rubiaceae	Tree	Undragonda
653.	10306	*Knoxia sumatrensis* (Retz.) DC.	Rubiaceae	Herb	Dasmantpur

Contd...

Table 3.2–*Contd...*

Sl.No.	Field No.	Botanical Name	Family	Habit	Locality
654.	11125	*Meyna spinosa* Roxb. ex Link. var. *pubescens* Robyns	Rubiaceae	Tree	Chitapari
655.	10339	*Mitracarpus villosus* (Sw.) DC.	Rubiaceae	Herb	Govindapally
656.	10540	*Mitragyna parvifolia* (Roxb.) Korth.	Rubiaceae	Tree	Balimela
657.	11091	*Morinda pubescens* Sm.	Rubiaceae	Tree	Satiguda
658.	12195	*Paederia foetida* L.	Rubiaceae	Climber	Tamasapalli
659.	11345	*Pavetta crassicaulis* Bremek.	Rubiaceae	Shrub	Goiparbat
660.	10350	*Richardia scabra* L.	Rubiaceae	Herb	Govindapally
661.	11353	*Spermacoce articularis* L.f.	Rubiaceae	Herb	Goiparbat
662.	11000	*Spermadictyon suaveolens* Roxb.	Rubiaceae	Shrub	Chitrakonda
663.	10993	*Tamilnadia uliginosa* (Retz.) Tirveng. & Sastre	Rubiaceae	Tree	Chitrakonda
664.	11010	*Wendlandia heynei* (Roem. & Schult.) Sant. & Merch.	Rubiaceae	Tree	Chitrakonda
665.	10960	*Aegle marmelos* (L.) Corr.	Rutaceae	Tree	Chitapari
666.	12514	*Citrus aurantium* L.	Rutaceae	Tree	Satiguda
667.	12173	*Citrus limon* (L.) Burm.f.	Rutaceae	Shrub	Tamasapalli
668.	12294	*Murraya koenigii* (L.) Spreng.	Rutaceae	Shrub	Tunibeda
669.	10965	*Murraya paniculata* (L.) Jack	Rutaceae	Shrub	Chitrakonda
670.	12506	*Santalum album* L.	Santalaceae	Tree	Satiguda
671.	11745	*Cardiospermum halicacabum* L.	Sapindaceae	Climber	Khirkoliguda
672.	11168	*Lepisanthes rubiginosa* (Roxb.) Leenh.	Sapindaceae	Shrub	Disariguda
673.	11718	*Sapindus emarginata* Vahl	Sapindaceae	Tree	Motu
674.	11116	*Schleichera oleosa* (Lour.) Oken	Sapindaceae	Tree	Chitapari
675.	10587	*Madhuca indica* Gmel.	Sapotaceae	Tree	Nalagunthi

Contd...

Table 3.2–*Contd...*

Sl.No.	Field No.	Botanical Name	Family	Habit	Locality
676.	10290	*Manilkara hexandra* (Roxb.) Dub.	Sapotaceae	Tree	Balimela
677.	10698	*Manilkara zapota* (L.) P. Royen	Sapotaceae	Tree	Chitrakonda
678.	12537	*Mimusops elengi* L.	Sapotaceae	Tree	Malkangiri
679.	11723	*Adenosma indianum* (Lour.) Merr.	Scrophulariaceae	Herb	Bhubanpalli
680.	11407	*Centranthera tranquebarica* (Spreng.) Merr.	Scrophulariaceae	Herb	Kolipari
681.	12516	*Dopatrium junceum* (Roxb.) Buch.-Ham. ex Benth.	Scrophulariaceae	Herb	Way 2 Jayapuriaguda
682.	10641	*Limnophila aromatica* (Lam.) Merr.	Scrophulariaceae	Herb	Undragonda
683.	10358	*Limnophila heterophylla* (Roxb.) Benth.	Scrophulariaceae	Herb	Dasmantpur
684.	11727	*Limnophila indica* (L.) Druce	Scrophulariaceae	Herb	Bhubanpalli
685.	10658	*Limnophila repens* (Benth.) Benth.	Scrophulariaceae	Herb	Balimela
686.	10265	*Limnophila rugosa* (Roth) Merr.	Scrophulariaceae	Herb	Biralaxmanpur
687.	12260	*Lindernia anagallis* (Burm.f.) Pennell	Scrophulariaceae	Herb	Chitapari
688.	11146	*Lindernia antipoda* (L.) Alston.	Scrophulariaceae	Herb	Chitapari
689.	11408	*Lindernia ciliata* (Colsm.) Pennell	Scrophulariaceae	Herb	Kolipari
690.	10629	*Lindernia procumbens* (Krock.) Philcox	Scrophulariaceae	Herb	Undragonda
691.	11406	*Lindernia viscosa* (Horn.) Bold.	Scrophulariaceae	Herb	Kolipari
692.	10700	*Mecardonia procumbens* (Mill.) Small	Scrophulariaceae	Herb	Chitrakonda
693.	11149	*Scoparia dulcis* L.	Scrophulariaceae	Herb	Nakamamudi
694.	11414	*Sopubia delphinifolia* (L.) G. Don	Scrophulariaceae	Herb	Kolipari
695.	11726	*Striga angustifolia* (D.Don) Saldanha	Scrophulariaceae	Herb	Bhubanpalli
696.	11393	*Torenia cordifolia* Roxb.	Scrophulariaceae	Herb	Goiparbat
697.	11131	*Selaginella repanda* (Desv. ex Poir.) Spring	Selaginellaceae	Herb	Chitapari

Contd...

Table 3.2–*Contd...*

Sl.No.	Field No.	Botanical Name	Family	Habit	Locality
698.	12188	*Ailanthus excelsa* Roxb.	Simaroubaceae	Tree	Sorishamal
699.	12191	*Simarouba glauca* DC.	Simaroubaceae	Tree	Tamasapalli
700.	10308	*Smilax zeylanica* L.	Smilacaceae	Climber	Dasmantpur
701.	11722	*Datura metel* L.	Solanaceae	Shrub	Bhubanpalii
702.	10559	*Datura stramonium* L.	Solanaceae	Shrub	Kalimela
703.	11098	*Physalis minima* L.	Solanaceae	Herb	Satiguda
704.	10460	*Solanum erianthum* D. Don	Solanaceae	Herb	Mudulipara
705.	10340	*Solanum torvum* Sw.	Solanaceae	Herb	Govindapally
706.	10632	*Solanum viarum* Dunal	Solanaceae	Herb	Undragonda
707.	10269	*Solanum violaceum* Orteg.	Solanaceae	Herb	Biralaxmanpur
708.	11040	*Mitreola petiolata* (Gmel.) Torr. & A. Gray	Spigeliaceae	Herb	Chitapari
709.	11367	*Byttneria herbacea* Roxb.	Sterculiaceae	Herb	Gojparbat
710.	11128	*Firmiana colorata* (Roxb.) R. Br.	Sterculiaceae	Herb	Chitapari
711.	10287	*Helicteres isora* Linn.	Sterculiaceae	Shrub	Balimela
712.	10408	*Melochia corchorifolia* L.	Sterculiaceae	Herb	Mudulipara
713.	11413	*Pterospermum xylocarpum* (Gaertn.) Sant & Wagh	Sterculiaceae	Tree	Koilipari
714.	10597	*Sterculia urens* Roxb.	Sterculiaceae	Tree	Kalimela
715.	11157	*Sterculia villosa* Roxb. ex DC.	Sterculiaceae	Tree	Disariguda
716.	10444	*Waltheria indica* L. var. *indica*	Sterculiaceae	Shrub	Mudulipara
717.	11473	*Strychnos nux-vomica* L.	Strychnaceae	Tree	Pulkelkonda
718.	10542	*Tacca leontopetaloides* (L.) Kuntze	Taccaceae	Herb	Balimela
719.	12389	*Pronephrium nudatum* (Roxb. ex Griff.) Holttum	Thelypteridaceae	Herb	Dasmantpur

Contd...

Table 3.2–*Contd...*

Sl.No.	Field No.	Botanical Name	Family	Habit	Locality
720.	11375	*Corchorus aestuans* L.	Tiliaceae	Herb	Goiparbat
721.	11483	*Corchorus olitorius* L.	Tiliaceae	Herb	Motu
722.	12502	*Grewia disperma* Rottl.	Tiliaceae	Shrub	Satiguda
723.	10516	*Grewia hirsuta* Vahl	Tiliaceae	Shrub	Bandhaguda
724.	10252	*Grewia rothii* DC.	Tiliaceae	Shrub	Biralaxmanpur
725.	11368	*Grewia tiliifolia* Vahl	Tiliaceae	Shrub	Goiparbat
726.	11469	*Triumfetta annua* L.	Tiliaceae	Herb	Pulkelkonda
727.	11421	*Triumfetta pentandra* A. Rich	Tiliaceae	Herb	Kolipari
728.	11464	*Triumfetta rhomboidea* Jacq.	Tiliaceae	Herb	Pulkelkonda
729.	11752	*Trapa natans* L. var. *bispinosa* (Roxb.) Makino	Trapaceae	Herb	Balisagar
730.	12180	*Turnera ulmifolia* L.	Turneraceae	Herb	Tamasapalli
731.	10551	*Holoptelia integrifolia* (Roxb.) Planch.	Ulmaceae	Tree	Kalimela
732.	10416	*Trema orientalis* (L.) Bl.	Ulmaceae	Tree	Mudulipara
733.	11386	*Elatostema cuneatum* Wight	Urticaceae	Herb	Goiparbat
734.	10459	*Girardinia diversifolia* (Link.) Friis	Urticaceae	Herb	Mudulipada
735.	11381	*Laportea interrupta* (L.) Chew	Urticaceae	Herb	Goiparbat
736.	11376	*Neodistemon indicum* (Wedd.) Babu & Henry	Urticaceae	Herb	Goiparbat
737.	11409	*Pouzolzia auriculata* Wight	Urticaceae	Herb	Kolipari
738.	11163	*Callicarpa tomentosa* (L.) Murr.	Verbenaceae	Tree	Disariguda
739.	11720	*Clerodendrum inerme* (L.) Gaertn.	Verbenaceae	Shrub	Motu
740.	10286	*Clerodendrum serratum* (L.) Moon	Verbenaceae	Shrub	Balimela
741.	11169	*Clerodendrum viscosum* Vent.	Verbenaceae	Shrub	Disariguda

Contd...

Table 3.2–*Contd...*

Sl.No.	Field No.	Botanical Name	Family	Habit	Locality
742.	12112	*Duranta repens* L.	Verbenaceae	Shrub	Chitrakonda
743.	11106	*Gmelina arborea* Roxb.	Verbenaceae	Tree	Tamasapalli
744.	10466	*Lantana camara* L. var. *aculeata* (L.) Mold.	Verbenaceae	Shrub	Mudulipara
745.	10298	*Lippia javanica* (Burm.f.) Spreng	Verbenaceae	Shrub	Dasmantpur
746.	11072	*Phyla nodiflora* (L.) Greene	Verbenaceae	Herb	Satiguda
747.	10476	*Stachytarpheta jamaicensis* (L.) Vahl	Verbenaceae	Herb	Kamalapadar
748.	10696	*Tectona grandis* L.f.	Verbenaceae	Tree	Chitrakonda
749.	12161	*Vitex negundo* L.	Verbenaceae	Shrub	Tamasapalli
750.	10637	*Hybanthus enneaspermus* (L.) F.v. Muell.	Violaceae	Herb	Undragonda
751.	11250	*Ampelocissus latifolia* (Roxb.) Planch	Vitaceae	Climber	Kamalapadar
752.	10580	*Ampelocissus tomentosa* (Roth) Planch.	Vitaceae	Climber	Nalagunthi
753.	11734	*Cayratia auriculata* (Wall.) Gamble	Vitaceae	Climber	Khirkoliguda
754.	12116	*Cayratia pedata* (Lour.) Juss. ex Gagnep.	Vitaceae	Shrub	Chitrakonda
755.	10569	*Cayratia trifolia* (L.) Domin.	Vitaceae	Climber	Nalagunthi
756.	11248	*Leea asiatica* (L.) Ridsd.	Vitaceae	Shrub	Kamalapadar
757.	12279	*Leea indica* (Burm.f.) Merr.	Vitaceae	Shrub	Chitrakonda
758.	10347	*Leea macrophyla* Roxb. ex Hornem.	Vitaceae	Shrub	Govindapally
759.	10395	*Xyris pauciflora* Willd.	Xyridaceae	Herb	Kamalapadar
760.	10309	*Costus speciosus* (Koenig.) Sm.	Zingiberaceae	Herb	Dasmantpur
761.	11366	*Curcuma angustifolia* Roxb.	Zingiberaceae	Herb	Goiparbat
762.	11777	*Globba marantina* L.	Zingiberaceae	Herb	Dasmantpur
763.	11261	*Globba racemosa* Sm.	Zingiberaceae	Herb	Dasmantpur

and *Stereospermum chelonoides* were collected from the district which needs immediate attention for conservation.

The three plant species namely *Blepharispermum subsessile, Saraca asoca* and *Stemona tuberosa* were enlisted as species of high conservation priority by Biswal and Nair (2008). The species like *Buddleja asiatica, Cochlospermum religiosum, Crotalaria ramosissima, Firmiana colorata, Girardinia diversifolia, Heterophragma quadriloculare, Mucuna nigricans, Pueraria tuberosa, Tephrosia villosa* and *Tinospora sinensis* were sited occasionally or very rarely. The species might not be threatened in the state context, but can certainly be included under the red listed category so far as the district is concerned.

Conclusion

The floristic study of Malkangiri district of Orissa enumerates 763 different plant species belonging to 505 genera under 147 families. The collection includes as many as 745 angiospermic species (Dicots- 596 species and Monocots- 149 species), 16 species of vascular cryptogams and only 2 gymnospermic species. This indicates the floristic richness of the district which can still be explored thoroughly without the fear of naxalites and other anti-social elements for new discoveries. The study is the first noble attempt to botanise the district for a small but worthy contribution to the flora of the state as well as the country. The small piece of work may definitely be considered as a baseline data for further research and planning works in the district on floristics, systematics, ethnobotany and biodiversity conservation.

Acknowledgements

The Director, CSIR-IMMT, Bhubaneswar is gratefully acknowledged for providing lab facilities to carry on the present work. The financial support of National Medicinal Plant Board (NMPB) is highly acknowledged. Authors dedicate the entire research findings to late Dr M. Brahmam who has taken a lot of pain on his shoulder to bring the project from the funding agency as well as providing full guidance till his last breath. Thanks are due to the DFO, Malkangiri for allowing to conduct the exploration work in the forests of Malkangiri and provide necessary assistance whenever required. The tribals of the locality are also thanked for accompanying through the hilly forests of the district and provide necessary information and support during the work.

References

Aminuddin and Girach, R. D. 1991. Ethnobotanical studies on Bonda tribes of District Koraput, Orissa, India. *Ethnobotany* 3:15-19.

Biswal, A. K. and Nair, M. V. 2008. Threatened plants of Orissa and priority species for conservation. In *ENVIS: Special Habitats and Threatened Plants of India*. 11 (1):175-186.

Champion, H.G. and Seth, S.K. 1968. *A Revised Survey of Forest Types of India*. Govt. of India Press, New Delhi.

Gamble, J. S. and Fischer C. E. C. 1915-36. *Flora of Presidency of Madras*. Adlard and Son Ltd, London.

Good, R. 1956. *Features of evolution in the flowering plants.* London.

Haines, H. H. 1921-25. *The Botany of Bihar and Orissa.* Adlard and Son and West Newman Ltd, London.

Hemadri, K. and Rao, S. S. B. 1989-91. Folklore claims of Koraput and Phulbani district of Orissa. *Indian Medicine* 1:11-13, 2:4-6, 3:10-14.

Hooker, J. D. 1872-97. *The Flora of British India.* London.

Mishra, M. K. and Das, P. K. 1988. Some ethnobotanical plants of Koraput districts of Orissa. *Ancient Science of Life* 8:60-67.

Mooney, H. F. 1950. *Supplement to the Botany of Bihar and Orisssa.* Ranchi.

Pattnaik C., Reddy, C. S., Murthy, M. S. R. and Reddy, P. M. 2006. Ethnomedicinal observations among the tribal people of Koraput District, Orissa, India. *Res. J. Bot.* 1(3): 125-128.

Pattnaik, C., Reddy, C. S., Das, R. and Reddy, P. M. 2007. Traditional medicinal practices among the tribal people of Malkangiri district, Orissa, India. *Nat. Prod. Rad.* 6 (5): 430-435.

Pattnaik, C., Reddy, C. S. and Murthy, M. S. R. 2008. An ethnobotanical survey of medicinal plants used by the Didayi tribe of Malkangiri district of Orissa, India. *Fitoterapia.* 79 (1):67-71.

Prusty, A. B. 2007. Plants used as ethnomedicine by Bonda tribe of Malkangiri district of Orissa. *Ethnobotany* 19:105-110.

Prusty, A. B. and Behera, K. K. 2007. Ethnobotanical exploration of Malkangiri district of Orissa, India. *Ethnobotanical Leaflets* 11:122-140.

Saxena, H. O. and Brahmam, M. 1994-96. *The Flora of Orissa.* OFDC, Bhubaneswar.

Teki Surayya, Mishra, M. and Mishra, R. P. 2003. Marketing of selected NTFPs: A Case Study of Koraput, Malkangiri and Rayagada districts, Orissa, *India. J. Non-Timber Forest Prod.* 10 (3-4): 186-194.

Tzvelen, N. K. 1989. The systems of grasses and their evolution. *Bot. Rev.* **55**:141-203.

Ved, D. K., Kinhal, G. A., Ravikumar, K., Vijayasankar, R., Sumathi, R., Mahapatra, A.K. and Panda, P.C. 2007. *Conservation Assessment and Management Prioritisation for Medicinal plants of Orissa, India.* RPRC, Bhubaneswar and FRLHT, Bangalore.

Chapter 4

The Epiphyllous Liverworts of the Southern Western Ghats: An Overview

A.E.D. Daniels

Bryology Laboratory, Department of Botany and Research Centre,
Scott Christian College (Autonomous), Nagercoil – 629 003, Tamil Nadu

Introduction

Reports on epiphyllous (folicolous) liverworts date back to the late 18th century. According to Gradstein (1997), it was Swartz (1788) who first reported the liverwort *Jungermannia flava* Sw. [now *Lejeunea flava* (Sw.) Nees] growing as an epiphyll on old fern fronds in Jamaica (*vide* Zhu and So, 2001). Since then, these elusive plants have attracted the attention of numerous bryologists and ecologists world-wide. Many epiphyllous species have been described by several hepaticologists from various parts of the world but studies on epiphyllous liverworts started to appear only in the early 20th century (Herzog, 1926; Schiffner, 1929; Richards, 1932; Allorge and Allorge, 1938; Kamimura, 1939; Pandé and Misra, 1943; Pandé and *al.*, 1957; Mizutani, 1966, 1975; Olarinmoye, 1974, 1975, 1977; Pócs, 1975, 1980, 1984, 1985; Piippo, 1994; Tixier, 1995; Gradstein, 1997; Pócs and Streimann, 1999; Daniels and Daniel, 2004, 2009; Dey and Singh, 2012).

In India, studies on epiphyllous liverworts were first initiated on the Western Ghats by Pandé and Misra (1943) who reported 5 species falling under 3 genera based on Rev. I. Pfleiderer's collections made in 1924 from Kudremukh in the Western Ghats of Karnataka, H.S. Rao in 1934 from Gersoppa falls (Jog falls) in Shimoga

District in Karnataka and also by S.N.D. Gupta from Mungpoo in Darjeeling, Eastern Himalaya. This kindled their interest on these plants which led to further collections in the W. Ghats of Karnataka, E. Himalaya and also Sri Lanka, and was published by Pandé & *al.* (1957). Meanwhile, Kachroo (1951) reported 2 epiphyllous liverworts from Assam. More recent studies include Udar and Srivastava's (1983, 1985) in Karnataka from where they described two new epiphyllous species of *Cololejeunea* Schiffn. *viz., C. mizutaniana* and *C. kashyapii* respectively, Udar & *al.* (1985) who reported the occurrence of *C. furcilobulata* (Berrie & E.W. Jones) R.M. Schust. in Karnataka, Lal (1979, 2003) in Manipur and India respectively, Daniels and Daniel (2004, 2009), Daniels & *al.* (2010) in the Southern Western Ghats, Dey and Singh (2007, 2009, 2010, 2011a,b, 2012), and Dey & *al.* (2008a,b) in E. Himalaya and Chavan (2010) in Andaman Islands. Most of the current novelties and discoveries from India and China are epiphyllous forms. In spite of such rigorous surveys carried out in various regions of the world, the exact number of epiphyllous liverworts world-wide is not known, since several floristic regions await investigations. On the other hand some epiphyllous genera need taxonomic revisions (Zhu and So, 2001).

Pócs (1996) provided a world-wide data bank for 1000 epiphyllous liverworts belonging to 22 genera, including 504 Asian and 375 American species. Lucking (1997) gave a total estimation of 535 typical epiphyllous bryophyte species world-wide which included several mosses also based on the epiphyllous bryophyte diversity of Neotropics, tropical Africa and Asia as discussed by Pócs (1978) (*vide* Zhu and So, 2001).

Factors Influencing the Growth of Epiphyllous Liverworts

According to Wu & *al.* (1987), light, temperature and relative humidity are certainly the primary ecological factors which affect the growth and distribution of epipyllous liverworts. But some species are also tolerant to extremely low temperatures. Shirasaki (1997) reported that *Cololejeunea nakajimae* S. Hatt., an epiphyllous liverwort, endemic to Japan, grew well in the winter snow-covered district of central Japan. Longevity, texture, and cuticular properties of host leaves appear to affect the growth and distribution of epiphyllous liverworts which usually prefer evergreen, coriaceous leaves in Chinese subtropical forests (Zhu and So, 2001). The evergreen leaves perhaps have a sufficiently long life for a successful colonization by liverworts. In the SW. Ghats too epiphylls collected so far were on evergreen species with coriaceous leaves. Most epiphyllous liverworts appear to avoid hairy leaves, whose surface makes it difficult for them to adhere. In China, hairy leaves, however, are sometimes occupied by several very common epiphyllous liverworts, such as *Cololejeunea goebelii* (Gottsche ex K.I. Goebel) Schiffn., *C. spinosa* (Horik.) Pandé & R.N. Misra, *Leptolejeunea elliptica* (Lehm. & Lindenb.) Schiffn. and *Radula acuminata* Steph. (Zhu and So, 2001). However, in the SW. Ghats no collection has been made so far from hairy leaves, not even the common ones such as *L. elliptica*; also Dey and Singh (2012) in E. Himalaya.

Epiphyllous Liverworts and their Adaptive Features

The phyllosphere of vascular plants in the temperate and mediterranean regions is an ephemeral environment and only special communities of some fungi and algae can occur (Zhu and So, 2001). However, in tropical and subtropical regions where the forests are constantly moist and humid, rich epiphyllous communities composed of cyanobacteria, algae, lichens, liverworts and mosses frequently develop. Of these, liverworts generally dominate with most of them belonging to the Lejeuneaceae, the largest family in Marchantiophyta (Hepaticae). These epiphyllous liverworts show remarkable morphological adaptations to the phyllosphere which usually has a relatively higher humidity and temperature due to the loss of water in the form of vapour and energy in the form of heat through the stomata.

The characteristic features of epiphyllous liverworts are as follows:

Growth Pattern

Epiphyllous liverworts of the SW. Ghats grow strongly appressed to the phylloplane (adaxial surface of leaf) in order to obtain sufficient moisture and to evade desiccation since water hardly sticks to coriaceous leaf surfaces on which they mostly grow (Plate 4.1A). Thus, water often becomes a limiting factor to the growth of epiphyllous liverworts although water is abundant in rainforests. Therefore, they either grow in patches or ramified strands that help to stop the free flow of water down the leaf surface. Such growth patterns have been observed by Magdefrau (1982), Richards (1984) and Daniels (1998) in other epiphytic forms also. However, *Microlejeunea ulicina* (Taylor) A. Evans occasionally grows decumbent and loosely appressed to the phylloplane.

Morphological and Anatomical Adaptations

The stem tends to be strongly reduced and flattened in most typical epiphyllous liverworts (Plate 4.1B). In the epiphyllous genus *Cololejeunea* (Spruce) Schiffn., with the largest number of species, the stem consists of one row of cortical cells surrounding only one or a few medullary cells. In species like *Cololejeunea lanciloba* Steph., the stem as said earlier, is roughly flattened. Reduced investment in stem structure might enable a more rapid growth and an increase of the reproductive rate (Thiers, 1988).

The leaves are usually imbricate and closely appressed to the substratum in most epiphyllous liverworts. When plenty of moisture is available, they are open and absorb moisture, but close rapidly to prevent water loss as soon as the atmosphere becomes dry (Zhu and So, 2001). Morphologically, the leaf is 2-lobed, forming a dorsal lobe and a ventral lobule in most epiphyllous leafy liverworts (a character also seen in most leafy liverworts). The leaf lobule in most species is strongly inflated, with a usually incurved or inrolled free lateral margin. Such a leaf lobule is also called a 'water sac' (Plate 4.1C). The storage of water is considered a basic function of water sacs and they also provide some symbiotic cyanobacteria with a moist habitat (Chen and Wu, 1964). The thick wall of peptidoglycan surrounding a gummy sheath in these cyanobacteria might help to maintain a moist environment for the epiphylls. It has been shown by Edmisten (1970) and Thiers (1988) that more nitrogen is fixed on leaves with a covering of epiphyllous liverworts than on leaves without them.

In many epiphyllous liverworts, the leaf cells bear various dorsal protrusions (Plate 4.1D), an adaptation to changes in humidity. A good example of one such species in the SW. Ghats is *Cololejeunea spinosa* (Horik.) Pandé & R.N. Misra which

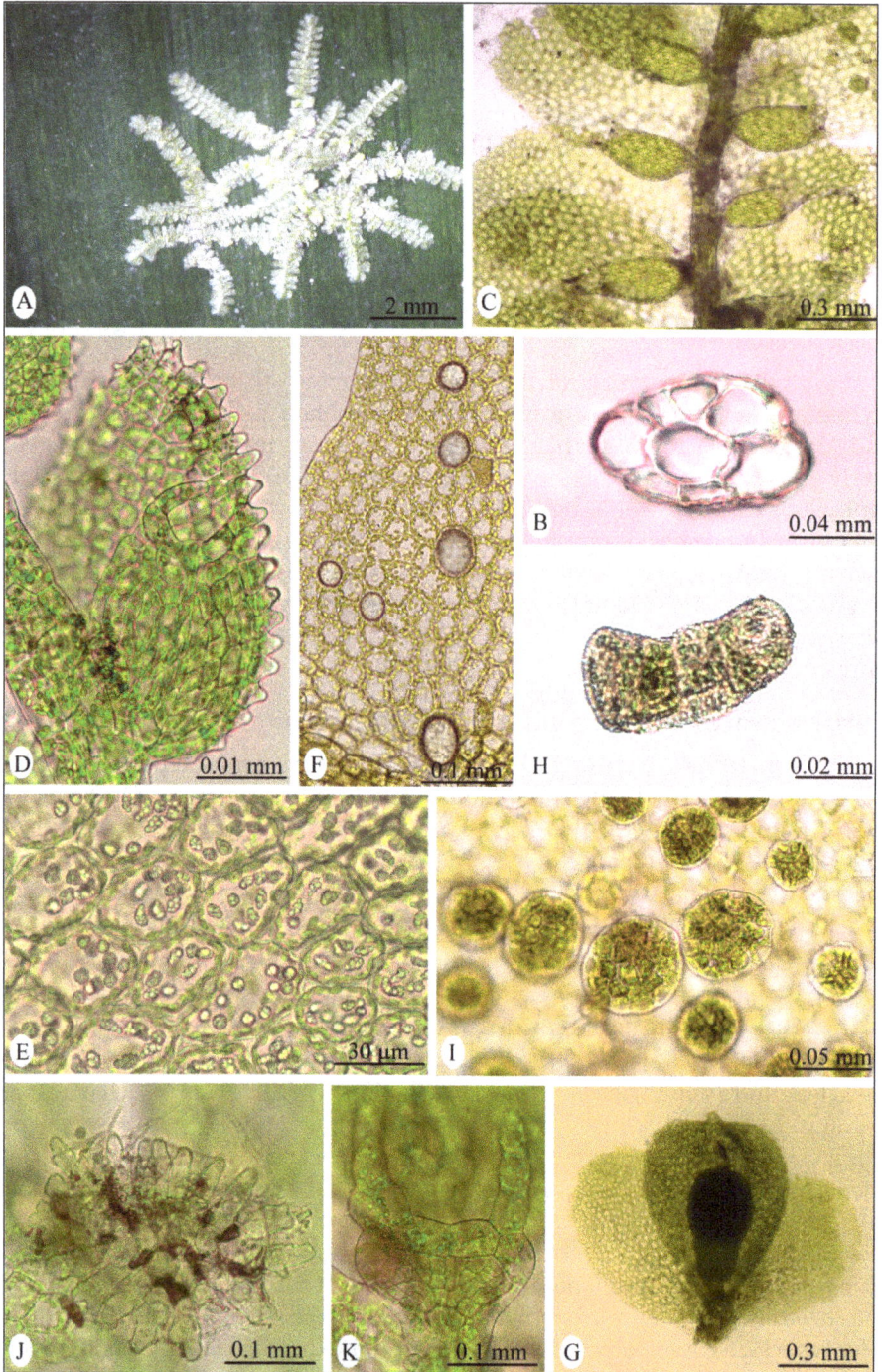

**Plate 4.1: A. Habit; B. Cross section of stem; C. Leaf lobule (water sac);
D. Dorsal protrusions; E. Oil bodies; F. Ocelli; G. Perianth; H. Spore; I. Gemmae;
J. Fasciculate rhizoids; K. Rhizoidal disc.**

shows diverse dorsal protrusions and leaf margin denticulations. When dry, dorsal protrusions start to contract, thus reducing the surface area of evaporation, and help prevent excessive desiccation. Undoubtedly when plenty of moisture is present, the dorsal protrusions function like capillaries to absorb water. In some epiphyllous species, marginal cells are different from inner cells of leaf lobe in size and physiological function. Marginal cells, which are very large and hyaline, form a partial or entire hyaline border (example *Cololejeunea lanciloba* Steph.). These cells may arise from the long-term epiphyllous habitat and by loss of chloroplasts, because they, undoubtedly, are more prone to water loss. They have a greater volume than adjacent cells because of the absence of cytoplasm, and can absorb more water from the substrate than inner living cells (Daniels, 1998; Zhu and So, 2001). According to Thiers (1988), they may serve as water reservoirs for the photosynthetic cells of the leaf. They not only have a greater storage capacity of water when plenty of moisture is present, but can also be more tightly appressed to the substrate to prevent water loss from the adjacent, living cells.

The presence of oil bodies (Plate 4.1E) and ocelli (Plate 4.1F) are features of taxonomic interest but whether they help in the absorption and storage of water is not known since according to Asakawa (1997) oil bodies mainly contain sesqui- and diterpenoids and lipophilic aromatic compounds which are hydrophobic. Similarly, ocelli also contain oils.

Most of the epiphyllous liverworts possess dorsiventrally flattened perianths which may be more adaptive to their habitat (Plate 4.1G).

The spores of epiphyllous liverworts are generally large and irregularly-shaped (Plate 4.1H) which can undergo precocious germination before the capsules dehisce. According to Fulford (1951), spores of some Lejeuneaceae are not adapted to long distance dispersal and can get desiccated within an hour resulting in loss of viability. But, they are suitable for rapid colonization on leaf surfaces.

Asexual reproduction by means of gemmae is a common feature in most epiphylls found in SW. Ghats. Gemmae (Plate 4.1I) are abundant on leaves during monsoon showers and have considerable advantages in short distance dispersal. Moreover, they are quick to germinate. This may be one of the reasons why epiphyllous liverworts grow well in moist tropical and subtropical forests. Asexual reproduction is one of the most important reproductive strategies of epiphyllous liverworts.

The epiphyllous liverworts so far found in the SW. Ghats produce hyaline to faintly brown unicellular rhizoids, which are generally fasciculate (Plate 4.1J). These fasciculate rhizoids some times fuse to form a large "adhesive disc" that helps the stem to be tightly adhered to the leaf surface. Some epiphyllous liverworts possess rhizoids whose apices are digitate. A rhizoid disc (paramphigastrium) is a special structure arising from the underleaf cells and is usually well-developed in the genus *Leptolejeunea* (Spruce) Schiffn (Plate 4.1K). These discs secrete adhesive exudates that help to adhere the epiphylls firmly onto the leaf surface.

Types of Epiphyllous Liverworts

Epiphyllous liverworts are certainly not a natural group which have been treated differently by different authors (Zhu and So, 2001). According to Chen and Wu (1964) all liverworts that can grow on living leaves of vascular plants can be called "epiphyllous liverworts". This view has been followed by most bryologists. However, Gradstein (1997) proposed a term "typical epiphyllous liverworts" which distinguishes it from those which occasionally occur on living leaves. According to Pócs (1996), almost all liverworts and many mosses can occur on living leaves in very wet conditions in super humid rain forests found in Choco, North Madagascar or New Guinea and regions where rainfall exceeds 3000 mm annually. Zhu and *al.* (1998) recognized three types of epiphyllous liverworts, based on their occurrence on living leaves: Type 1. those that grow exclusively on living leaves - "obligate epiphyllous liverworts", Type 2. those that grow predominantly on living leaves and only rarely on other substrates - "facultative, common epiphyllous liverworts" and Type 3. those that grow occasionally on living leaves, but predominantly on other substrates - "occasional epiphyllous liverworts". However, Thiers (1988) and Pócs (1996) doubt whether any liverwort is an obligate epiphyll. But, in the SW. Ghats, *Cololejeunea distalopapillata* (E.W. Jones) R.M. Schust., *C. floccosa* (Lehm. & Lindenb.) Schiffn., *C. vidaliana* Tixier, *Lejeunea himalayensis* (Pandé & R.N. Misra) A.E.D. Daniels & P. Daniel, *Leptolejeunea elliptica* (Lehm. & Lindenb.) Schiffn., *L. epiphylla* (Mitt.) Steph., *L. maculata* (Mitt.) Schiffn. & *L. sikkimensis* Udar & U.S. Awasthi have been found only as epiphylls. But, *Cololejeunea lanciloba* Steph., *C. minutissima* (Sm. & J.C. Sowerby) Schiffn., *C. spinosa* (Horik.) Pandé & R.N. Misra and *Microlejeunea ulicina* (Taylor) A. Evans have been found growing on other substrates too.

Host Specificity

The leaves of certain vascular plants appear to be preferred by certain epiphyllous liverworts (Zhu and So, 2001). This tendency has been observed by many bryologists and ecologists (Berrie and Eze, 1975; Richards, 1984; Dey and Singh, 2012). However, the reasons are still poorly known (Richards, 1984). In the SW. Ghats also, epiphyllous forms tend to show a preference towards coriaceous leaves. Epiphyllous collections made so far in the SW. Ghats have been from angiosperms or pteridophytes with coriaceous leaves. For example, *Leptolejeunea balansae* Steph. was collected from *Elaeocarpus venustus* Bedd. (Elaeocarpaceae), an evergreen tree endemic to the SW. Ghats with coriaceous leaves (Daniels and Daniel, 2004). Furthermore, *Cololejeunea distalopapillata* (E.W. Jones) R.M. Schust., was found on *Syzygium rama-varmae* (Bourd.) Chithra (Myrtaceae), another evergreen tree endemic to the SW. Ghats bearing coriaceous leaves, *Leptolejeunea epiphylla* (Mitt.) Steph. on *Arenga wightii* Griff. (Arecaceae), an endemic, under canopy palm with coriaceous leaves commonly seen in the evergreen and riparian forests of the SW. Ghats, *L. sikkimensis* Udar & U.S. Awasthi on *Garcinia rubro-echinata* Kosterm. (Clusiaceae) another tree with coriaceous leaves endemic to the SW. Ghats and the neighbouring Sri Lanka. However, *Cololejeunea vidaliana* Tixier was found growing on *Syzygium rama-varmae* as well as *Agrostystachys borneensis* Becc. (Euphorbiaceae), another coriaceous-leaved species but not endemic to the SW. Ghats (Daniels and Daniel, 2009). Similarly, *Leptolejeunea maculata* (Mitt.) Schiffn. was found growing on *Elaeocarpus venustus* Bedd. and *Syzygium rama-varmae* (Bourd.) Chithra (Daniels and Daniel, 2013). All the three tree species form a community in the cloud forests of Tirunelveli-Travancore Hills in the SW. Ghats. E.W. Jones found *C. distalopapillata* (as *Leptocolea distalopapillata*) and *C. vidaliana* (as *Leptocolea punctata*, now a synonym of *C. vidaliana*) growing on large leaves of a Commelinaceae member in Africa. E.H. Man found *Leptolejeunea balansae* Steph. on the frond pinnules of *Angiopteris evecta* (Forst.) Hoffm., a pteridophyte and on the leaves of *Heritiera littoralis* Dryand (Sterculiaceae), a large tree in Andaman Islands, and Chavan (2010) reported its occurrence on bamboo leaves and also as a corticolous form but without mentioning the host plant. *Cololejeunea lanciloba* Steph. was found as a corticolous form on *Terminalia paniculata* Roth (Combretaceae), a lofty tree and an epiphyll on *Calamus thwaitesii* Becc. (Arecaceae), a palm endemic to the Western Ghats and Sri Lanka and also on leaves of *Arenga wightii* Griff. (Arecaceae), and on cultivated *Camellia chinensis* Kuntz (Theaceae) and *Coffea arabica* L. (Rubiaceae), and *Drynaria quercifolia* (L.) J. Sm. (Drynariaceae), a pteridophyte growing on rocks adjacent to water sources or epiphytic on trees. *Cololejeunea minutissima* (Sm. & J.C. Sowerby) Schiffn. was found growing as an epiphyll on *Syzygium rama-varmae* (Bourd.) Chithra, as a corticolous form on *Nothopegia* sp. (Anacardiaceae) and *Areca catechu* L. (Arecaceae).

Olarinrnoye (1975) found that in Nigeria epiphylls appear to avoid dissected leaves; probably they offer less landing surface for falling gemmae. But in northwestern Yunnan, a lot of epiphyllous species were found on ferns with dissected fronds (Zhu and So, 2001). However, in China, epiphyllous liverworts seem to avoid the leaves of gymnosperms. Only two specimens have been so far reported to grow on gymnosperm needles *viz.*, *Cololejeunea macounii* (Spruce ex Underw.) A. Evans and *Metzgeria furcata* (L.) Corda which occurred on the needles of a seedling of *Torreya grandis* Fortune ex Lindl. This may be due to the small surface area of the needles and the rare occurrence of gymnosperms in moist and warm habitats.

Dey and Singh (2012), in their study on the epiphyllous liverworts of E. Himalaya found angiosperm leaves to be the most preferred ones hosting 82 epiphylls. Of these, 30 were exclusive to angiosperms. On pteridophytes, 58 taxa have been reported of which 6 are exclusive to pteridophytes. On the other hand 9 species were found to grow on gymnosperm needles. Forty four species were found growing on angiosperms and pteridophytes and 8 taxa on all the three hosts.

According to Zhu and So (2001), in several provinces of eastern China, *Indocalamus tessellates* (Munro) Keng f. was found to be a favoured host where about 60 per cent of the local epiphyllous liverworts were found. Whereas in Hong Kong, the most preferred host seemed to be *Pronephrium simplex* (Hook.) Holttum, a very common species along river banks in moist forests. The number of hosts is much more than that of the epiphylls. A total of about 400 species belonging to over 155 genera of hosts were recorded in China. Thus, it is evident that epiphyllous liverworts do not show any host specificity.

Diversity of Epiphyllous Liverworts

Based on the studies carried out so far on the SW. Ghats, there are 13 epiphyllous species falling under 4 genera. They may be *Cololejeunea distalopapillata* (E.W. Jones) R.M. Schust., *C. floccosa* (Lehm. & Lindenb.) Schiffn., *C. lanciloba* Steph., *C. minutissima* (Sm. & J.C. Sowerby) Schiffn., *C. spinosa* (Horik.) Pandé & R.N. Misra, *C. vidaliana* Tixier, *Lejeunea himalayensis* (Pandé & R.N. Misra) A.E.D. Daniels & P. Daniel, *Leptolejeunea balansae* Steph., *L. elliptica* (Lehm. & Lindenb.) Schiffn., *L. epiphylla* (Mitt.)

Steph., *L. maculata* (Mitt.) Schiffn., *L. sikkimensis* Udar & U.S. Awasthi and *Microlejeunea ulicina* (Taylor) A. Evans. *Cololejeunea* with 6 species followed by *Leptolejeunea* with 5 and *Lejeunea* and *Microlejeunea* with one each. More studies will certainly add many more genera/species to the existing list.

Ecology and Distribution of Epiphyllous Liverworts

The epiphyllous liverworts have always been found on hosts growing in riparian forests or cloud forests at altitudes above 1000 m. Similar observations have been made by Zhu and So (2001) in China. Up to 4 species have been found growing on a single leaf of *Syzygium rama-varmae* (Bourd.) Chithra, an endemic and critically endangered tree distributed in the cloud forests of the SW. Ghats, and a homogeneous population of 33 plants of *Leptolejeunea balansae* Steph. were found growing on a leaf of *Elaeocarpus venustus* Bedd. (Daniels and Daniel, 2004), another endemic and critically endangered tree found in the same habitat.

Conservation of Epiphyllous Liverworts

The phyllosphere of vascular plants is a very fragile habitat (Zhu and So, 2001). According to Pócs (1997), epiphyllous liverworts are very good environmental indicators and are probably the most fragile group of plants that respond immediately to the slightest change caused in their microclimates. Therefore, microclimatic niches are a matter of concern since they are very important for their survival and distribution in tropical forests, and any slight change in the habitat is very sensitively reflected on their composition resulting in the disappearance of some species. Liu and *al.* (1988) found that the dominant species changed immediately after a slight change in habitat caused due to laying of a road for slipping logs. Pócs (1996) states that any impact on the structure of the canopy or underlying storeys of a forest habitat causes serious impoverishment or total loss of epiphyllous communities. According to Zhu & *al.* (1994), in China, the main threats towards epiphyllous liverworts are over-felling of natural forests, ill-planned public construction works and inappropriate conservation practices. The conservation of epiphyllous liverworts relies greatly on the conservation of related forests.

Acknowledgements

The author thanks the Tamil Nadu and Kerala State Forest Departments for permission to explore the study area, D.G. Long (E), R.L. Zhu (HSNU), G. Winter, Senckenberg Natural History Museum, Germany, M.J. Wigginton (Peterborough), G. Hardy (E) and D.K. Singh (CAL), for help with literature and the Principal, Scott Christian College, for facilities.

References

Allorge, V. and Allorge, P. 1938. Sur la répartition et l'écologie des hépatiques epiphylls aux Açores. *Bol. Soc. Brot.* **13**: 211 - 231.

Asakawa, Y. 1997. Biologically active compounds from bryophytes. *IAB Symposium on 2000's Bryology Abstracts:* 86 - 90. Beijing.

Berrie, G.K. and Eze, J.M.O. 1975. The relationship between an epiphyllous liverwort and host leaves. *Ann. Bot.* **39**: 955 - 963.

Chavan, S.J. 2010. Studies on the family Lejeuneaceae from Andaman Islands, India. *Bioinfolet* **7**: 4 - 8.

Chen, P.-C. and Wu, P.-C. 1964. Studies on epiphyllous of China. I. *Acta Phytotax. Sin.* **9**: 213 - 276.

Daniels, A.E.D. 1998. Ecological adaptations of some bryophytes of the Western Ghats. *J. Ecobiol.* **10**: 261 - 270.

Daniels, A.E.D. and Daniel, P. 2004. *Leptolejeunea balansae* (Hepaticae: Jungermanniales) - a new record of Bryoflora from the Indian mainland. *J. Bombay Nat. Hist. Soc.* **101**: 333 - 334.

Daniels, A.E.D. and Daniel, P. 2009. *Cololejeunea distallopapillata* (E.W. Jones) R.M. Schust. and *C. vidaliana* Tixier (Lejeuneaceae) - New to the liverwort flora of India. *Acta Bot. Hung.* **51**: 61 - 66.

Daniels, A.E.D. and P. Daniel 2013. *The Bryoflora of the Southernmost Western Ghats, India*. Bishen Singh Mahendra Pal Singh, Dehra Dun.

Daniels, A.E.D., Kariyappa, K.C. and P. Daniel, P. 2010. Circumscription of the Polymorphic *Cololejeunea lanciloba* Steph. (Lejeuneaceae - Hepaticae) and species falling within it. *Acta Bot. Hung.* **52**: 287 - 295.

Dey, M., Singh, D. and Singh, D.K. 2008a. A new species of *Cololejeunea* (Hepaticae: Lejeuneaceae) from Eastern Himalaya, India. *Taiwania* **53**: 258 - 263.

Dey, M. and Singh, D.K. 2007. *Cololejeunea yipii* R.L. Zhu (Hepaticae: Lejeuneaceae) - new to Indian bryoflora from West Siang, Arunachal Pradesh. *Phytotaxonomy* **7**: 35 - 37.

Dey, M. and Singh, D.K. 2009. *Caudalejeunea lehmanniana* (Gottsche) A. Evans (Hepaticae: Lejeuneaceae) - a new record for Indian bryoflora from Lohit district, Arunachal Pradesh. *Indian J. Forest.* **32:** 653 - 656.

Dey, M. and Singh, D.K. 2010. Two new epiphyllous *Leptolejeunea* (Hepaticae: *Lejeuneaceae*) from Eastern Himalaya, India. *Taiwania* **55**: 355 - 362.

Dey, M. and Singh, D.K. 2011a. Four foliicolous species of *Cololejeunea* (Spruce) Schiffn. (Marchantiophyta: Lejeuneaceae) new to India. *J. Bryol.* **33**: 163 - 167.

Dey, M. and Singh, D.K. 2011b. A new *Lopholejeunea* (Spruce) Schiffn. (Hepaticae: *Lejeuneaceae*) from India. *Nelumbo* **53**: 197 - 200.

Dey, M. and Singh, D.K. 2012. *Epiphyllous liverworts of Eastern Himalaya*. Botanical Survey of India, Kolkata.

Dey, M., Singh, D.K and Singh, D. 2008b. Two new species of *Lejeunea* Lib. (Hepaticae: Lejeuneaceae) from Sikkim, India. *J. Bryol.* **30**: 126 - 130.

Edmisten, J. 1970. Preliminary studies of the nitrogen budget of a tropical rain forest. In: Odum, H.T. and Pigeon, R.F. (ed.), *A tropical rainforest, a study of irradiation and ecology at E1 Verde, Puerto Rico*, H-21: 211 - 215. U.S. Atomic Energy Commission, Washington, D.C.

Fulford, M. 1951. Distribution patterns of the genera of leafy Hepaticae of South America. *Evolution* **5**: 243 - 264.

Gradstein, S.R. 1997. The taxonomic diversity of epiphyllous bryophytes. *Abstr. Bot.* **21**: 15 - 19.

Herzog, T. 1926. *Geographie der Moose.* Jena.

Kachroo, P. 1951. Studies in Assam hepaticae. I. On some epiphyllous liverworts of Assam. *J. Univ. Gauhati* **2**: 31 - 39.

Kamimura, M. 1939. Studies on the epiphyllous Hepaticae and its attached plants in Sikoku, Japan. *J. Jap. Bot.* **15**: 63 - 83.

Lal, J. 1979. Epiphyllous bryophytes of Manipur. Metzgeriaceae. *Proc. 66ᵗʰ Indian Sci. Congr. Assoc.* **3**: 42.

Lal, J. 2003. Studies in epiphyllous liverworts of India - its present position. *Bull. Bot. Surv. India* **45**: 1 - 4.

Liu, Z.-L., Guo, X.-H. and Hu, R.-L. 1988. Investigation on the epiphyllous liverworts from southern part of Anhui Province. *J. East China Normal Univ. (Nat. Sci.)* **4**: 89 - 96.

Lücking, A. 1997. Diversity and distribution of epiphyllous bryophytes in a tropical rainforest in Costa Rica. *Abstr. Bot.* **21**: 79 - 87.

Magdefrau, K. 1982. Life forms of bryophytes. In: Smith, A.J.E. (ed.), *Bryophyte Ecology*: 45 - 58. Chapman and Hall, London.

Mizutani, M. 1966. Epiphyllous species of Lejeuneaceae from Sabah (North Borneo). *J. Hattori Bot. Lab.* **29**: 154 - 170.

Mizutani, M. 1975. Epiphyllous species of Lejeuneaceae from the Philippines. *J. Hattori Bot. Lab.* **39**: 255 - 262.

Olarinmoye, S.O. 1974. Ecology of epiphyllous liverworts: growth in three natural habitats in Western Nigeria. *J. Bryol.* **8**: 275 - 289.

Olarinmoye, S.O. 1975. Ecological studies of epiphyllous liverworts in Western Nigeria. II. Notes on competition and successional change. *Rev. Bryol. Lichénol.* **41**: 457 - 463.

Olarinmoye, S.O. 1977. Studies on epiphyllous liverwort-phorophyte relationship. *Nova Hedwigia* **27**: 647 - 654.

Pandé, S.K. and Misra, R.N. 1943. Studies in Indian hepaticae. II. On the epiphyllous liverworts of India and Ceylon. I. *J. Indian Bot. Soc.* **22**: 159 - 169.

Pandé, S.K., Srivastava, K.P. and Ahmad, S. 1957. Epiphyllous liverworts of India and Ceylon. II. *J. Indian Bot. Soc.* **36**: 335 - 347.

Piippo, S. 1994. On the biogeography of western Melanesian Lejeuneaceae, with comments on their epiphyllous occurrence. *Trop. Bryol.* **9**: 43 - 57.

Pócs, T. 1975. New or little-known epiphyllous liverworts. I. *Cololejeunea* from tropical Africa. *Acta Bot. Acad. Sci. Hung.* **21**: 353 - 375.

Pócs, T. 1978. Epiphyllous communities and their distribution in East Africa. *Bryophyt. Biblioth.* **13**: 681 - 713.

Pócs, T. 1980. New or little-known epiphyllous liverworts. II. Three new *Cololejeunea* from East Africa. *J. Hattori Bot. Lab.* **48**: 305 - 320.

Pócs, T. 1984. New or little-known epiphyllous liverworts. III. The genus *Aphanolejeunea* Evans in Tropical Africa. *Cryptog. Bryol. Lichénol.* **5**: 239 - 267.

Pócs, T. 1985. East African bryophytes. VII. The Hepaticae of the Usambara rain forest project expedition, 1982. *Acta Bot. Hung.* **31**: 113 - 133.

Pócs, T. 1996. Epiphyllous liverwort diversity at worldwide level and its threat and conservation. *Anales Inst. Bioi. Ubiv. Nac. Auton. Mexico, Ser. Bot.* **67**: 109 - 127.

Pócs, T. 1997. Taxonomy and ecology of the Cololejeuneae (an overview). IAB Symposium on 2000's bryology abstracts (suppl.). Beijing.

Pócs, T. and Streimann, H. 1999. Epiphyllous liverworts from Queensland, Australia. *Bryobrothera* **5**: 165 - 172.

Richards, P.W. 1932. Ecology. In: Verdoorn, F. (ed.), *Manual of Bryology*: 367 - 395. The Hague.

Richards, P.W. 1984. The ecology of tropical forest bryophytes. In: Schuster, R.M. (ed.), *New Manual of Bryology* **2**: 1223 - 1270. *The Hattori Botanical Laboratory, Nichinan.*

Schiffner, V. 1929. Über epiphylle Lebermoose aus Japan nebst einigen Beobachtungen über Rhizoiden, Elateren und Brutkörper. *Ann. Bryol.* **2**: 87 - 106.

Shirasaki, H. 1997. Distribution and ·ecology of the epiphyllous liverwort *Cololejeunea nakajimae* in the winter snow-covered district of Niigata Prefecture and its adjacent regions, central Japan. *Bryol. Res.* **7**: 1 - 7.

Swartz, O. 1788. Nova genera and species plantarum seu prodromus. *Bibliopoliis Acad. M. Swederi, Homiae, Upsaliae and Aboae.*

Thiers, B.M. 1988. Morphological adaptations of the Jungermanniales (Hepaticae) to the tropical rainforest habitat. *J. Hattori Bot. Lab.* **64**: 5 - 14.

Tixier, P. 1995. Résultats taxonomiques de 1' éxpédition Bryotrop au Zaire et Rwanda. 30. Bryophytes épiphylles (récoltes de E. Fischer). *Trop. Bryol.* **11**: 11 - 76.

Udar, R. and Srivastava, G. 1983. A new *Cololejeunea* from India. *Misc. Bryol. Lichénol.* **9**: 137 - 139.

Udar, R. and Srivastava, G. 1985. *Cololejeunea* (*Pedinolejeunea*) *kashyapii* sp. nov. from Karnataka, India. *Geophytology* **15**: 64 - 66.

Udar, R., Srivastava, G. and Srivastava, S.C. 1985. *Cololejeunea* (*Pedinolejeunea*) *furcilobulata* (Berrie et Jones) Schuster: new to Asia. *Proc. Indian Acad. Sci. (Pl. Sci.)* **95**: 303 - 307.

Wu, P.-C., Li, D.-K. and Gao, C.-H. 1987. Light and epiphyllous liverworts in the subtropical evergreen forests of Southeast China. *Symp. Bio. Hung.* **35**: 27 - 32.

Zhu, R.-L. and So, M.L. 2001. Epiphyllous liverworts of China. *Nova Hedwigia* **121**: 1 - 418.

Zhu, R-L., Hu, R.-L. and Ma, Y.-J. 1994. Some comments on rare and endangered liverworts in mainland China. *Arctoa* **3:** 7 - 12.

Zhu, R.-L., So, M.L. and Ye, Y.-X. 1998. A synopsis of the hepatic flora of Zhejiang, China. *J. Hattori Bot. Lab.* **84**: 159 - 174.

Chapter 5

Macrofungi in Coastal Sand Dunes and Mangroves of Southwest India

*Sudeep D. Ghate and Kandikere R. Sridhar**

Department of Biosciences, Mangalore University,
Mangalagangotri, Mangalore – 574 199, Karnataka

ABSTRACT

Although coastal sand dunes (CSDs) and mangroves are important biogeographic ecological zones in India, they are threatened due to human interference (encroachment, urbanization, sand mining and fishery activities). This chapter projects occurrence of macrofungi in CSDs and mangroves based on surveys in Dakshina Kannada (Karnataka State) during monsoon season (June–October). A brief description of fourteen macrofungi based on field and laboratory observations along with their extent of occurrence, substrate preference and commercial value is given. Some macrofungi occurring on the CSDs and mangroves are edible and dependent on sandy/silty soils, leaf litter, woody litter and herbivore dung (*e.g.*, *Coprinus plicatilis*, *Lycoperdon decipiens* and *Termitomyces schimperi*). Some are ectomycorrhizal in CSDs or in mangroves (*Collybia fusipes*, *Inocybe petchii* and *Lycoperdon decipiens*). This study as well as earlier observations revealed occurrence of several edible, medicinal and mycorrhizal macrofungi in

* Corresponding Author: E-mail: kandikere@gmail.com

CSDs and mangroves. Currently, CSDs and mangroves are highly neglected and it is necessary to safeguard, promote and utilize macrofungal resources as social and cultural heritage of coastal dwellers.

Keywords: Coastal sand dunes, Mangroves, Diversity, Macrofungi, Mushrooms, Ectomycorrhizae.

Introduction

Fungi have been recognized as one of the major diverse group of organisms next to arthropods. They occupy a key position due to their involvement in decomposition of organic matter, uptake of nutrients by mutualistic association with plants and cause diseases in plants or animals as pathogens. Several fungi have been recognized as source of nutritional (edible or improve nutritional quality), medicinal (antibiotics, toxins and nutraceuticals) and industrial (fermentation and probiotics) importance. Thus, interest on exploring fungal diversity and their conservation has become an issue of priority especially after Biodiversity Convention in Rio de Janeiro in 1992. The first important step to document, understand diversity and utilize fungi is through inventorying and monitoring a wide range of natural ecosystems. Macrofungi as non-timber forest product represent a major economic resource worldwide due to their importance in food, pharmaceutical and industries. There are several biomes untouched or partially inventoried to explore the diversity of macrofungi.

Indian subcontinent represents 12 mega-biodiversity centres of the world encompassing 10 biogeographic zones with 25 biogeographic provinces consisting over 400 biomes (Rodgers and Panwar, 1988; Mehta, 2000). The coastal region represents one of the 10 biogeographic zones (west coast, east coast, Andaman, Nicobar and Lakshadweep: 13,000 km^2) with 26 National Parks/Wildlife Sanctuaries. Coastal region consists of two important habitats of special significance include coastal sand dunes (CSDs) and mangroves. A variety of psammophytic (saline-tolerant) plant species have adapted to saline habitats of CSDs and mangroves (Rao and Meher-Homji, 1985; Rao and Sherieff, 2002; Suresh and Rao, 2001). Most of the diversity studies pertaining to macrofungi of Peninsular India are confined to the Western Ghats (*e.g.*, Bhagwat *et al.*, 2005; Manoharachary *et al.*, 2005; Natarajan *et al.*, 2005a, 2005b; Brown *et al.*, 2006; Swapna *et al.*, 2008; Mohanan, 2011; Karun *et al.*, 2014; Usha and Janardhana, 2014), while coastal regions especially CSDs and mangroves are ignored. Similar to other fungi, macrofungal occurrence and diversity is also dependent on the detritus generated by the psammophytes in CSDs and mangroves. Usually CSD plant species have xeric adaptation and withstand harsh conditions especially high temperature, wind and sand abrasion. Relatively studies on macrofungal resource of CSDs and mangroves of the Indian coast are scanty (*e.g.*, Dutta *et al.*, 2013; Ghate *et al.*, 2014). Coastal dwellers are dependent on CSDs and mangroves mainly for seafood. Besides, CSD and mangrove vegetation also provide a wide range of nutritional, medicinal, agricultural and industrial benefits (Arun *et al.*, 1999; Kathiresan and Bingham, 2001; Sridhar and Bhagya, 2007; Sridhar, 2009). One of the interesting aspects of CSDs and mangroves is occurrence of several macrofungi beneficial as source of food, medicine and ectomycorrhizal (Dutta *et al.*, 2013; Ghate *et al.*, 2014). The current study envisaged to evaluate selected CSDs and mangroves of the Southwest coast of India to follow macrofungal occurrence, distribution and substrate preference with brief description of diagnostic characteristics.

Study Area, Sampling and Assessment

Five coastal sand dunes (CSDs) chosen for macrofungal survey on the Southwest India include Someshwara (12°47′N, 74°51′E), Thannir Bavi (12°53′N, 74°48′E), Surathkal (12°99′N, 74°79′E), Padubidri (13°12′N, 74°76′E) and Kaup (13°12′N, 74°44′E) located at a total distance of 51 km (from Someshwara to Kaup). Five mangroves selected for survey include: Paduhithlu (13°7′N, 74°45′E), Nadikudru (13°5′N, 74°46′E), Sasihithlu (13°4′N, 74°46′E), Thokottu (12°49′N, 74°51′E) and Batapady (12°45′N, 74°51′E) spanned at a distance of 42 km (Paduhithlu to Batapady). Survey of macrofungi on the CSDs and mangroves was conducted during monsoon season (June through October, 2013).

Opportunistic sampling was carried out by transect method in chosen locations. About 50 m distance in each CSDs (on mid- and hind-dunes) and easily accessible part of mangroves was surveyed for occurrence of macrofungi on soil/sand, tree bases, litter dumps and other autochthonous and allochthonous debris. Essential features of each fungus were documented in the field which includes colour, extent of fruit bodies and substrate preference (soil, leaf litter, woody litter and other substrates). Those occurring near the tree bases were assessed for their intimate association with root system as ecotomycorrhizae. Representative specimens were carried to the laboratory for microscopic examination and preservation. Fragile mushrooms after measurements were fixed in lactophenol in vials on the spot. Fruit bodies of macrofungi were fixed in formalin-ethanol-water (14:5:1) in the laboratory.

Macrofungi were indentified based on overall general and specific characteristics using descriptions in monographs matching with microscopic observations (Jordan, 2004; Phillips, 2006; Mohanan, 2011). Dimensions of basidiomata (pileus and stipe) were based on 5-25 random samples using vernier calipers, while spore dimensions were based on 25 random spores using high power microscope (Olympus CX41RF, USA). Minimum and maximum dimensions for each fungus are presented in parenthesis and the rest outside the parenthesis as range.

Macrofungi in Coastal Sand Dunes

A total of nine macrofungi was recovered from the CSDs surveyed (Table 5.1; Figures 5.1–5.2). Out of them, three were edible (*Coprinus plicatilis*, *Lycoperdon decipiens* and *Termitomyces schimperi*) and two were ectomycorrhizal in association with exotic tree species (*Collybia fusipes* with *Acacia auriculiformis* and *Casuarina equisetifolia*; *L. decipiens* with *C. equisetifolia*). Detailed descriptions of eight species are given below.

Collybia fusipes (Bull.) Quél. (Tricholomataceae – Basidiomycotina) (# MUBSSDGKRSMF-001) (Figures 5.1a, b)

Medium size agaric, with reddish brown cap, cap surface dull in appearance, pale brownish gills, medium length stipe, occurs in tufts at the base of dune vegetation, common annual and poisonous.

Table 5.1: Occurrence and Commercial Value of Macrofungi in Coastal Sand Dunes and Mangroves of the Southwest India

Taxon	Soil	Leaf^v	Wood	Other	Commercial Value
			Litter		
Coastal sand dunes					
Collybia fusipes (Figures 5.1a and b)**	Base: *Acacia* and *Casuarina*	–	–	–	Poisonous and mycorrhizal
Conocybe apala (Figures 5.1c and d)***	Humus	–	Twigs	Lawn	
Coprinus plicatilis (Figures 5.1e)**	Soil/sand with litter and humus	–	–	Cow dung	Edible with least flesh
Crepidotus uber (Figures 5.1f and g)**	–	–	Twigs and Branches	–	
Lentinus polychrous (Figures 5.1h)*	–	–	Logs of *Acacia*	–	Lignocellulase (bioremediation)
Lepiota clypeolaria (Figures 5.2a and b)*	Base of *Casuarina*	–	–	–	Poisonous
*Lycoperdon decipiens***	–	–	–	Roots: *Casuarina*	Edible (young) and mycorrhizal
Marasmius kisangensis (Figures 5.2c and d)***	–	*Acacia* and *Casuarina*	–	–	
Termitomyces schimperi (Figures 5.2e-i)*	Termite mounds	–	–	–	Highly edible
Mangroves					
*Coprinus plicatilis***	Soil with litter	–	–	–	Edible with least flesh
Hygrocybe nivosa var. *nivosa* (Figures 5.3a)*	Soil	–	–	–	
Inocybe petchii Boedijn (Figures 5.3b and c)*	Soil	–	–	Roots: *Rhizophora*	Mycorrhizal

Contd...

Table 5.1-*Contd...*

Taxon	Soil	Litter			Commercial Value
		Leaf[Ψ]	Wood	Other	
Leucocoprinus brebissonii (Figures 5.3d and e)**	Soil	Acacia	–	–	–
Lycoperdon decipiens (Figures 5.3f)**	–	–	–	Roots: *Acacia* and *Rhizophora*	Edible (young) and mycorrhizal
Marasmiellus stenophyllus (Figures 5.3g and h)***	–	–	Twigs, branches and logs	–	
*Marasmius kisangensis****	–	*Rhizophora*	–	–	
Marasmius confertus (Figures 5.3i and j)**	Soil	Acacia	–	–	

*: Rare; **: Common; ***: Frequent; –: Not found.

Acacia: *A. auriculiformis*; *Casuarina*: *C. equisetifolia*; *Rhizophora*: *R. mucronata.*

Ψ, dead phyllodes of *A. auriculiformis*, dead needles of *C. equisetifolia* and leaf litter of *R. mucronata.*

Figure 5.1: Macrofungi in Coastal Sand Dunes: *Collybia fusipes* (a, b), *Conocybe apala* (c, d), *Coprinus plicatilis* (e), *Crepidotus uber* (f, g) and *Lentinus polychrous* (h).

Basidiomata: Pileus (2.7) 3–6.5 (7) cm diam. (n=25), conical to convex becoming expanded, with a broad umbo and wavy margin; pileal surface sepia to dark reddish brown, smooth, hygrophanous and slightly viscid; context of pileus reddish brown, thick and fleshy. *Lamellae* at first creamish turning to pale brown, free or emarginated and crowded with short gills of 2–3 lengths. *Stipe* (4) 4.3–8.1 (8.8) × (0.6) 0.7–1.4 (1.5) cm (n=25), cylindrical, swollen at the centre, pale at the apex becoming dark brown at the base, sometimes twisted, grooved and fused at the base with several others. *Annulus* absent. *Spore print* creamish white. *Spores* (3.7) 4–6 (6.3) × (1.7) 2–3.5 (3.8) µm (n=25), hyaline and ellipsoid to almond-shaped with thick wall.

Substrate and distribution: Occurs gregariously in clusters on soil in the base of *Acacia auriculiformis* and *Casuarina equisetifolia* trees as mycorrhizal in coastal sand dunes of Padubidri, Thannir Bhavi and Surathkal during June and July.

Conocybe apala (Fr.) Arnolds (Bolbitaceae – Basidimycotina) (# MUBSSDGKRSMF-002) (Figures 5.1c, d)

Small size agaric with whitish to light ochreous cap, bell-shaped cap with a small pale centre, pale creamish gills, medium length stipe, occurs in tufts amongst roadside (grasses, meadows and litter), frequent, annual, odour and taste not distinct and inedible.

Basidiomata: Pileus (1) 1.5–2 (2.3) cm diam. (n=8), conical to narrowly bell-shaped with straight and entire margin; context of pileus creamish, smooth and very thin. *Lamellae* white turning yellowish, free and distant with short gills of 2 lengths. *Stipe* (7.5) 8–11.3 (11.7) × (0.1) 0.15–0.2 (0.25) cm (n=8), central, slender, hollow with slightly bulbous base, whitish, powdery and smooth. *Annulus* absent. *Spores* (10.5) 11–14.3 (15) × (6.4) 7–9.6 (10) µm (n=25), ovoid to ellipsoid with a germ-pore and yellowish brown.

Substrate and distribution: Occurs gregariously in small groups on roadside lawns near sand dunes of Kaup and Padubidri during July.

Coprinus plicatilis (Curtis) Fr. (Agaricaceae – Basidimycotina) (# MUBSSDGKRSMF-003) (Figure 5.1e)

Small size agaric with pale brown cap, which is almost flat with a small central depression, pale grey gills, small to medium length stipe, occurs singularly or in small troops (on soil, humus and dung), common, annual and edible but possess less flesh.

Basidiomata: Pileus (0.7) 1–2.6 (3) cm diam. (n=20), ovoid turning into companulate and finally applanate like a parasol and margin radially plicate-sulcate, fragile and withering early and surface pale brown with a cinnamon-coloured disk at the centre; context of pileus pale brown, thin and delicate. *Lamellae* light grey, sometimes with a pink undertone turning black and distant with short gills of 2–3 lengths. *Stipe* (2.5) 3–6.3 (7) × (0.5) 0.8–2 (2.3) cm (n=20), central, cylindrical, hollow with a slightly swollen base, buff and smooth. *Annulus* absent. *Spores* (9.5) 10–12 (12.5) × (6.9) 7.5–9.3 (10) µm (n=25), black, teardrop-shaped or almond-shaped with thick wall.

Substrate and distribution: Occurs solitary or dispersed in small numbers on soil in lawns, on humus and also on dung on the coastal dunes of Padubidri, Surathkal and Thannir Bhavi during June to September and also in mangrove soils with leaf litter in Sasihithlu and Nadikudru during October. This fungus was also found in the grassy fields of Sundarban mangroves of the West Bengal (Dutta *et al.*, 2013).

Crepidotus uber (Berk. & M.A. Curtis) Sacc. (Crepidotaceae – Basidimycotina) (# MUBSSDGKRSMF-004) (Figures 5.1f, g)

Small size agaric with pearl white to creamish cap, pale pinkish gills, devoid of stipe, occurs in overlapping groups on wood (branches and twigs), common, perennial and inedible.

Basidiomata: Pileus (0.6) 0.7–2 (2.2) cm diam. (n=20), semicircular or curved to reniform shape with a wavy and entire margin; pileal surface pale pink when damp and turning to creamish when dry, smooth, viscid when damp; context of pileus creamish and thin. *Lamellae* pale pink to light yellow, excentric to more or less decurrent and crowded with short gills of 3 lengths. *Stipe* absent. *Spores* (4.3) 5–7.7 (8.1) × (3.7) 4–5 (5.2) μm (n=25), hyaline, broadly ellipsoid to fusiform, with a thin buff coloured wall and carrying a refractile guttule.

Substrate and distribution: Occurs in small overlapping troops on the surface of twigs and branches in the coastal sand dunes of Padubidri and Someshwara during June-October.

Lentinus polychrous Lév. (Polyporaceae – Basidimycotina) (# MUBSSDGKRSMF-005) (Figure 5.1h)

Small to medium size polypore, with creamish to brownish cap, cocoa brown gills, medium length central to eccentric or sidelong stipe, occurs in clusters of small numbers on decaying wood, rare, perennial, odour not distinct and inedible.

Basidiomata: Pileus (4.5) 5–7.6 (8) cm diam. (n=10), tough, coriaceous when fresh, convex becoming fan-like and shammy to dusky brown in the centre turning to pale brown at the margin; pileal surface shrouded by fine hairs along with recurved squamules at the centre becoming glabrescent at the periphery; context of pileus ochreous brown, thick, tough and leathery. *Lamellae* coffee brown to pale brown, deeply decurrent and fairly crowded with short gills of 4–5 lengths. *Stipe* (3.8) 4–6 (6.3) × (0.4) 0.5–0.8 (0.9) cm (n=10), central to excentric or lateral, cylindrical, solid, pale creamish, with presence of small scales and swollen dark base. *Annulus* absent. *Spores* (7.8) 8–9.5 (10.2) × (4) 4.3–5.2 (6) μm (n=25), translucent, ellipsoidal to narrowly cylindrical and thin-walled.

Substrate and distribution: Occurs gregariously or in large clumps on the logs of *Acacia auriculiformis* in coastal sand dunes of Thannir Bhavi during August.

Lepiota clypeolaria (Bull.) P. Kumm. (Agaricaceae – Basidimycotina) (# MUBSSDGKRSMF-006) (Figures 5.2 a, b)

Small to medium size agaric with pale umber brown cap, cap somewhat conical to convex becoming flattened with obtuse umbo at the centre, pale creamish to light

Figure 5.2: Macrofungi in Coastal Sand Dunes: *Lepiota clypeolaria* (a, b), *Marasmius kisangensis* (c, d) and *Termitomyces schimperi* (e–f).

yellowish gills, medium length stipe, occurs in clusters on coastal dune litter, rare, annual, odour and taste not distinctive, and inedible.

Basidiomata: Pileus (2) 2.3– 4.2 (4.6) cm diam. (n=10), broadly bell-shaped with a ochreous cocoa brown centre and creamish to pale yellow on the periphery; pilial surface quickly breaking into concentrically arranged woolly squamules revealing creamish flesh underneath with a abrased margin; context of pileus creamish, thin and fuzzy. *Lamellae* white turning yellowish, free and crowded with short gills of 2 lengths. *Stipe* (4) 4.5– 6 (6.3) × (0.2) 0.25–0.3 (0.35) cm (n=10), central, cylindrical, fragile, creamish at the top becoming light brown and covered by woolly powdery scales in the lower region. *Annulus* white, woolly and short-lived. *Spore print* white. *Spores* (10.5) 11–20 (21) × (3.4) 4–4.5 (4.8) µm (n=25), ovoid to fusiform and hyaline with thin wall.

Substrate and distribution: Occurs solitarily or gregariously on floor of *Casuarina equisetifolia* Thannir Bhavi during June.

Marasmius kisangensis Singer (Marasmiaceae – Basidimycotina) (# MUBSSDGKRSMF-007) (Figures 2c, d)

Small size agaric with reddish to ochreous cap with a dark centre, convex, very small size, hair-like fragile stipe, pale creamish gills, medium length stipe, occurs in dense clusters on leaf litter, frequent, annual, and taste and odour not distinctive.

Basidiomata: Pileus (0.7) 0.8–1.8 (2) cm diam. (n=25), convex expanding to conico-campanulate; pileus surface marked with many ridges and grooves, and radially wrinkled with straight entire margin; context pileus of reddish, very thin and smooth. *Lamellae* pearl white, adnate to decurrent and distant with short gills of 2 lengths. *Stipe* (3) 3.3–5.3 (5.6) × (0.1) 0.15–0.2 (0.23) cm diam. (n=25), central, slender, cylindrical, surface dark brown to pale brown, powdery and emerging from leaf and associated with dark horsehair-like mycelium. *Annulus* absent. *Spore print* white. *Spores* (6.3) 7–8.3 (8.6) × (4) 4.3–5 (5.3) µm (n=25), ovoid to ellipsoid, hyaline, thin-walled and smooth with presence of a few guttules.

Substrate and distribution: It grows frequently on phyllodes *Acacia auriculiformis* and needles of *Casuarina equisetifolia* in CSDs of Kaup and Someshwara during June, August and September. It was also frequent on leaf litter of *Rhizophora mucronata* in mangroves of Nadikudru (July) and Paduhithlu (July and October).

Termitomyces schimperi (Pat.) R. Heim (Lypphyllaceae – Basidimycotina) (# MUBSSDGKRSMF-008) (Figures 2e–i)

Large size agaric with creamy whitish cap with large, thick, persistent velar scales (which form a large plate-like covering constitutes main identifying feature) give the appearance of large plate-like reddish brown covering on the disk (umbrinous disk), occurs singly or in small troops near coastal dune vegetation having with termite activities, rare, annual, pleasant smell, excellent taste and highly edible.

Basidiomata: Pileus (4.8) 6.6–8.3 (8.8) cm diam. (n=18), sub-globose at first, enlarging to convexo-applanate, missing a pointed perforatorium and covered by orangish to rust brown coloured squamules on surface; pileus surface with entire, straight or slightly incurved margin, slimy when damp and otherwise dry; context of pileus white, thick and fleshy. *Lamellae* at first white, becoming creamish with age, free to adnexed, crowded with lamellulae of different lengths. *Stipe* (7.5) 8–16 (17.7) × (0.8) 1–2.2 (2.4) cm (n=18), solid, central and cylindrical, long, swollen at the base near the soil surface, with spindle-shaped pseudorrhiza (9) 12–25 (27) cm long (n=18), tapering downwards, whitish, hollow and fibrous. *Annulus* membranous, thin and present at centre of the stipe. *Spore print* creamish. *Spores* (5.6) 6.5–10 (11) × (3.5) 4–7 (8) μm (n=25), broadly ellipsoid, thin walled and hyaline.

Substrate and distribution: Rare but gregarious or in large troops on termite mounds in Kaup beach. It is highly edible to coastal dwellers.

Macrofungi in Mangroves

Eight species of macrofungi were recovered from the mangroves surveyed (Table 5.1; Figure 5.3). Among them, two each were edible (*Coprinus plicatilis* and *Lycoperdon decipiens*) and ectomycorrhizal (*Inocybe petchii* with *Rhizophora mucronata* and *L. decipiens* with *Acacia auriculiformis* and *R. mucronata*). Detailed descriptions of six species are given below.

Hygrocybe nivosa var. *nivosa* (Berk. & Broome) Leelav., Manim. & Arnolds
(Hygrophoraceae – Basidimycotina) (# MUBSSDGKRSMF-040) (Figure 3a)

Small size agaric, whitish cap with a small central depression, creamish gills, short central stipe, occurs in small troops on mangrove soil, rare, annual, odour and taste not distinctive, and inedible.

Basidiomata: Pileus (1.2) 1.3–2 (2.2) cm diam. (n=5), convex with a small depression at the centre and rolled margin; pileal surface pearl white, semi-transparent and smooth when damp; context of pileus white, very thin and semi-transparent. *Lamellae* creamish, decurrent and moderately crowded with short gills of 3–4 lengths. *Stipe* (1.5) 1.8–2.8 (3.2) × (0.8) 1–2 (2.1) cm (n=5), central, cylindrical but often becoming flattened or compressed, with a bulbous apex, smooth and white. *Annulus* absent. *Spore print* white. *Spores* (5) 5.4–7.5(8.3) × (3.3) 3.7–4.8 (5.3) μm (n=25) and ovoid to ellipsoid with a thin wall.

Substrate and distribution: It was found growing commonly in mangrove soils of Nadikudru region during August.

Inocybe petchii Boedijn (Inocybaceae – Basidimycotina) (# MUBSSDGKRSMF-041)
(Figures 5.3b, c)

Small to medium size agaric with brownish cap, abruptly sharp or tapering umbo, pale brownish gills, small to medium length stipe, occurs singly or in small troops on soils and base of *Rhizophora mucronata*, rare, annual, odour and taste not distinct and inedible.

Basidioimata. Pileus (2) 2.5–4.5 (4.7) cm diam. (n=17), conical to convex with a sharp pointed umbo and straight and entire margin; pileal surface smooth, silky and fibrillose; context of pileus yellowish brown, thin and fleshy. *Lamellae* yellowish

Figure 5.3: Macrofungi in Mangroves: *Hygrocybe nivosa* var. *nivosa* (a), *Inocybe petchii* (b, c), *Leucocoprinus brebissonii* (d, e), *Lycoperdon decipiens* (f: upper inset, opened young basidiomata; lower inset, basidiomata with rhizomorph), *Marasmiellus stenophyllus* (g, h) and *Marasmius confertus* (i, j).

brown to brown, adnate to adnexed and crowded with short gills of 2–3 lengths. *Stipe* (5) 6–10 (10.5) × (0.15) 0.2–0.35 (0.4) cm (n=17), central, cylindrical, slender, with a somewhat enlarged base, reddish brown, smooth and solid. *Annulus* absent. *Spore print* brown. *Spores* (9.5) 10–14 (14.5) × (10.4) 11–12.6 (13) µm (n=25), light brown, globose to subglobose and with thick wall decorated by numerous spines giving a fairly star-like appearance under the microscope.

Substrate and distribution: It is rare and growing in the mangrove soils with roots of *Rhizophora mucronata* of Batapady during July.

Leucocoprinus brebissonii (Godey) Locq. (Agaricaceae – Basidimycotina) (# MUBSSDGKRSMF-042) (Figures 5.3d, e)

Small size agaric with creamish cap, pale creamish gills, small to medium length central stipe, occurs singly or in small numbers on mangrove litter, common, annual, and taste and odour not distinct.

Basidiomata: Pileus (2) 2.5–3.5 (3.7) cm diam. (n=16), conical turning into applanate and a groovy serrated margin; pileal surface pinkish buff in the centre, creamish towards the edges and decorated by small greyish brown scales, context of pileus creamish, thin and delicate. *Lamellae* whitish, free, slightly crowded with short gills of 3 lengths. *Stipe* (2.5) 3–6.4 (6.8) × (0.15) 0.2–0.7 (0.85) cm (n=16), central, cylindrical, base slightly swollen, creamish, grainy and hairy. *Annulus* present, white, central and ephemeral. *Spore print* creamish white. *Spores* (6)7–9.3 (10) × (4) 4.7–6.6 (7) µm (n=25), hyaline, tear-drop or almond-shaped with a thick wall and containing a small germ pore.

Substrate and distribution: Common on the soil and decaying phyllodes of *Acacia auriculiformis* in mangroves of Paduhithlu during September.

Lycoperdon decipiens Durieu & Mont. (Agaricaceae – Basidimycotina) (# MUBSSDGKRSMF-043) (Figure 5.3f)

Gastroid basidome, small to medium size, spiny surface, attached to the substrate by white rhizoids, occurs in gregarious clusters on mangrove soils, mycorrhizal in *Acacia auriculiformis* and *Rhizophora mucronata* trees, common, annual, odour light, taste excellent and edible when young.

Basidiomata: Globose to subglobose (2) 2.5–4 (4.3) cm across, (1.8) 2.2–3.8 (4) cm in height (n=25), whitish to pale creamish, outer layer scurfy with whitish, delicate spines, looking somewhat starry scaled appearance and inner wall light brown. *Gleba* whitish turning to light brown. *Spores* (3) 3.5–6 (6.3) µm diam. (n=25), spherical, warty and light brown.

Substrate and distribution: Common ectomycorrhizae on soil with decaying phyllodes of *Acacia auriculiformis* and leaf litter of *Rhizophora mucronata* in mangroves of Nadikudru during October, Paduhithlu during September. It was also common ectomycorrhizae in *Casuarina equisetifolia* base of CSD of Padubidri, Thannir Bhavi and Someshwara during June-September.

Marasmiellus stenophyllus (Mont.) Singer (Auriculariaceae – Basidimycotina) (# MUBSSDGKRSMF-044) (Figures 5.3g, h)

Small size agaric with hygrophanous whitish cap, pearl-coloured gills, small central to eccentric stipe, occurs in gregarious clusters on surface of decaying mangrove wood and on forest litter, frequent, annual, taste and odour not distinctive, and inedible.

Basidiomata: Pileus (2.3) 2.5–3.3 (3.6) cm diam. (n=25), convex becoming applanate and central depression with wavy margin; pileal surface whitish, hygrophanous, smooth with a pale wine red centre; context of pileus whitish thin and fleshy. *Lamellae* pearl white, adnate to decurrent and fairly distant with short gills of 2–3 lengths. *Stipe* (5) 5.4–8 (8.2) × (0.05) 0.07–0.25 (0.3) cm (n=25), central, cylindrical, slender, insidious, of creamish and solid. *Annulus* absent. *Spore print* white. *Spores* (8.3) 9–11.6 (12) × (4) 4.3–5.3 (5.6) µm (n=25), hyaline, spindle to tear shaped, thin walled with presence of guttules.

Substrate and distribution: It was frequent on fallen twigs, branches and logs and twigs in all mangroves studied (Paduhithlu, Nadikudru, Sasihithlu, Thokottu and Batapady) during July and August.

Marasmius confertus Berk. & Broome (Marasmiaceae – Basidimycotina) (# MUBSSDGKRSMF-045) (Figures 5.3i, j)

Small size agaric with light creamish brown cap, small central depression, creamish brown gills, small to medium central stipe, occurs in clusters on mangrove litter, common, annual, taste and odour not distinct, and inedible.

Basidiomata: Pileus (1.3) 1.5–2.5 (3.2) cm diam. (n=25), conical to convex, expanding to form a applanate structure with a central tiny depression and striated margin; pileal surface ochreous to light creamish brown and smooth; context of pileus ochreous brown, thin and fleshy. *Lamellae* at first creamish turning to pale brown, free to adnate and moderately crowded with short gills of 2 lengths. *Stipe* (3.5) 3.8–5.8 (6.1) × (0.05) 0.1–0.3 (0.32) cm (n=25), central, cylindrical, with a somewhat enlarged base, surface pale brown, smooth and hairless. *Annulus* absent. *Spore print* creamish. *Spores* (8.3) 8.5–10 (10.3) × (4) 4.3–5.6 (6) µm (n=25), teardrop-shaped and hyaline with a thin wall and containing guttules.

Substrate and distribution: It was found growing commonly on the phyllodes of *Acacia auriculiformis* in mangroves of Paduhithlu and Sasihithlu during September and October.

Discussion and Outlook

Altogether the present study revealed occurrence of 14 macrofungi in CSDs and mangroves with three overlapping species (*Coprinus plicatilis, Lycoperdon decipeins* and *Marasmius kisangensis*). *Coprinus plicatilis* occurred on soil/sand with litter, humus and cow dung on the dunes, while in mangroves on soil with leaf litter. *Lycoperdon decipeins* was ectomycorrhizal in dune *Casuarina equisetifolia*, while in *Acacia auriculiforms* and *Rhizophora mucronata* in mangroves. *Marasmius kisangensis* grew on dead phyllode of *A. auriculiformis* and dead needles of *C. equisetifolia* in dunes, while in mangroves on leaf litter of *R. mucronata*. Similar to our study, *C. plicatilis* was found in grass fields of the Sundarban mangroves of West Bengal with 2.6 per cent frequency occurrence (Dutta *et al.*, 2013). *Crepedotus uber* was recorded on bark, deadwood, rotting aerial roots and upright trunks of *Rhizophora mangle* in the mangroves of Southwestern Puerto Rico (Nieves-Rivera *et al.*, 2005).

Macrofungi are substrate-dependent and need canopy cover and accumulation of leaf/woody litter for their growth especially in CSDs. Many macrofungi grew on dead phyllode of *Acacia* and dead needles of *Casuarina* on dunes. Besides, dead grass shreds on the dunes are of special significance for growth of many macrofungi (*e.g.*, *Coprinus* and *Conocybe*) possibly due to their moisture retention ability. Accumulation and burial of woody debris on the dunes serve as durable lignocellulosic source for macrofungal perpetuation. In addition, animal activities leads to accumulation of dung (*e.g.*, cow dung and pellets of sheep/goat/rabbit), which also support selected macrofungi (*e.g.*, *Coprinus plicatilis* on cow dung). In mangroves, some macrofungi occur on elevated regions like dykes and some are found near the crab tunnels in high tide levels with accumulation of leaf litter.

In this study, three species are edible (*Coprinus plicatilis, Lycoperdon decipiens* and *Termitomyces schimperi*). *Termitomyces clypeatus* and *T. microcarpus* were reported on the termite mounds of Sundarban mangroves of West Bengal (Dutta *et al.*, 2013). Many edible (*Amanita* sp. *Dacryopinax spathularia, Lentinus squarrosulus, Lycoperdon utriforme, Macrolepiota rachoides Pleurotus flabellatus* and *Termitomyces umkowaan*), medicinal (*Collybia dryophila* var. *extuberans* and *Ganoderma lucidum*) and ectomycorrhizal (*Pisolithus albus* and *Scleroderma citrinum*) macrofungi have been reported from the CSDs of Southwest India (Ghate *et al.*, 2014). Edible *Amanita* sp. is a delicacy of coastal dwellers and it is likely an important ectomycorrhizal fungus in dune tree species. In CSDs and mangroves, unlike other habitats size of basidiocarp of *L. decipiens* was smaller and in tender stage they are eaten by the coastal dwellers. A three year study of macrofungi in the Sundarban mangroves yielded only one ectomycorrhizal fungus *Pisolithus arhizus* associated with *A. auriculiformis* and *Eucalyptus globulus* (Dutta *et al.*, 2013). Woodrot fungi in CSDs and mangroves are of special interest especially on wood logs and tree stubs. Polypores are host-specific and 81-88 per cent of collections showed host specificity in mangroves (Gilbert and Sousa, 2002; Gilbert *et al.*, 2008). Polypores preferred old-growth mangrove forests and xylophilous basidiomycetes were common in mangroves (Baltzar *et al.*, 2009; Hattori *et al.*, 2012), thus deserves special attention of exploration for their use in pharmaceutical products.

Psammophytic tree species on CSDs are occupied by *A. auriculiformis, C. equisetifolia* and other plantations of commercial interest. Tree species especially *A. auriculiformis* and *C. equisetifolia* will be harvested and remaining stubs serve as good substrate for growth of perennial fungi (*e.g.*, polypores). Some mid- and hind-dunes are occupied by xerophytic plants as fence (*e.g.*, cacti). There are several threats for growth of macrofungi in coastal habitats especially due to wood extraction and sand mining. Besides, occasional fires to clear the vegetation for beach sports and vehicular movements are also highly detrimental. Mangroves are also facing threats like sand mining, clear cutting, exotic plantations and wood seasoning.

Symbiotic endomycorrhizas and rhizobia are important catalysts for nitrogen and phosphorus cycles in dunes as well as mangroves (Beena *et al.,* 2000; Sridhar and Beena, 2001; Sridhar, 2009; Arun and Sridhar, 2005; Sridhar *et al.,* 2011). Besides psammophytic tree species, other vegetation (Asteraceae, Cyperaceae, Fabaceae and Poaceae) have major distribution in tropical CSDs (Sridhar and Bhagya, 2007). Conservation measures need to be focused to prevent destruction of CSD landscapes and mangroves for sustainable coastal management to safeguard the social and cultural heritage of coastal tribes. It is necessary to study the impact of replacement of psammophytic vegetation with exotic vegetation on macrofungi. It is also important to follow whether any macrofungi is CSD- and or mangrove-dependent. Detailed macrofungal inventories in CSDs and mangroves will add up many more into existing checklist and helpful to decide whether any of them are vulnerable or threatened or red-listed.

Acknowledgements

Authors are thankful to Mangalore University for permission to carry out this study in the Department of Biosciences. SDG acknowledges the award of an INSPIRE Fellowship by the Department of Science and Technology, New Delhi, Government of India DST/INSPIRE Fellowship/2013/132: Award # IF130237). KRS acknowledges the award and financial assistance under UGC-BSR Faculty Fellowship by the University Grants Commission, New Delhi.

References

Arun, A.B. and Sridhar, K.R. 2005. Growth tolerance of rhizobia isolated from sand dune legumes of southwest coast of India. *Engineering in Life Sciences* **5**: 134-138.

Arun, A.B., Beena, K.R., Raviraja, N.S. and Sridhar, K.R. 1999. Coastal sand dunes - A neglected ecosystem. *Curr. Sci.* **77**: 19-21.

Baltazar, J.M., Trierveiler-Pereira, L. and Loguercio-Leite, C. 2009. A checklist of xylophilous basidiomycetes (Basidiomycota) in mangroves. *Mycotaxon* **107**: 221–224.

Beena, K.R., Raviraja, N.S., Arun, A.B. and Sridhar, K.R., 2000. Diversity of arbuscular mycorrhizal fungi on the coastal sand dunes of the west coast of India. *Curr. Sci.* **79**: 1459-1466.

Bhagwat, S., Kushalappa, C., Williams, P. and Brown, N. 2005. The role of informal protected areas in maintaining biodiversity in the Western Ghats of India. Ecology and Society **10**: Article # 8, 1-40: http://www.ecologyandsociety.org/vol10/iss1/art8/

Brown, N., Bhagwat, S. and Watkinson, S. 2006. Macrofungal diversity in fragmented and disturbed forests of the Western Ghats of India. *J. Appl. Ecol.* **43**: 11-17.

Dutta, A.K., Pradhan, P., Basu, S.K. and Acharya, K. 2013. Macrofungal diversity and ecology of the mangrove ecosystem in the Indian part of Sundarbans. *Biodiversity* **14**: 196–206.

Ghate, S.D., Sridhar, K.R. and Karun, N.C. 2014. Macrofungi on the coastal sand dunes of South-western India. *Mycosphere* **5**: 144-151.

Gilbert, G.S. and Sousa, W.P. 2002. Host specialization among wood-decay polypore fungi in a Caribbean Mangrove forest. *Biotropica* **34**: 396–404.

Gilbert, G.S., Gorospe, J. and Ryvarden, L. 2008. Host and habitat preferences of polypore fungi in Micronesian tropical flooded forests. *Mycol. Res.* **112**: 674-680.

Hattori, T., Yamashita, S. and Lee, S.-S. 2012. Diversity and conservation of wood-inhabiting polypores and other aphyllophoraceous fungi in Malaysia. *Biodiversity and Conservation* **21**: 2375-2396.

Jordan, M. 2004. *The Encyclopedia of Fungi of Britain and Europe.* Francis Lincoln Publishers Ltd., London.

Karun, N.C., Sridhar, K.R. and Appaiah, K.A.A. 2014. Diversity and distribution of macrofungi in Kodagu region (Western Ghats): A preliminary account. In: *Biodiversity in India, Volume 7* (eds. Pullaiah, T., Karuppusamy, S. and Rani, S.). Regency Publications, New Delhi, pp.73-96.

Kathiresan, K. and Bingham, B.L. 2001. Biology of mangrove and mangrove ecosystems. *Adv. Mar. Biol.* **40**: 81–251.

Manoharachary, C., Sridhar, K.R., Singh, R., Adholeya, A., Suryanarayanan, T.S., Rawat, S. and Johri, B.N. 2005. Fungal biodiversity: Distribution, conservation and prospecting of fungi from India. *Curr. Sci.* **89**: 58–71.

Mehta, R. 2000. WWF – India. In: *Setting biodiversity conservation priorities for India* (eds. Singh, S., Sastry, A.R.K., Mehta, R. and Uppal, V.). Wildlife Institute of India, Dehradun, 245–266.

Mohanan, C. 2011. *Macrofungi of Kerala. Kerala, Handbook # 27.* Kerala Forest Research Institute, Peechi, India.

Natarajan, K., Narayanan, K., Ravindran, C. and Kumaresan, V. 2005a. Biodiversity of agarics from Nilgiri Biosphere Reserve, Western Ghats, India. *Curr. Sci.* **88**: 1890–1893.

Natarajan, K., Senthilarasu, G., Kumaresan, V. and Riviere, T. 2005b. Diversity in ectomycorrhizal fungi of a dipterocarp forest in Western Ghats. *Curr. Sci.* **88**: 1893–1895.

Nievas-Rivera, Á.M., Tattar, T.A. and Ryvarden, L. 2005. Manglicolous basidiomycetes of southwestern Puerto Rico and Southwestern Florida (U.S.A). *Hoehnea* **32**: 49-57.

Phillips, R. 2006. *Mushrooms.* Pan Macmillan, London.

Rao, T.A. and Meher-Homji, V,M. 1985. Strand plant communities of the Indian sub-continent. *Proc. Indian Acad. Sci. (Pl. Sci.)* **94**: 505-523.

Rao, T.A. and Sherieff, A.N. 2002. *Coastal Ecosystem of the Karnataka State, India II - Beaches.* Karnataka Association for the Advancement of Science, Bangalore.

Rodgers, W.A. and Panwar, H.S. 1988. *Planning a Wildlife Protected Area Network in India: Volumes 1 and 2*. Wildlife Institute of India, Dehradun.

Sridhar, K.R. 2009. Bioresources of coastal sand dunes - are they neglected? In: *Coastal Environments: Problems and Perspectives* (eds. Jayappa, K.S. and Narayana, A.C.). IK International Publishing House, New Delhi, 53–76.

Sridhar, K.R. and Beena, K.R. 2001. Arbuscular mycorrhizal research in coastal sand dunes – A review. *Proc. Nat. Acad. Sci. India* **71**: 179-205.

Sridhar, K.R. and Bhagya, B. 2007. Coastal sand dune vegetation: a potential source of food, fodder and pharmaceuticals. Livestock Research for Rural Development **19**: Article # 84: http://www.cipav.org.co/lrrd/lrrd19/6/srid19084.htm

Sridhar, K.R., Roy, S. and Sudheep N.M. 2011. Assemblage and diversity of arbuscular mycorrhizal fungi in mangrove plant species of the southwest coast of India. In: *Mangroves: Ecology, Biology and Taxonomy* (ed. Metras, J.N.). Nova Science Publishers Inc., New York, 257-274.

Suresh, P.V. and Rao, A. 2001. *Coastal Ecosystems of Karnataka State, India. I. Mangroves*. Karnataka Association for the Advancement of Science, Bangalore.

Swapna, S., Abrar, S. and Krishnappa, M. 2008. Diversity of macrofungi in semi-evergreen and moist deciduous forest of Shimoga District - Karnataka, India. *J. Mycol. Pl. Pathol.* **38**: 21–26.

Usha, N. and Janardhana, G.R. 2014. Diversity of macrofungi in the Western Ghats of Karnataka (India). *Indian Forester* **140**: 531-536.

Chapter 6

An Appraisal of Lichen Biota in Chittoor District of Andhra Pradesh, India

Satish Mohabe[1], B. Anjali Devi[1], A. Madhusudhana Reddy[1]*,
Sanjeeva Nayaka[2] and P. Chandramati Shankar[3]

[1]*Department of Botany, Yogi Vemana University, Vemanapuram,*
Kadapa – 516 003, Andhra Pradesh
[2]*Lichenology Laboratory, CSIR-National Botanical Research Institute,*
Rana Pratap Marg, Lucknow – 226 001, Uttar Pradesh
[3]*Department of Biotechnology, Yogi Vemana University,*
Vemanapuram, Kadapa – 516 003, Andhra Pradesh

ABSTRACT

The present investigation revealed the occurrence of 75 species of lichens belonging to 31 genera under 17 families in Chittoor district of Andhra Pradesh. Among the different growth forms, the foliose lichens exhibit the maximum diversity represented by 45 species followed by 24 species of crustose, 3 fruticose, 2 squamulose and a single species of leprose lichen. Based on their habitat preferences the corticolous lichens exhibited the maximum diversity represented by 42 species followed by 20 species of saxicolous while 13 species found growing both on bark and rock. A total of 18 species are new records to Andhra Pradesh; 31 species are new additions to Chittoor district and four species are endemic to India. The members of lichen families Parmeliaceae and Physciaceae are the most

* Corresponding Author: E-mail: grassed@yahoo.com

dominat taxa in the district with 24 and 22 species respectively. The lichen genus *Parmotrema* exhibited the luxuriant growth represented by 17 species. A most common foliose lichen *Parmotrema praesorediosum* found growing luxuriantly in 8 localities. Out of the 11 localities surveyed in the district, Horsley Hills and Tirumala hills exhibited the maximum diversity of lichens represented by 40 and 39 species respectively. The present data on lichens from the district will be useful for future biomonitoring studies and medicinaly important lichens can be utilized for bioprospection studies.

Key words: Eastern Ghats, Seshachalam hill, Lichen biota, Taxonomy, Biodiversity.

Introduction

Chittoor district is a part of Rayalaseema region in Andhra Pradesh and lies between 12°372 and 14°82 North latitudes and 78°32 and 79°552 East longitudes. The district occupies an area of 15,359 km² and bounded by Anantapur district to the Northwest, YSR (Kadapa) district to the north, Nellore district to the northeast, Tamil Nadu to the south and Karnataka state to the west. Thirty percent of the total land area is covered by forests in the district. Major forest types in the district comprises of dry, south Indian deciduous mixed forests, southern cutch thorn forests and tropical evergreen dry forest, which provides range of habitats and wide altitudinal gradient for the growth of plants.

Phytogeographically the region is enriched with a diverse flora with a large degree of endemism. The world famous Hindu piligrim centre, Tirumala hills which are the part of Seshachalam hill ranges of Eastern Ghats is situated in this district is the floristic hotspot harbouring many endemic and rare plants. Varied micro-climatological conditions, undulating topographical and interesting ecology prevailing in the district provide a very conducive environment for the growth of variety of plants. The district is well explored by several researchers for floristic wealth, medicinal and ethnobotanical plants. Vedavathy *et al.* (1997) surveyed the district and documented 202 medicinal plant used in tribal medicine. Chetty *et al.* (2008) enumerated nearly 1756 species of flowering plants belonging to 870 genera under 176 families from the district. Recently, Miria *et al.* (2012) reported 13 species of Orchids in Talakona Sacred Grove from Chittoor district. However, it is evident from the literature survey that studies on cryptogams are scarce for such an ecologically interesting region. The cryptogams, especially lichens are sensitive to changes in the microclimatic conditions and air pollution. The lichens are considered as an excellent bioindicators and biomonitors of air pollution as well as climate change. Therefore to generate a baseline data for future biomonitoring studies lichens of Chittoor district is studied in detail. In the earlier investigations a total of 44 species has been reported from different parts of Chittoor district such as Horsley hills, Talakona and Tirumala hills (Reddy *et al.*, 2011, Nayaka *et al.*, 2013, Mohabe *et al.*, 2014). These studies indicate that Chittoor district has good diversity of lichen biota. In the present study we further update the checklist of lichens from Chittoor district with 75 species (Table 6.1). Also we present here the brief description to the species and key for their identifications.

Table 6.1: List of Lichens from Chittoor District of Andhra Pradesh

Sl.No.	Families/Species Name	GF	Habit	Localities										
				1	2	3	4	5	6	7	8	9	10	11
	Candelariaceae													
1.	Candelaria concolor (Dicks.) Stein*	Fl	C	–	+	+	–	–	–	–	–	–	–	–
	Chrysothricaceae													
2.	Chrysothrix candelaris (L.) J.R. Laundon	Cr	C	–	+	–	–	–	–	–	–	–	–	–
3.	C. chlorina (Ach.) J.R. Laundon	Cr	C	+	–	–	–	–	–	–	+	–	–	–
	Collemataceae													
4.	Collema nigrescens (Huds.) DC.	Fl	C	+	–	–	–	–	–	–	–	–	–	–
5.	Leptogium denticulatum Nyl.*	Fl	C	+	–	–	–	–	–	–	–	–	–	–
	Graphidaceae													
6.	Diorygma junghuhnii (Mont. & Bosch) Kalb	Cr	C	+	–	–	–	–	+	–	–	–	–	–
	Lecanoraceae													
7.	Lecanora achora Ach.	Cr	C	–	+	+	–	–	+	–	–	–	+	+
8.	L. alba Lumbsch	Cr	C	–	+	–	–	–	–	–	–	–	–	–
9.	L. chlarotera Nyl.	Cr	C	–	–	–	–	–	+	–	+	–	–	–
10.	L. helva Stizenb.	Cr	C	+	–	–	–	–	–	–	–	–	–	–
11.	L. interjecta Müll. Arg.	Cr	C	–	+	+	–	–	+	–	–	–	–	–
12.	L. tropica Zahlbr.*	Cr	C	–	+	–	–	–	–	–	–	–	–	–
	Letrouitiaceae													
13.	Letrouitia transgressa (Malme) Hafellner & Bellem.	Cr	C	–	–	+	–	–	–	–	–	–	–	–

Contd...

Table 6.1–*Contd...*

Sl.No.	Families/Species Name	GF	Habit	\[Localities\] 1	2	3	4	5	6	7	8	9	10	11
	Ochrolechiaceae													
14.	*Ochrolechia subpallescens* Verseghy*	Cr	C	–	+	–	–	–	–	–	–	–	–	–
	Parmeliaceae													
15.	*Bulbothrix isidiza* (Nyl.) Hale	Fl	C	+	+	–	–	–	–	–	–	–	–	–
16.	*B. tabacina* (Mont. & Bosch) Hale*	Fl	C	–	–	–	–	–	–	–	–	–	–	–
17.	*Canoparmelia texana* (Tuck.) Elix & Hale	Fl	C	+	+	–	+	–	–	–	–	–	–	–
18.	*Parmelia sulcata* Taylor in J. Mackay	Fl	S, C	+	–	–	–	–	–	–	+	–	–	–
19.	*Parmelinella wallichiana* (Taylor.) Elix & Hale	Fl	C	–	+	–	–	–	–	–	–	+	+	+
20.	*Parmotrema andinum* (Mull.Arg.) Hale	Fl	S, C	+	+	–	+	–	+	–	+	–	+	+
21.	*P. austrosinense* (Zahlbr.) Hale	Fl	S, C	+	–	–	+	–	–	–	+	–	–	–
22.	*P. ravum* (Krog & Swinscow) Swisn. *	Fl	C	–	+	–	–	–	–	–	–	–	–	–
23.	*P. crinitoides* J.C. Wei	Fl	C	–	–	–	–	–	–	–	–	+	–	–
24.	*P. cristiferum* (Taylor) Hale	Fl	C	+	–	–	+	–	–	–	–	–	–	–
25.	*P. defectum* (Hale) Hale	Fl	C	–	+	–	+	–	–	–	+	–	–	–
26.	*P. eunetum* (Stirt.) Hale*	Fl	S	–	–	–	–	–	–	–	+	–	–	–
27.	*P. grayanum* (Hue) Hale	Fl	S, C	–	–	–	–	–	–	–	+	–	+	+
28.	*P. latissimum* (Fée) Hale*	Fl	S, C	–	–	–	–	–	–	–	+	–	+	+
29.	*P. melanothrix* (Mont.) Hale	Fl	S, C	+	–	–	+	–	–	–	+	–	+	–
30.	*P. mesotropum* (Müll. Arg.) Hale	Fl	S, C	+	+	–	+	–	+	–	+	+	+	+
31.	*P. nilgherrense* (Nyl.) Hale	Fl	S, C	+	+	–	–	–	–	–	+	+	+	–
32.	*P. praesorediosum* (Nyl.) Hale	Fl	S, C	+	+	–	+	–	+	+	+	+	–	+

Contd...

Table 6..1–Contd.

Sl.No.	Families/Species Name	GF	Habit	Localities										
				1	2	3	4	5	6	7	8	9	10	11
33.	P. reticulatum (Taylor) Choisy	Fl	S, C	+	–	–	+	+	–	–	+	–	+	+
34.	P. saccatilobum (Taylor) Hale*	Fl	C	–	–	–	+	–	–	–	–	–	–	–
35.	P. stuppeum (Taylor) Hale	Fl	C	+	–	–	–	–	–	–	–	–	–	+
36.	P. tinctorum (Despr. ex Nyl.) Hale	Fl	S, C	+	+	–	+	+	+	–	–	–	+	+
37.	Xanthoparmelia pseudocongensis Hale	Fl	S	–	–	–	–	–	+	–	–	–	–	–
38.	Usnea baileyi (Stirt.) Zahlbr.*	Fr	C	–	–	–	–	–	–	–	+	–	–	–
	Peltulaceae													
39.	Peltula farinosa Büdel	Sq	S	+	–	–	–	–	–	–	–	–	–	–
40.	P. placodizans (Zahlbr.) Wetmore	Sq	S	+	–	–	–	–	–	–	–	–	–	–
	Pertusariaceae													
41.	Pertusaria leucosora Nyl.	Cr	S	+	–	–	–	–	–	–	+	–	–	–
42.	P. melastomella Nyl.	Cr	C	+	–	–	–	–	–	–	–	–	–	–
43.	P. pustulata (Ach.) Duby*	Cr	S	–	–	–	–	–	–	–	–	–	–	+
44.	P. quassiae (Fée) Nyl.	Cr	C	+	–	–	–	–	–	–	–	–	–	–
	Physciaceae													
45.	Buellia hemispherica S.R. Singh & D.D. Awasthi*#	Cr	S	–	–	–	–	–	–	–	+	–	–	–
46.	Dirinaria aegialita (Afz. in Ach.) Moore	Fl	C	+	–	–	–	–	–	–	–	–	–	–
47.	D. applanata (Fée) D.D. Awasthi	Fl	C	–	+	–	–	–	–	–	–	–	–	–
48.	Hafellia curatellae (Malme) Marbach	Cr	C	+	+	–	–	–	–	–	–	–	–	–
49.	Heterodermia albicans (Pers.) Swinsc. & Krog	Fl	C	+	–	–	–	–	–	–	–	–	–	–

Contd...

Table 6..1–*Contd...*

Sl.No.	Families/Species Name	GF	Habit	1	2	3	4	5	6	7	8	9	10	11
									Localities					
50.	*H. diademata* (Taylor) D.D. Awasthi	Fl	C	+	–	–	–	+	–	–	–	–	–	–
51.	*H. dissecta* (Kurok.) D. D. Awasthi	Fl	S	–	–	–	–	–	–	–	–	–	–	+
52.	*H. isidiophora* (Vain.) D.D. Awasthi	Fl	S, C	+	–	–	+	+	+	–	+	–	+	+
53.	*H. obscurata* (Nyl.) Trevis*	Fl	C	–	–	–	–	–	–	–	–	–	+	–
54.	*H. pseudospeciosa* (Kurok.) W.L. Culb.	Fl	C	–	–	–	–	–	–	–	–	+	–	+
55.	*Hyperphyscia adglutinata* (Flörke) H. Mayrhofer & Poelt*	Fl	S, C	+	+	–	–	–	–	–	–	–	–	–
56.	*H. adglutinata* var. *pyrithrocardia* (Müll. Arg.) D.D. Awasthi*	Fl	C	–	+	–	–	–	–	–	–	–	–	–
57.	*Phaeophyscia hispidula* (Ach.) Moberg	Fl	C	+		–	–	–	–	–	–	–	–	–
58.	*Physcia* sp.	Fl	C	+	–	–	–	–	–	–	–	–	–	–
59.	*P. abuensis* D.D. Awasthi & S.R. Singh#	Fl	S	–	–	–	–	–	–	–	+	–	–	–
60.	*P. tribacoides* Nyl.	Fl	C	+	–	–	–	–	–	–	–	–	–	–
61.	*Pyxine minuta* Vain.*	Fl	S	+	–	–	–	–	–	–	–	+	–	–
62.	*P. petricola* var. *pallida* Swinscow & Krog, Norweg	Fl	S, C	–	+	–	–	–	–	–	–	–	–	–
63.	*P. petricola* var. *petricola* Nyl.	Fl	S, C	–	+	–	–	–	–	–	–	–	–	–
64.	*P. punensis* Nayaka & Upreti#	Fl	S, C	+	–	–	–	–	–	–	–	–	–	–
65.	*P. subcinerea* Stirton	Fl	S	–	+	–	–	–	–	–	–	–	–	–
66.	*Rinodina badiella* (Nyl.) Th. Fr.*	Cr	s	–	–	–	–	–	–	–	+	–	–	–
	Porinaceae													
67.	*Porina tetracerae* (Afz.) Müll. Arg.	Cr	C	–	–	–	–	–	+	–	–	–	–	–

Contd...

Table 6..1–*Contd...*

Sl.No.	Families/Species Name	GF	Habit	1	2	3	4	5	6	7	8	9	10	11
									Localities					
	Ramalinaceae													
68.	*Ramalina conduplicans* Vain.	Fr	C	–	+	–	–	–	–	–	+	–	–	–
	Stereocaulaceae													
69.	*Lepraria coriensis* (Hue) Sipman	LP	S	+	–	–	–	–	–	+	–	–	–	–
	Roccellaceae													
70.	*Roccella montagnei* Bél	Fr	S	+	–	–	–	–	–	–	–	–	–	–
	Teloschistaceae													
71.	*Caloplaca bassiae* (Willd. ex Ach.) Zahlbr.	Cr	C	+	+	–	–	–	–	–	–	–	–	–
72.	*C. cinnabarina* (Ach.) Zahlbr.	Cr	S	+	–	–	–	–	–	–	+	–	–	–
73.	*C. poliotera* (Nyl.) Stein	Cr	S	–	–	–	–	–	+	–	–	–	–	–
74.	*C. tropica* Y. Joshi & Upreti#	Cr	S	+	–	+	–	–	–	–	–	–	–	–
	Thelotremataceae													
75.	*Diploschistes actinostomus* (Pers. ex Ach.) Zahlbr.*	Cr	S	+	–	–	–	–	–	–	+	–	–	–
	Totals	40	26	6	12	4	12	2	23	4	11	13		

Abbreviations: GF: Growth Forms; Cr: Crustose, Fl: Foliose; Sq: Squamulose; LP: Leprose; C: Corticolous; S: Saxicolous; +: Present –: Absent; *: New additions for state; #: Endemic to India.

Name of Localities: 1: Horsley Hills; **2:** Talakona; **3:** Thambalapalli hills; **4:** Tirumala Hills; **5:** TH Dharmagiri; **6:** TH Japali Anjaneya Swami Temple; **7:** TH Papavinasanam; **8:** TH Shilathoranam; **9:** TH Srivari mettu; **10:** TH Srivari padalu; **11:** TH Vedapatashala.

Materials and Methods

The present study is based on observation of 326 lichen specimens collected during the year 2011-2014 from different forest localities of Chittoor district namely Horsley Hills, Talakona, Tirumala hills (Dharmagiri, Japali Anjaneya Swami Temple, Papavinasnam, Shilathoranam, Vedapathasala) and Mallaiah Konda hills of Thambalapalli (Plate 6.1 and Figure 6.1). The lichens were growing over tree trunks and exposed rocks. The morphological features of lichen thallus and ascomata were observed under Magnüs MS 24/13 and Leica S8AP0 stereozoom microscopes. Spot test for colour reaction were carried out by 10 per cent aqueous solution of potassium hydroxide (K), Steiner's stable para-henylenediamine solution (PD) and calcium hypochlorite solution (C). For anatomical investigation of fruiting bodies light microscope of ZEISS Axiostar plus and Leica DM500 compound microscope were used. All the measurements of anatomical structures were taken in water. The lichen substances were identified with Thin Layer Chromatography in solvent system 'A' following White and James (1985) and Orange *et al.* (2001). The following literatures were referred for identification of lichen samples; Awasthi (1991, 2007), Divakar and Upreti (2005), Joshi (2008), Mayrhofer *et al.* (1996), Nayaka (2005), Laundon (1981) and Upreti *et al.* (2010). The nomenclature and classification of lichens were updated following Lumbsch and Huhndorf (2007). A key to the genera are provided separately and the species key are provided within the genera. A brief description of 75 species along with specimen citation is also provided and species under each genus are arranged alphabetically. The identified specimens were labelled, documented, digitalized and preserved in the Herbarium of Department of Botany, Yogi Vemana University (YVUH), Kadapa, Andhra Pradesh.

Results and Discussion

The present investigation revealed the occurrence of 75 species belonging to 31 genera under 17 families. Among the different growth forms, the foliose lichens exhibited the maximum diversity represented by 45 (60 per cent) species followed by 24 (32 per cent) species of crustose, 3 (4 per cent) fruticose, 2 (3 per cent) squamulose and a single (1 per cent) species of leprose lichen (Figure 6.2). Based on their habitat the corticolous lichens exhibited the maximum diversity represented by 42 (56 per cent) species followed by 20 (27 per cent) species of saxicolous while 13 (17 per cent) species are common growing both on bark and rock (Figure 6.3). Out of these 18 species are new records to Andhra Pradesh and 31 species are new additions to Chittoor district. *Buellia hemispherica, Caloplaca tropica, Pyxine punensis* and *Physcia abuensis* are the endemic lichen taxa recorded from the district to India (Singh and Sinha, 2010). The members of lichen families Parmeliaceae and Physciaceae are the most dominant in the district with 24 and 22 species under 7 and 8 genera followed by Lecanoraceae with 6 species, Teloschistaceae and Pertusariaceae with 4 species each (Figure 6.4). The lichen genus *Parmotrema* exhibited the luxuriant growth represented by 17 species followed by 6 species each of *Heterodermia* and *Lecanora*. A large number genera (20 nos.) are represented by only single species under them. The foliose lichen *Parmotrema praesorediosum* found growing luxuriantly was represented by 8 localities followed by *P. andinum, P. mesotropum, P. tinctorum, Heterodermia*

Plate 6.1: Lichen Rich Sites in Chittoor District.

A. Beautiful view of rock slopes with luxuriant growth of lichens in Horsley hills, B. A view of Shilathoranam in Tirumala hills, C. Different growth forms of lichens on exposed rock in Thamballapalli, Mallaiah Konda hills, D. *Diorygma* on tree trunk near Japali Anjaneya Swami Temple, Tirumala hills, E. Luxuriant growth of *Pyxine* on *Mangifera indica* tree branch in Talakona, F. Beautiful patches of *Parmotrema* on tree base in Thambalapalli hills.

MAP OF CHITTOOR DISTRICT

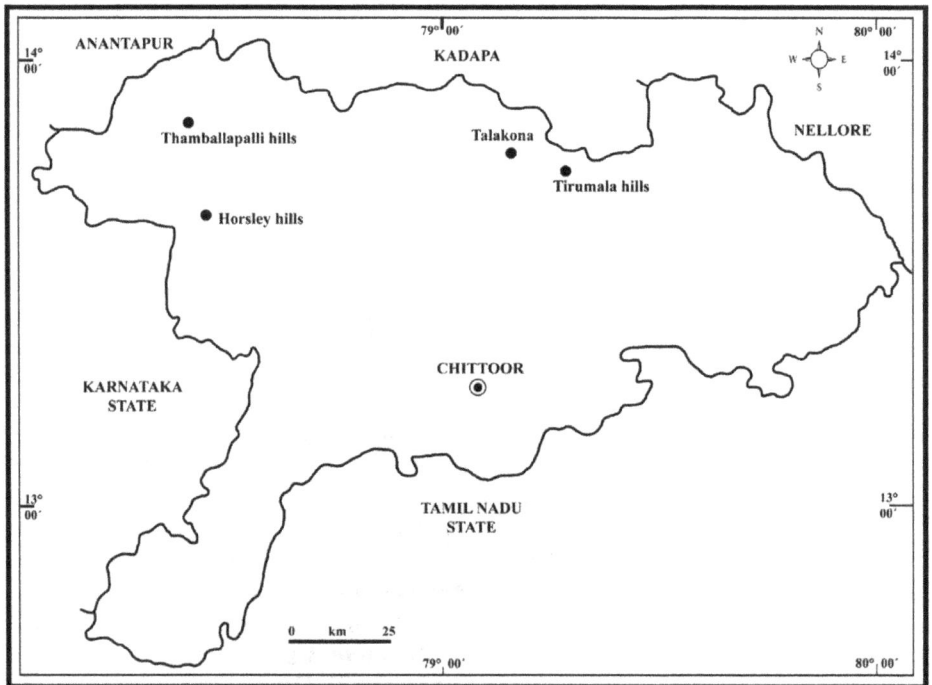

Figure 6.1: Map Showing Explored Areas in Chittoor District.

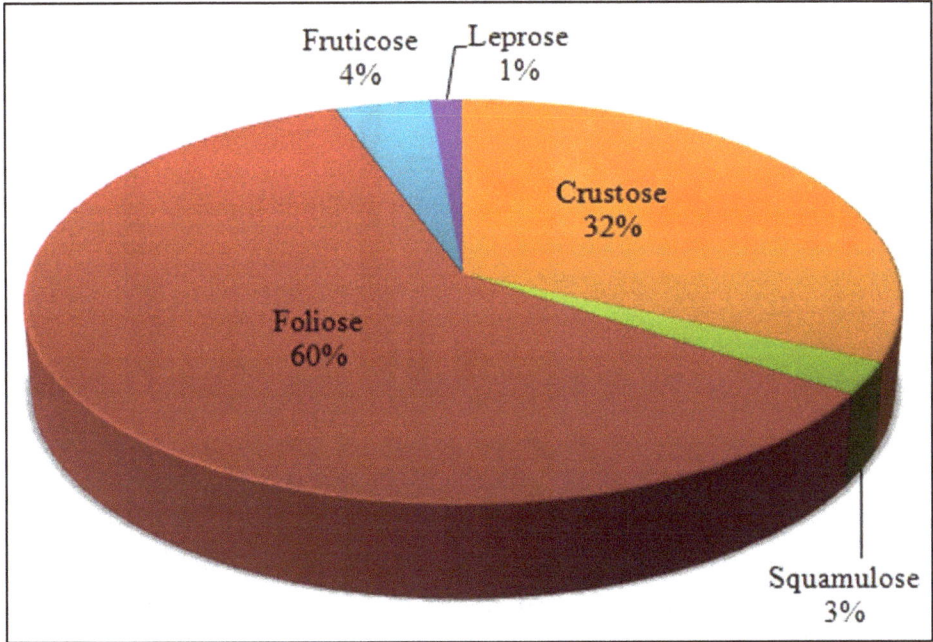

Figure 6.2: Proportion of Various Growth Forms Represented in Chittoor District.

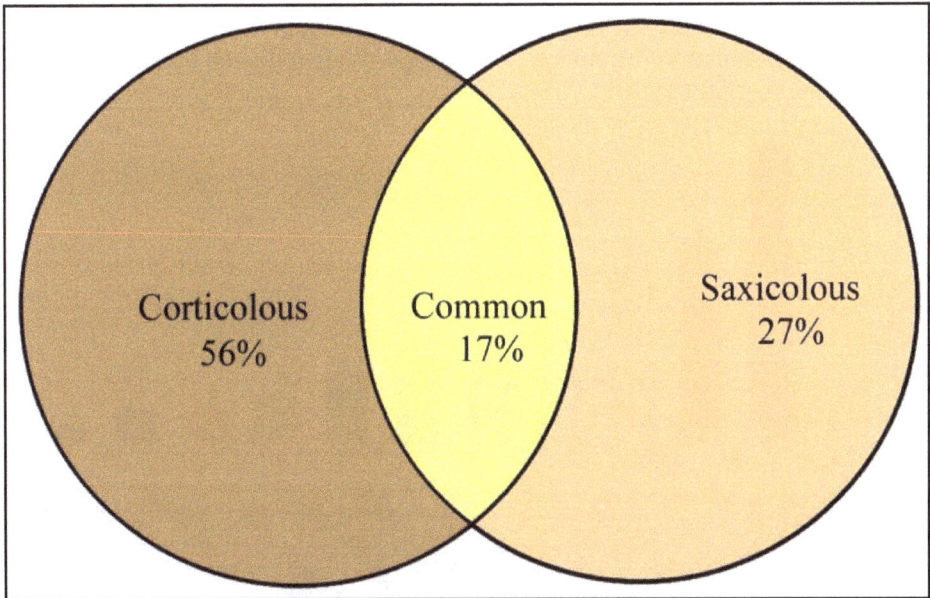

Figure 6.3: Habit Preference of Lichens in Chittoor District.

isidiophora were found in 7 localities each while, *P. reticulatum* represented in 6 localities. It is interesting to note that, all these species are medicinally important especially for antimicrobial properties. These foliose species were found luxuriantly

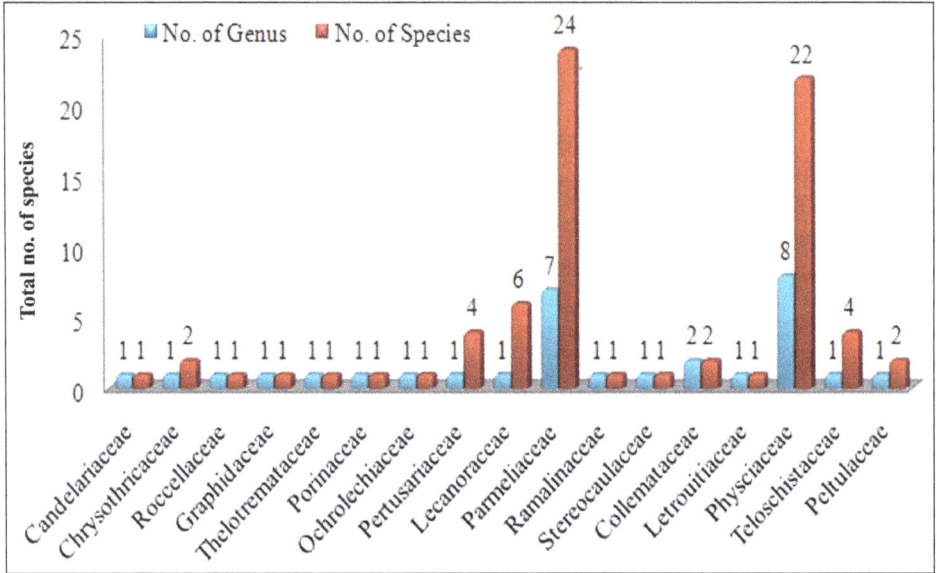

Figure 6.4: Representation of different Families in Chittoor District.

growing both on tree trunks and on exposed rocks. It indicates that the forest areas are well preserved. Out of 11 localities surveyed in the present study, Horsley Hills and Tirumala hills (including adjacent localities) exhibited the maximum diversity of lichens represented by 40 and 39 species respectively followed by Talakona with

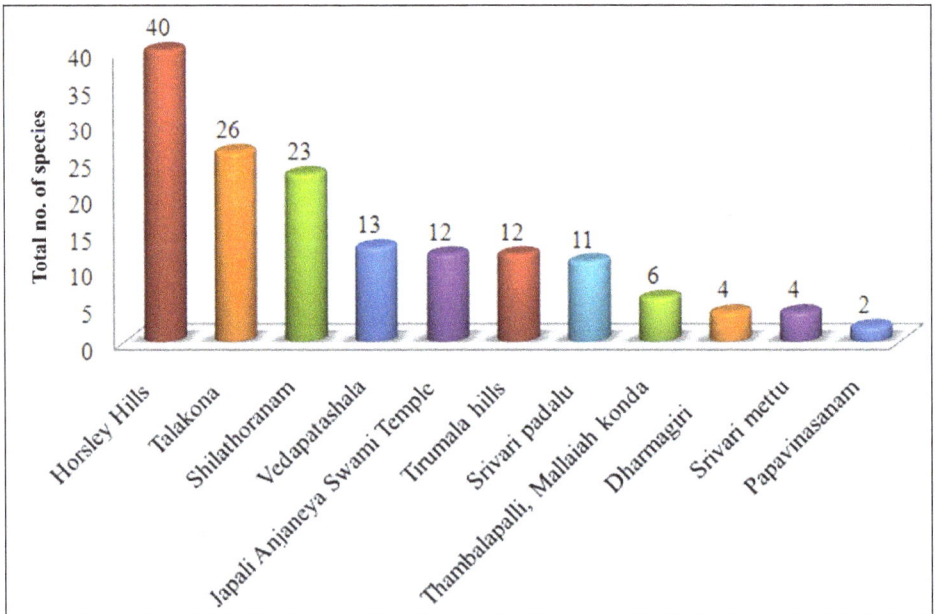

Figure 6.5: Lichen Diversity in different Localities of Chittoor District.

26 species, while other localities in the district shows poor or scare growth of lichens (Figure 6.5). Probable reasons for the poor diversity of lichens in these areas may be due to frequent human interference as many of these are famous pilgrim centres.

Conclusion

The present investigation on lichens clearly indicates that the presence of high lichen diversity in the Chittoor district. This provides a baseline data on lichens of Andhra Pradesh. Devi *et al.* (2013) enumerated 46 species from YSR district with 28 new additions to the state. So far a total of 107 species of lichens are known from the Andhra Pradesh State. After the addition of 18 species from the present study the total number of lichens would rise to 125 species. It is interesting to note that out of 26 species recorded from Talakona 12 species were found on *Mangifera indica* which is most suitable host tree for the growth of lichens. The present data on lichens from the district will be useful for future biomonitoring studies while medicinaly important lichens will be useful for bioprospection studies. The present investigation on lichens in Rayalaseema region is still in progress and it is expected that more species would be added in the present list.

SYSTEMATIC ENUMERATION

Key to the Genera

1. Thallus leprose, crustose or squamulose 2

1a. Thallus foliose or fruticose 15

2. Thallus completely squamulose, containing bluegreen alga .... *Peltula*

2a. Thallus leprose or crustose, containing green alga 3

3. Thallus leprose 4

3a Thallus crustose 5

4. Thallus greenish-grey to grey, powdery crust or granular .... *Lepraria* (*L. coriensis*)

4a. Thallus greenish-yellow to yellowish green, powdery crust or granular .... *Chrysothrix*

5. Ascocarp elongate or lirellate, white .... *Diorygma* (*D. junghuhnii*)

5a. Ascocarp apothecia, round 6

6. Apothecia perithecioid, disc opening by pore 7

6a. Apothecia apothecioid, disc widened open 9

7. Ascospores muriform, brown, saxicolous .... *Diploschistes* (*D. actinostomus*)

7a. Ascospores simple or transversely septate, colourless, saxicolous or corticolous 8

8. Ascospores simple, large .... *Pertusaria*

8a. Ascospores transversely septate, fusiform .... *Porina* (*P. tetracerae*)

9. Ascospores simple or transversely septate, colourless 10

9a. Ascospores transversely septate or submuriform, brown 12

10 Apothecia lecidine or biatorine, ascospores polaribilocular .... *Caloplaca*

10a. Apothecia lecanorine, ascospores simple, large.. 11

11. Ascospores large, more than 30 μm .... *Ochrolechia (O. subpallescens)*

11a. Ascospores smaller, less than 30 μm .... *Lecanora*

12. Apothecia lecanorine or biatorine 13

12a. Apothecia lecidine 14

13 Apothecia lecanorine, ascospore1-septate,
disc brown to black .... *Rinodina (R. badiella)*

13a. Apothecia biatorine, ascospore submuriform,
disc yellow orange .... *Letrouitia (L. transgressa)*

14. Thallus saxicolous, ascospores brown,
smooth on surface .... *Buellia (B. hemispherica)*

14a. Thallus corticolous, ascospores brown,
ornamented on surface .... *Hafellia (H. curtellae)*

15. Thallus foliose.. 16

15a. Thallus fruticose 30

16. Thallus containing green alga 17

16a. Thallus containing blue green alga 29

17. Ascospores brown, septate 18

17a. Ascospores colourless, simple or ellipsoid, rarely 2-celled 23

18. Lower surface without rhizines 19

18a. Lower surface with rhizines.. 20

19. Lobes dichotomously to pinnately divided,
atranorin in upper cortex .... *Dirinaria*

19a. Lobes usually small and narrow, lichen substances absent.......... *Hyperphyscia*

20. Thallus UV+ yellow .... *Pyxine*

20b. Thallus UV 21

21 Atranorin always present in upper cortex 22

21a. Atranorin always absent in upper cortex .... *Phaeophyscia (P. hispidula)*

22. Thallus dichotomously or irregularly branched ....*Heterodermia*

22a. Thallus radially lobate ....*Physcia*

23. Thallus minutely lobate, greenish yellow
to yellowish green .... *Candelaria (C. concolor)*

23a. Thallus broadly lobate, greenish grey to grey 24

24 Lobes with cilia 25

24a. Lobes without cilia 27

25. Lobes with bulbate cilia, rhizines dichotomousely branched .... *Bulbothrix*

25a Lobes with simple cilia, rhizines simple or simple to branched 26

26. Cilia restricted in central part of lobes,
     lacking pseudocyphalae .... *Parmelinella* (*P. wallichiana*)

26a. Lobes with or without cilia, pseudocyphalae
      present or absent .... *Parmotrema*

27. Thallus with pseudocyphalae, rhizines simple to
     dichotomously branched .... *Parmelia* (*P. sulcata*)

27a. Thallus lacking pseudocyphalate, rhizines simple 28

28. Lobes sublinear to irregular, atranorin present in
     upper cortex .... *Canoparmelia* (*C. texana*)

28a. Lobes branched and black rimmed, usnic acid
      present in upper cortex .... *Xanthoparmelia* (*X. pseudocongensis*)

29. Thallus ecorticated or with pseudocortex .... *Collema* (*C.nighrescens*)

29a. Thallus corticated on both sides .... *Leptogium* (*L. denticulatum*)

30. Thallus cylindrical, subcylindrical, rarely angular,
     ascospores simple .... *Usnea* (*U. baileyi*)

30a. Thallus flat to strap shaped, ascospores 2-celled or
      transversely septate 31

31. Thallus with chondroid tissue beneath cortex,
     C -ve .... *Ramalina* (*R. conduplicans*)

31a. Thallus lacking chondroid tissue beneath cortex,
      C+ pink .... *Roccella* (*R. montegnei*)

CANDELARIACEAE

Candelaria A. Massal.

1. Candelaria concolor (Dicks.) Stein in Cohn-Krypt. Fl. Schles. 2(2): 84. 1879. Plate
   6.2F

It is a foliose lichen, found growing on bark of trees and rock, greenish yellow to deep yellow in colour, 2.0 cm across, minutely lobate, lobes up to 2 mm long, 0.5 mm wide with granular soredia at lobe ends, rhizinate, apothecia rare, rounded, adnate to sessile, 0.1 - 0.3 mm wide, lecanorine, disc yellowish brown, ascus 8 spored, ascospores simple, hyaline, contains calycin and pulvinic acid.

Distribution in India: Himachal Pradesh, Jammu and Kashmir, Karnataka, Madhya Pradesh, Nagaland, Sikkim and Uttarakhand. It is a new record for Andhra Pradesh and new addition for Chittoor district.

Plate 6.2: Some Prominent Lichens of Chittoor District.
A. *Buellia hemispherica*, B. *Bulbothrix isidiza*, C. *Caloplaca bassiae*, D. *C. cinnabarina*,
E. *C. tropica*, F. *Candelaria concolor*.

Specimens examined: Andhra Pradesh, Chittoor district, Talakona, before jungle thrills, N 14°28.698´ E 078°42.749´ alt. 539 m, on bark of *Mangifera indica*, 16.03.2013, Anjali Devi B. and Satish Mohabe 3700 (YVUH); Tirumala Hills, Shilathornam, on Rock, N 13°41.282´ E079°20.418´ alt. 937 m, 07.02.2013, Anjali Devi B. and Satish Mohabe 3735 (YVUH); Thamballapalli, Mallaiah Konda hills, Temple premises, N 13°60.025´ E 078°25.424´ alt. 856 m, on bark, 05.01.2013, A. Madhusudhana Reddy and Satish Mohabe 2796 (YVUH).

CHRYSOTHRICACEAE

Chrysothrix Mont.

1. Thallus yellow to slight orange, containing calycin and
    pinastrict acid .... *C. candelaris*

1a. Thallus lemon yellow, containing calycin and vulpinic acid ....*C. clorina*

2. Chrysothrix candelaris (L.) Laudon-Lichenologist 13(2): 110. 1981. Plate 6.3B

It is a crustose, leprose lichen, found growing on bark of trees, greenish-yellow to yellowish green, sometimes slightly orange in colour, granular, forming uniformly thick or pulverulent mass, contains calycin and pinastric acid.

Distribution in India: Andhra Pradesh, Himachal Pradesh, Jammu and Kashmir, Sikkim and Tamil Nadu. It is a new addition for Chittoor district.

Specimens examined: Andhra Pradesh, Chittoor district, Talakona, before jungle thrills, N 14°28.698´ E 078°42.749´ alt. 539 m, on bark of *Mangifera indica*, 16.03.2013, Anjali Devi B. and Satish Mohabe 3696 (YVUH).

3. Chrysothrix chlorina (Ach.) J.R. Laundon-Lichenologist 13(2): 106. 1981.

It is a crustose, leprose lichen, found growing on bark, lemon yellow in colour, forming uniformly thick or pulverulent mass, contains calycin and vulpinic acid.

Distribution in India: Andhra Pradesh, Himachal Pradesh, Madhya Pradesh, Jammu and Kashmir and Sikkim.

Specimens examined: Andhra Pradesh, Chittoor district, Horsley Hills, on stem, 04.03.2010, A. Madhusudhan Reddy and B. Ravi Prasad Rao 0812 (YVUH); Tirumala hills, Shilathoranam, on bark, 04.01.12, Anjali Devi B. 1434 (YVUH).

COLLEMATACEAE

Collema Weber ex F.H. Wigg.

4. Collema nighrescens (Huds.) DC. in Lam. & DC.-Fl. Franç., ed. 3, 2: 384. 1805.

It is a foliose lichen, found growing on tree bark, sterile, olive-green to brown in colour, thallus loosely attached to the substratum, 3.0 - 7.0 cm across, longitudinally densely ridged and postulate isidiate, isidia granular to oblong.

Distribution in India: Andhra Pradesh, Madhya Pradesh, Tamil Nadu and Uttarakhand.

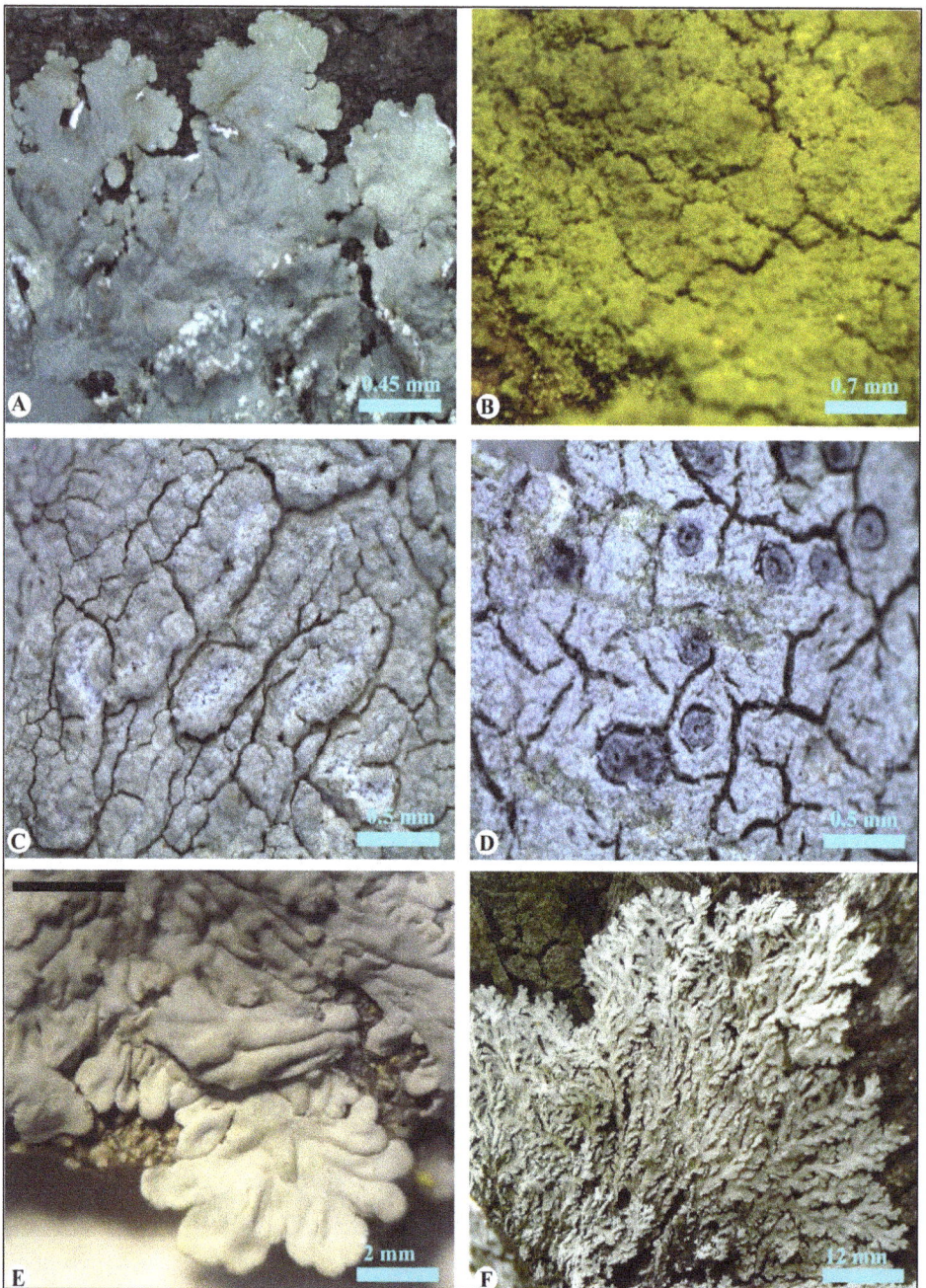

Plate 6.3: Some Prominent Lichens of Chittoor District.
A. *Canoparmelia texana,* **B.** *Chrysothrix candelaris,* **C.** *Diorygma junghuhnii,*
D. *Diploschistes actinostomus,* **E.** *Dirinaria applanata,* **F.** *Heterodermia pseudospeciosa.*

Specimens examined: Andhra Pradesh, Chittoor district, Horsley hills, on bark, 12.06.12, A. Madhusudhana Reddy, Sanjeeva Nayaka and Anjali Devi B. 1704 (YVUH).

Leptogium (Ach.) Gray

5. Leptogium denticulatum Nyl.-Ann. Sci. Nat., Bot., ser. 5, 7: 302. 1867. Plate 6.5B

It is a foliose lichen, found growing on tree bark, lead grey to darker in colour, adante, 4 cm across, slightly wrinkled, isidiate, isidia squamuliform, lower side paler, etomentose, apothecia rare.

Distribution in India: Andaman and Nicobar Islands, Arunachal Pradesh, Goa, Karnataka, Kerala, Madhya Pradesh, Maharashtra, Manipur, Meghalaya, Nagaland, Sikkim, Tamil Nadu and West Bengal –hills. It is a new record for Andhra Pradesh and new addition for Chittoor district.

Specimens examined: Andhra Pradesh, Chittoor district, Horsley Hills, on Bark, 12.06.12, A. Madhusudhana Redd, Sanjeeva Nayaka and Anjali Devi B. 1712 (YVUH).

GRAPHIDACEAE

Diorygma Eschw.

6. Diorygma junghuhnii (Mont. & Bosch) Kalb, Staiger & Elix- Symb. Bot. Upsal. 34(1): 157. 2004. Plate 6.3C

It is a crustose lichen, found growing on bark, whitish to pale grey in colour, ascocarp oval to elongate and irregularly branched, immersed in thallus, disc wide, covered by whitish pruina, exciple pale, poorly developed, ascus 1 – 2 spored, ascospores hyaline, muriform, contains norstictic acid.

Distribution in India: Andaman and Nicobar Islands, Assam, Karnataka, Kerala, Maharashtra, Manipur, Meghalaya, Nagaland, Sikkim and Tamil Nadu.

Specimens examined: Andhra Pradesh, Chittoor district, Horsley hills, alt. 895 m, on bark, 12.06.12, A. Madhusudhana Reddy, Sanjeeva Nayaka and Anjali Devi B. 1837, 2034, 2035 (YVUH); on bark, 04.03.2010, A. Madhusudhana Reddy 0811B (YVUH); Tirumala hills, Japali Anjaneya Swamy temple, alt. 746.5 m, on bark, 13.06.12, A. Madhusudhana Reddy and Sanjeeva Nayaka 1820, 1825 (YVUH).

LECANORACEAE

Lecanora Ach.

1.   Apothecial disc red brown, epihymenium not dissolving in KOH 2

1a. Apothecial disc yellow, orange, brown, epihymenium dissolving in KOH 3

2.   Thallus containing usnic acid, amphithecium with large crystals .... *L. alba*

2a. Thallus containing zeorin, amphithecium with large
     and small crystals .... *L. tropica*

3.   Thallus containing 22-methyl-perlatolic acid, disc pale orange 4

3a. Thallus lacking 22-methyl-perlatolic acid, disc pale orange to brown 5

4.   Thallus contains usnic acid .... *L. achroa*

4a.   Thallus lacking usnic acid .... *L. helva*

5.   Disc pale brown to orange brown, contains atranorin
      and usnic acid .... *L. interjecta*

5a.   Disc pale yellow to brownish yellow or orange,
      contains atranorin and zeorin .... *L. chlarotera*

7. Lecanora achroa Nyl. in Cromb.- J. Bot. 14: 263. 1876. Plate 6.4B

It is a crustose lichen, found growing on bark of trees, grey to yellowish grey in colour, continuous to rimose areolate, slightly verruculose, lacking isidia and soredia, apothecia sessile, 0.2 – 1.0 mm in diam, disc yellowish orange to pale brown, epruinose, margin smooth to verruculose, sometimes excluded at maturity, exciple with large crystals, epihymenium yellowish brown, granular, with coarse crystals, dissolving in KOH, contains atranorin, 22-*O*-methylperlatolic, usnic and arthothelin.

Distribution in India: Andhra Pradesh, Himachal Pradesh, Madhya Pradesh, Manipur, Sikkim and Uttar Pradesh.

Specimens examined: Andhra Pradesh, Chittoor district, Japali Anjaneya Swamy temple, on bark, 08.02.2013, Anjali Devi B. and Satish Mohabe 3568 (YVUH); Talakona, on bark, 16.03.2013, Anjali Devi B. and Satish Mohabe 3714, 3744 (YVUH); Thamballapalli, Mallaiah Konda hills, Temple premises, N 13°60.025´ E 078°25.424´ alt. 856 m, on bark, 05.01.2013, Satish Mohabe and A. Madhusudhana Reddy 2789, 2794 (YVUH); Tirumala Hills, Srivari Padalu, on bark, N 13°40.658' E 079°19.924', alt. ca. 1076 m, Satish Mohabe and Anjali Devi B. 3522 (YVUH).

8. Lecanora alba Lumbsch-Bryologist 98: 565. 1995.

It is a crustose lichen, found growing on bark, yellowish grey to greenish white or whitish grey in colour, continuous to verruculose, apothecia numerous, sessile 0.5 - 0.9 mm in diam., disc red brown to brown, epruinose, margin smooth to verruculose, exciple with large crystals, epihymenium red brown, egranular, lacking crystals, not dissolving in K, contains atranorin, arthothelin and usnic, acids.

Distribution in India: Andhra Pradesh, Arunachal Pradesh, Himachal Pradesh and Uttarakhand. It is a new addition for Chittoor district.

Specimens examined: Andhra Pradesh, Chittoor district, Talakona, before jungle thrills, N 14°28.698' E 078°42.749' alt. 539 m, on bark of *Mangifera indica*, 16.03.2013, Anjali Devi B. and Satish Mohabe 3658 (YVUH).

9. Lecanora chlarotera Nyl.-Bull. Soc. Linn. Normandie, ser. 2, 6: 274. 1872. Plate 6.4C

It is a crustose lichen, found growing on bark of trees, greenish grey to grey, verruculose to verrucose, apothecia numerous, crowded, flat to concave, sometimes flexuose, 0.2 – 0.9 mm in diam., margin smooth to verruculose, crenulated, disc pale orange to orange brown or reddish brown, epruinose, exciple with large crystals and

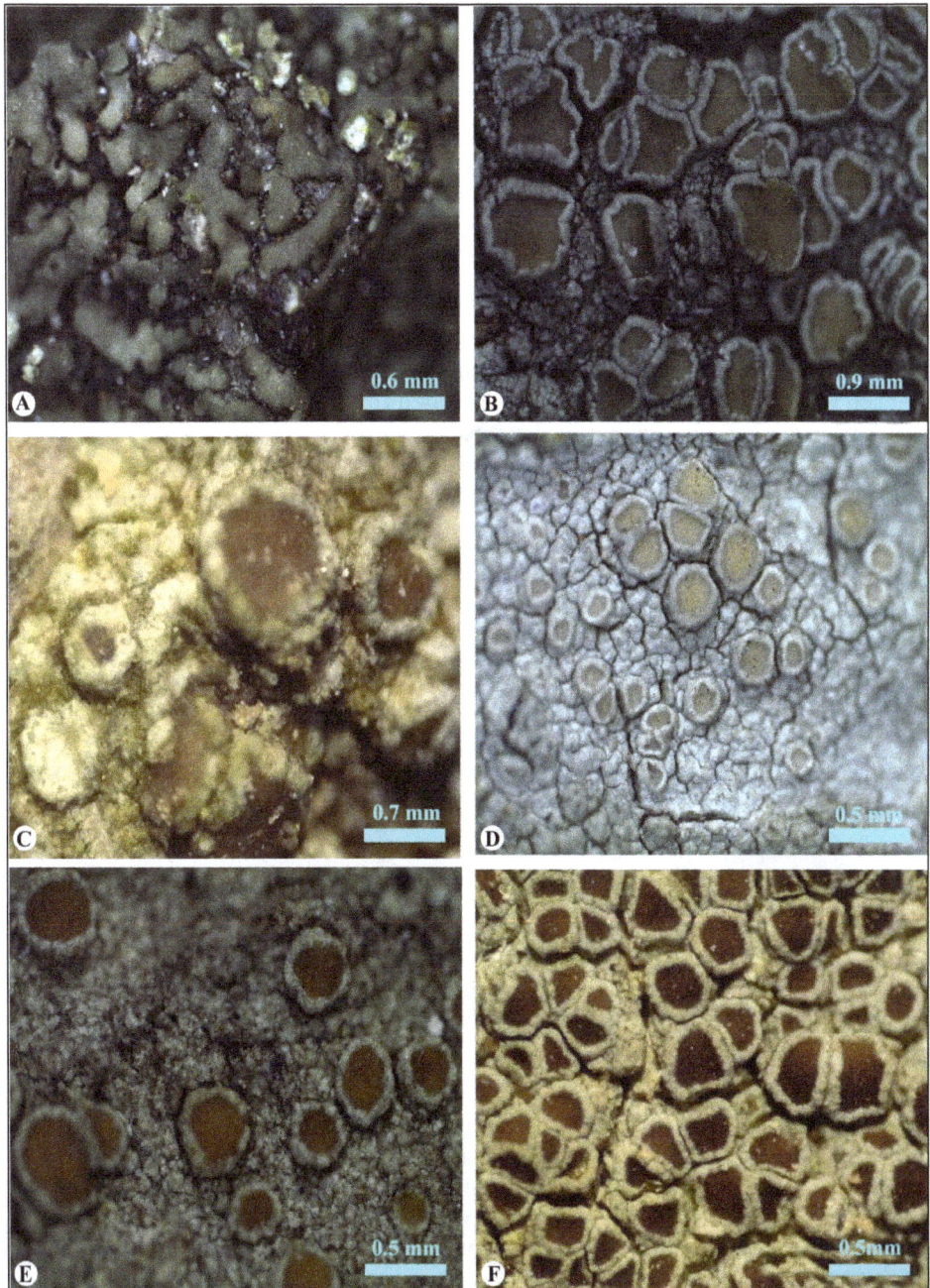

Plate 6.4: Some Prominent Lichens of Chittoor District.

A. *Hyperphyscia adglutinata,* **B.** *Lecanora achroa, C. L. chlarotera,* **D.** *L. helva,*
E. *L. interjecta,* **F.** *L. tropica.*

algal cells, epihymenium yellowish to brownish, egranular, dissolving in K, contains atranorin and zeorin.

Distribution in India: Jammu and Kashmir, Karnataka, Maharashtra, Manipur, Nagaland, Rajasthan, Tamil Nadu, Uttarakhand and West Bengal.

Specimens examined: Andhra Pradesh, Chittoor district, Shilathoranam, on the backside of the arch 10 ft away, on bark, 13.06.2012, A. Madhusudhana Reddy and Sanjeeva Nayaka 1806/B (YVUH); Japali Anjaneya Swamy temple, on bark,13.06.2012, A. Madhusudhana Reddy and Sanjeeva Nayaka 1822 (YVUH).

10. Lecanora helva Stizenb.-Ber. Thätigk. St. Gallischen Naturwiss. Ges. 1888/1889: 218. 1890. Plate 6.4D

It is a crustose lichen, found growing on bark of tree, greenish-grey to grey, smooth to verruculose, ecorticated, apothecia numerous, rounded, 0.2 – 1.0 mm in diam., margin smooth, sometimes flexuose, lecanorine, disc pale yellow to brownish yellow or orange, epruinose or slightly pruinose, flat to concave, exciple with large crystals and algal cells, epihymenium yellowish to brownish, dissolving in KOH, contains 22 methyl-perlatolic acid and atranorin.

Distribution in India: Assam, Goa, Himachal Pradesh, Kerala, Madhya Pradesh, Maharashtra and Tamil Nadu.

Speciemens examined: Andhra Pradesh, Chittoor district, Horsley Hills, on bark, 12.06.2012, Madhusudhana Reddy, Sanjeeva Nayaka and Anjali Devi B. 1896 (YVUH).

11. Lecanora interjecta Müll. Arg.- Nuovo Giorn. Bot. Ital. 23: 390. 1891. Plate 6.4E

It is a crustose lichen, found growing on bark, greenish-grey in colour, verruculose to verrucose, apothecia numerous, sessile, 0.2 – 1.0 mm in diam, margin smooth to verruculose or crenulated, lecanorine, disc pale brown to orange brown, epruinose, plane to convex, exciple with large crystals and algal cells, epihymenium yellowish to pale brown, granular, pigmentation dissolving in KOH, contains atranorin and usnic acid.

Distribution in India: Arunachal Pradesh and Himachal Pradesh.

Specimens examined: Andhra Pradesh, Chittoor district, Japali Anjaneya Swamy temple alt. 746.5 m, on Bark, 13.06.2012, A. Madhusudhana Reddy and Sanjeeva Nayaka 1817, 1818, 1824, 1839 (YVUH); 08.02.013, A. Madhusudhana Reddy and Satish Mohabe 3603 (YVUH); Talakona, on bark, 16.03.13, Anjali Devi B. and Satish Mohabe 3730, 3732 (YVUH); Thamballapalli, Mallaiah Konda hills, Temple premises, N 13°60.025´ E 078°25.424´ alt. 856 m, on bark, 05.01.2013, Satish Mohabe and A. Madhusudhana Redddy 2825, 2832, 2891 (YVUH); Tirumala hills, on bark, 07.02.2013, Anjali Devi B. and Satish Mohabe 3464 (YVUH).

12. Lecanora tropica Zahlbr.-Cat. Lich. Univ. 5: 589. 1928. Plate 6.4F

It is a crustose lichen, found growing on bark, whitish grey to greenish grey, rough, verruculose, to verrucose, continuous to areolate, epruinose, apothecia numerous, crowded, sessile, 0.5 – 1.5 mm diam., disc pale to dark red brown, epruinose, plane to concave, margin thick, verrucose to verruculose, exciple with small and large crystals, epihymenium reddish brown, egranular, lacking crystals, not dissolving in K, thallus contains atranorin, chodatin and zeorin.

Distribution in India: Himachal Pradesh, Karnataka, Madhya Pradesh, Orissa, Sikkim, Tamil Nadu, Uttar Pradesh, Uttarakhand and West Bengal plains. It is a new record for Andhra Pradesh and new addition for Chittoor district.

Specimens examined: Andhra Pradesh, Chittoor district, Talakona, before jungle thrills, N 14°28.698' E 078°42.749' alt. 539 m, on bark of *Mangifera indica*, 16.03.2013, Anjali Devi B. and Satish Mohabe 3680 (YVUH).

LETROUITIACEAE
Letrouitia Hafellner and Bellem.

13. Letrouitia transgressa (Malme) Hafellner & Bellem. in Hafellner-Nova Hedwigia 35: 710. 1983. Plate 6.5C

It is a crustose lichen, found growing on bark of trees, greenish grey to greenish yellow in colour, apothecia scattered, rounded, sessile, constricted at the base, 0.3 - 1.0 mm wide, disc reddish brown to brown, plane to concave margin prominent, yellowish orange, K+ purple, ascus 8 spored, ascospores submuriform, ellipsoid, transversely 7 – 10 septate and vertically 3 - 4 septate.

Distribution in India: Andhra Pradesh, Arunachal Pradesh, Karnataka, Madhya Pradesh, Nagaland, Uttar Pradesh and West Bengal hills. It is a new addition for Chittoor district.

Specimens examined: Andhra Pradesh, Chittoor district, Thamballapalli, Mallaiah Konda hills, Temple premises, N 13°60.025´ E 078°25.424´ alt. 856 m, on bark, 05.01.2013, Satish Mohabe and A. Madhusudhana Reddy 2840 (YVUH).

OCHROLECHIACEAE
Ochrolechia A. Massal.

14. Ochrolechia subpallescens Vers.-Beih. Nova Hedwigia 1: 118. 1962. Plate 6.5D

It is a crustose lichen, found growing on bark of trees, greenish grey to whitish-grey in colour; verruculose to verrucose, central parts rimose cracked, apothecia rounded, sessile, wide, disc open, pale yellow to orange, epruinose, margin thick, lecanorine, ascus 6 - 8 spored, ascospores simple, large, colourless, contains Gyrophoric and lecanoric acid.

Distribution in India: Karnataka and Tamil Nadu. It is a new record for Andhra Pradesh and new addition for Chittoor district.

Specimens examined: Andhra Pradesh, Chittoor district, Talakona, before jungle thrills, N 14°28.698' E 078°42.749' alt. 539 m, on bark of *Mangifera indica*, 16.03.2013, Anjali Devi B. and Satish Mohabe 3635, 3760 (YVUH); Thamballapalli, Mallaiah Konda hills, Temple premises, N 13°60.025´ E 078°25.424´, alt. 900 m, on bark, 05.01.2013, Satish Mohabe and A. Madhusudhana Reddy 2830 (YVUH).

Plate 6.5: Some Prominent Lichens of Chittoor District.

A. *Lepraria coriensis,* **B.** *Leptogium denticulatum,* **C.** *Letrouitia transgressa,* **D.** *Ochrolechia subpallescens,* **E.** *Parmelia sulcata,* **F.** *Parmelinella wallichiana.*

PARMELIACEAE

Bulbothrix Hale

1. Lower side of the thallus pale brown to brown, simple to
    coralloid isidia .... *B. isidiza*

1a. Lower side of the thallus black, isidia simple .... *B. tabacina*

15. Bulbothrix isidiza (Nyl.) Hale-Phytologia 28: 480. 1974. Plate 6.2B

It is a foliose lichen, found growing on bark, greenish grey in colour, adnate, 0.5 - 10.0 cm across, isidiate, isidia simple to coralloid, lobes 0.2 - 0.6 mm wide, bulbate cilia along the margin, lower side pale brown to brown, contains salazinic acid.

Distribution in India: Andhra Pradesh, Arunachal Pradesh, Karnataka, Kerala, Madhya Pradesh, Manipur, Meghalaya, Maharashtra, Nagaland, Sikkim, Tamil Nadu, Uttar Pradesh, Uttarakhand and West Bengal-foot hills.

Specimens examined: Andhra Pradesh, Chittoor district, Horsley Hills, on bark, 24.09.11, Anjali Devi B. 1129, (YVUH); 06.12.11, Anjali Devi B. 1338 (YVUH); Talakona, before jungle thrills, N 14°28.698' E 078°42.749' alt. 539 m, on bark of *Mangifera indica*, 16.03.2013, Anjali Devi B. and Satish Mohabe 3734, 3760 (YVUH).

16. Bulbothrix tabacina (Mont. & Bosch) Hale-Phytologia 28: 481.1974.

It is a foliose lichen, found growing on bark, greyish-green to grey in colour, adnate up to 0.2 - 0.5 cm across, isidiate, isidia simple, lobes 0.2 - 0.3 mm wide, bulbate cilia along margin, lower side jet black, contains atranorin, salazinic and consalazinic acids.

Distribution in India: Nagaland. It is a new record for Andhra Pradesh and new addition for Chittoor district

Specimens examined: Andhra Pradesh, Chittoor district, Tirumala hills, down the way from Tirumala hills, on bark, 15.11.11, Anjali Devi B. 1195 (YVUH).

Canoparmelia Elix & Hale

17. Canoparmelia texana (Tuck.) Elix & Hale-Mycotaxon 27: 279. 1986. Plate 6.3A

It is a foliose lichen, found growing closely attached to the bark, grey to yellowish grey in colour, up to 10 cm in diam., lobes up to 4 mm wide, margin rotund, sorediate, contains atranorin and divaricatic acid.

Distribution in India: Chhattisgarh, Andhra Pradesh, Himachal Pradesh, Jammu and Kashmir, Karnataka, Kerala, Madhya Pradesh, Maharashtra, Manipur, Meghalaya, Nagaland, Sikkim, Tamil Nadu and Uttarakhand.

Specimens examined: Andhra Pradesh, Chittoor district, Horsley Hills, on stem, 04.03.2010, A Madhusudhana Reddy 0807 (YVUH); Talakona, before jungle thrills, N 14°28.698' E 078°42.749' alt. 539 m, on bark of *Mangifera indica*, 16.03.2013, Anjali Devi B. and Satish Mohabe 3672 (YVUH).

Parmelia Ach.

18. Parmelia sulcata Taylor in J. Mackay-Flora Hibern. 2: 145. 1836. Plate 6.5E

It is a foliose lichen, found growing on rock, greenish to dark grey in colour, adnate, 8.0 - 12.0 cm across, white maculate, maculae turning into pseudocyphellae, granular sorediate, soredia along the margins and ridges of pseudocyphalae, lobes sublinear to truncate, 1.5 - 3.5 mm wide, lower side black, rhizines squarrosely branched, apothecia substipate, margin sorediate, contains salazinic and protocetraric acid.

Distribution in India: Andhra Pradesh, Himachal Pradesh, Jammu and Kashmir, Sikkim, Tamil Nadu and Uttarakhand.

Specimens examined: Andhra Pradesh, Chittoor district, Shilathoranam, on rock, 06.12.11, Anjali Devi B. 1336 (YVUH); Horsley Hills, on rock, 12.06.12, A. Madhusudhana Reddy and Sanjeeva Nayaka 2043 (YVUH).

Parmelinella Elix & Hale

19. Parmelinella wallichiana (Taylor.) Elix & Hale-Mycotaxon 29: 242. 1987. Plate 6.5F

It is a foliose lichen, found growing on bark, greenish-grey to dark grey in colour, closely adnate, up to 15 cm across, coriaceous, rugulose in central part densely isidiate, isidia simple to branched, brown to black tiped, ciliate, cilia restricted of lobes axils, lower side black, rhizines in central part, contains atranorin, consalazinic acid and salazinic acid.

Distribution in India: Arunachal Pradesh, Assam, Himachal Pradesh, Karnataka, Kerala, Madhya Pradesh, Manipur, Maharashtra, Meghalaya, Nagaland, Sikkim, Tamil Nadu, Uttarakhand and West Bengal-hills

Specimens examined: Andhra Pradesh, Chittoor district, Talakona, near water fall, on bark, 07.12.11, Anjali Devi B. 1479 (YVUH); Srivari padalu, on bark, 03.01.12, Anjali Devi. B. 1472 (YVUH); on the way to Srivari mettu, on bark, 05.12.11, Anjali Devi. B. 1328 (YVUH); Vedapatashala, on bark, 06.12.11, Anjali Devi B. 1290 (YVUH).

Parmotrema A. Massal.

1.   Thallus sorediate or isidiate 2

1a. Thallus lacking soredia and isidia 12

2.   Thallus isidiate 3

2a. Thallus sorediate 5

3.   Lobes up to 15 mm wide, containing lecanoric acid .... *P. tinctorum*

3a. Lobes up to 6.5 mm wide, lacking lecanoric acid 4

4.   Isidia simple to coralloid, stictic and constictic acid present .... *P. crinitoides*

4a. Isidia simple to filiform, protocetreric acid present .... *P. saccatilobum*

5.   Lobes margin ciliate 6

5a.   Lobes margin eciliate 8

6.   Upper surface maculate or maculae reticulatelly fissured .... *P. reticulatum*

6a.   Upper surface emaculate 7

7.   Soredia farinose, salazinic and consalazinic acid present .... *P. stuppeum*

7a.   Soredia granular, protolichesterinic acid present .... *P. grayanum*

8.   Medulla P- 9

8a.   Medulla P+ orange or orange red 11

9.   Medulla K-, C-, containing fatty acids only .... *P. praesorediosum*

9a.   Medulla K-, C+, containing lecanoric acid 10

10.   Thallus corticolous, loosely attached, large lobed .... *P. austrosinense*

10a. Thallus saxicolous, closely to strongly attached, smaller lobed .... *P. defectum*

11.   Medulla K + yellow turning red, salazinic acid present .... *P. cristiferum*

11a. Medula K-, protocetraric acid present .... *P. ravum*

12.   Lobe margins ciliate 13

12a. Lobe margins eciliate 15

13.   Medulla C+ rose red (Gyrophoric acid) .... *P. eunetum*

13a. Medulla C- 14

14.   Medulla KC-, Protolichesterinic acid present .... *P. melanothrix*

14a. Medulla KC+ pink or red, α-collatolic acids present .... *P. nilgherrense*

15.   Medulla C+ red (Lecanoric acid) .... *P. andinum*

15a. Medulla C- 16

16.   Medulla P+ orange .... *P. latissimum*

16a. Medulla P- .... *P. mesotropum*

20. Parmotrema andinum (Müll.Arg.) Hale-Phytologia 28: 334, 1974.

It is a foliose lichen, found growing on bark of tree and rocks, greenish grey to grey in colour, loosely adnate, 6.0 - 14.0 cm across, lobes ashy grey, 5.0 - 10.0 mm across, eciliate, contains atranorin, lecanoric acid.

Distribution in India: Andhra Pradesh, Himachal Pradesh, Jharkhand, Karnataka, Madhya Pradesh, Odisha, Tamil Nadu and Uttarakhand. It is a new addition for Chittoor district.

Specimens examined: Andhra Pradesh, Chittoor district, Horsley Hills, on rock, 24.09.11, Anjali Devi B. 1132 (YVUH); Talakona, near water fall, on bark, 07.12.11, Anjali Devi B. 1390, 1391 (YVUH); down the way from Tirumala hills, on rock, 09.02.12, Anjali Devi B. 1517, 1525 (YVUH); Tirumala hills, Shilathoranam, on rock, 17.11.11 Anjali Devi B. 1227 (YVUH); 06.12.11, Anjali Devi B. 1357 (YVUH); on rock, 04.01.12, Anjali Devi B. 1428, 1429, 1431, 1434, 1436, 1437, 1441, 1442 (YVUH); on rock, 08.02.12,

Anjali Devi B. 1566, 1567, 1569, 1571, 1573, 1582, 1591 (YVUH); Japali Anjaneya Swamy Temple, on bark, 16.11.11, Anjali Devi B. 1243, 1249, (YVUH); 05.12.11, Anjali Devi B. 1365 (YVUH); Srivari Padalu, on rock, 08.02.12, Anjali Devi B. 1560 (YVUH); Vedapatashala, on rock, 04.01.12, Anjali Devi B. 1449 (YVUH).

21. Parmotrema austrosinense (Zahlbr.) Hale-Phytologia 28: 335. 1974. Plate 6.6A

It is a foliose lichen found growing on bark and rocks, glaucose white to grey surface; loosely attached to the substratum, 0.5 - 10 cm across, lobes rotund 5.0 - 11 mm across, eciliate, soredia marginal, farinose, contains atranorin and lecanoric acid.

Distribution in India: Andhra Pradesh, Assam, Himachal Pradesh, Jammu and Kashmir, Karnataka, Kerala, Madhya Pradesh, Maharashtra, Manipur, Meghalaya, Nagaland, Tamil Nadu and Uttarakhand.

Specimens examined: Andhra Pradesh, Chittoor district, Horsley Hills, on bark, 24.09.11, Anjali Devi B. 1138 (YVUH); Tirumala hills, down the way from tirumala hills, on bark, 09.02.12, Anjali Devi B. 1526 (YVUH); Shilathoranam, on rock, 06.12.11, Anjali Devi B. 1332, 1342, 1346, 1354 (YVUH); 08.02.12, Anjali Devi B. 1581, 1592 (YVUH); on bark, 17.11.11, Anjali Devi B. 1230, 1231 (YVUH).

22. Parmotrema ravum (Krog & Swinscow) Sérus. in Vì zda-Lich. Sel Exs. Fasc. 75 No. 1857. 1983.

It is a foliose lichen found growing on tree trunk, pale yellowish grey in colour, adnate to appressed, 0.7 - 10.0 cm across, emaculate, lobes rotund, 5.0 - 9.0 mm across, eciliate, cracked in centre, lateral margins sorediate, soredia granular, pale grey, contains atranorin, protocetraric and usnic acid.

Distribution in India: Arunachal Pradesh, Kerala, Manipur, Nagaland, Tamil Nadu, Uttarakhand. It is a new record for Andhra Pradesh and new addition for Chittoor district.

Specimens examined: Andhra Pradesh, Chittoor district, Talakona, near water fall, on bark, 07.12.11, A. Madhusudhana Reddy 1481 (YVUH).

23. Parmotrema crinitoides J.C. Wei-Enum. Lich. China: 177. 1991.

It is a foliose lichen, found growing on bark of tree, pale grey in colour, adnate, 0.4 - 0.7 cm across, smooth, emaculate, isidiate, isidia laminal or marginal, simple to coralloid, lobes subirregular to imbricate, up to 5.0 mm wide, margin eciliate, contains atranorin, chloroatranorin, stictic and constictic acid.

Distribution in India: Andhra Pradesh, Himachal Pradesh, Jammu and Kashmir, Karnataka, Kerala, Tamil Nadu and Uttarakhand.

Specimens examined: Andhra Pradesh, Chittoor district, Tirumala hills, Srivari Mettu, on bark, Anjali Devi B. 1331 (YVUH).

Plate 6.6: Some Prominent Lichens of Chittoor District.

A. *Parmotrema austrosinense,* **B.** *P. praesorediosum,* **C.** *P. stuppeum,* **D.** *P. tinctorum,*
E. *Pertusaria melastomella,* **F.** *P. quassiae.*

24. Parmotrema cristiferum (Taylor) Hale-Phytologia 28: 335. 1974.

It is a foliose lichen, found growing loosely attached to bark, pale grey in colour, thallus large, up to 15 cm across, coriaceous, lobes rotund, up to 12 mm wide, sorediate, sorelia marginal to submarginal, lower surface black with shiny, brown, broad erhizinate marginal area, contains atranorin and salazinic acid.

Distribution in India: Andhra Pradesh, Arunachal Pradesh, Assam, Bihar, Jammu and Kashmir, Jharkhand, Karnataka, Kerala, Manipur, Meghalaya, Nagaland, Odisha and West Bengal.

Specimens examined: Andhra Pradesh, Chittoor district, Horsley Hills, on bark, 04.03.2010, A. Madhusudhana Reddy and S. Raja Gopal Reddy 0809 (YVUH); Tirumala hills, down the way from Tirumala hills, on bark, 09.02.12, Anjali Devi B. 1529 (YVUH).

25. Parmotrema defectum (Hale) Hale-Phytologia 28: 335. 1974.

It is a foliose lichen found growing on rocks, grey, shiny towards periphery, adnate and closely attached to the substratum, cracked in centre, up to 7.0 cm, sorediate, sorelia marginal, lobular to linear, revolute, ascending, granular, lobes rotund or irregularly incised, up to 5.0 mm wide eciliate, lower side centrally black, medulla white, contains atranorin, lecanoric acid.

Distribution in India: Andhra Pradesh and Tamil Nadu.

Specimens examined: Andhra Pradesh, Chittoor district, Talakona, on bark, 07/ 12/11, A. Madhusudhana Reddy and B. Anjali Devi 1259 (YVUH); Tirumala hills, down the way from Tirumala hills, on rock, 15.11.11, Anjali Devi B. 1193 (YVUH); Shilathoranam, on rock, 04.01.12, Anjali Devi B. 1440 (YVUH).

26. Parmotrema eunetum (Stirt.) Hale-Phytologia 28: 336. 1974.

It is a foliose lichen found growing on bark of tree trunk, mineral grey, loosely adnate to substratum, 0.4 - 0.8 cm across, smooth, cracked with age, without isidia, soredia and pustules, lobes rotund, margin entire to creanate, ciliate, cilia black, tapparing, medulla white, contains atranorin and gyrophoric acid.

Distribution in India: Kerala and Uttarakhand. It is a new record for Andhra Pradesh and new addition for Chittoor district.

Specimens examined: Andhra Pradesh, Chittoor district, Tirumala Hills, Shilathoranam, on bark, 08.02.12, Anjali Devi B. 1577 (YVUH).

27. Parmotrema grayanum (Hue) Hale-Phytologia 28: 336. 1974.

It is a foliose lichen, found growing on rock and bark, ashy grey to grey brown in colour, adnate to loosely attached to the substratum, 3.5 - 6.5 cm across, marginal soredia in the centre part of the thallus, granular, often with grey-brown tinge, lobes rotund, subimbricate to crowded, narrow, 4.5 - 5.5 mm wide, margin ciliate, cilia dense thick up to 1.5 mm long and medulla contains protolichesterinic and fatty acid.

Distribution in India: Andhra Pradesh, Karnataka, Kerala, Madhya Pradesh, Tamil Nadu and Uttarakhand.

Specimens examined: Andhra Pradesh, Chittoor district, Tirumala hills, Shilathoranam, on rock, 17.11.011, Anjali Devi B. 1228 (YVUH); Srivari Padalu, on Bark, 03.01.12, Anjali Devi B. 1422 (YVUH); 08.02.12, Anjali Devi B. 1549 (YVUH); Vedapatashala, on rock, 04.01.12, Anjali Devi B. 1456 (YVUH).

28. Parmotrema latissimum (Fée) Hale-Phytologia 28: 337. 1974.

It is a foliose lichen found growing on bark and rock, mineral grey to pale grey in colour, loosely attached to the substratum, 12 - 22 cm across, without isidia and soredia, lobes rotund 10 - 18 mm wide, margin eciliate, contains atranorin and salazanic acid.

Distribution in India: West Bengal plains. It is a new record for Andhra Pradesh and new addition for Chittoor district.

Specimens examined: Andhra Pradesh, Chittoor district, Tirumala hills, Srivari Padalu, on bark, 03.01.12, Anjali Devi B. 1423 (YVUH); Shilathoranam, on rock, 06.12.11, Anjali Devi B. 1343 (YVUH); Vedapatashala, on bark, 06.12.11, Anjali Devi B. 1286 (YVUH).

29. Parmotrema melanothrix (Mont.) Hale-Phytologia 28: 337. 1974.

It is a foliose lichen, found growing on bark of tree and rock, mineral grey, loosely attached to the substratum, 5.0 - 15.0 cm across, without isidia, soredia and pustules, lobes rotuned, 6.0 - 9.0 mm wide, margin crenate, short dentate to laciniate, densely ciliate, cilia short, lacking substances.

Distribution in India: Andhra Pradesh, Assam, Himachal Pradesh and Uttarakhand. It is a new addition for Chittoor district.

Specimens examined: Andhra Pradesh, Chittoor district, Horsley hills, on rock, 24.09.11, Anjali Devi B. 1123 (YVUH); Tirumala hills, down the way from Tirumala hills, on rock, 09.02.12, Anjali Devi B. 1523 (YVUH); Shilathoranam, on rock, 17.11.11, Anjali Devi B. 1233 (YVUH); on bark 06.12.11, Anjali Devi B. 1341, 1345, 1356, 1550 (YVUH); on rock, 04.01.12, Anjali Devi B. 1428, 1431, 1433, 1435, 1437, 1438, 1442 (YVUH): on rock, 08.02.12, Anjali Devi B. 1575, 1593, 1601 (YVUH); Srivari Padalu, on rock, 03.01.12, Anjali Devi B. 1419, 1555 (YVUH); 08.02.12, Anjali Devi B. 1553, 1558 (YVUH).

30. Parmotrema mesotropum (Müll. Arg.) Hale-Phytologia 28: 337. 1974.

It is a foliose lichen, found growing on rock, pale grey in colour, up to 6 cm in diam., lacking isidia, soredia and pustules, lobes rotund, up to 4 mm wide, lower surface erhizinate, margin brown, contains atranorin and caperatic acid.

Distribution in India: Andhra Pradesh, Arunachal Pradesh, Himachal Pradesh, Karnataka, Kerala, Madhya Pradesh and Uttarakhand.

Specimens examined: Andhra Pradesh, Chittoor district, Horsley Hills, on rock, 04.03.2010, A. Madhusudhana Reddy 0806A (YVUH); Talakona, near water fall, on bark, 07.12.11, Anjali Devi B. 1269 (YVUH); Tirumala hills, down the way from Tirumala hills, on rock, 15.11.11, Anjali Devi B. 1195 (YVUH); 09.02.12, A. Madhusudhana Reddy, 1509, 1510, 1511 (YVUH); Japali Anjaneya Swamy Temple, on rock, 16.11.11, Anjali Devi B. 1242 (YVUH); Shilathoranam, on rock, 08.02.12, Anjali Devi B. 1599 (YVUH); Srivari Mettu, on rock, 05.12.11, Anjali Devi B. 1310 (YVUH); Vedapatashala, on rock, 04.01.12, Anjali Devi B. 1444 (YVUH).

31. Parmotrema nilgherrense (Nyl.) Hale-Phytologia 28: 338. 1974.

It is a foliose lichen, found growing on bark and rock, mineral to ashy grey in colour, loosely attached to the substratum, coriaceous, 9.0 - 14.0 cm across, smooth sometimes rugose in older parts, maculate, lacking isidia, soredia and pustules, lobes plane to convoluted, rotund, 10 - 20 mm wide, margin ascending imbricate, entire to crenate-dentate, ciliate, cilia simple to furcate, 1.5 - 2.5 mm long, contains atranorin, alectoronic and alpha-collatolic acid.

Distribution in India: Andhra Pradesh, Assam, Himachal Pradesh, Jammu and Kashmir, Kerala, Manipur, Meghalaya, Nagaland, Sikkim, Tamil Nadu, Uttarakhand and West Bengal–hills.

Specimens examined: Andhra Pradesh, Chittoor district, Horsley hills, on bark, 24.09.11, Anjali Devi B. 1124 (YVUH); On the way near check post, 4th curve, on rock, 882 m, 12.06.12, A. Madhusudhana Reddy, Sanjeeva Nayaka and Anjali Devi B. 2036 (YVUH); Talakona, near water fall, on bark, 07.12.11, Anjali Devi B. 1479 (YVUH); Tirumala hills, Srivari padalu, on bark, 08.02.12, Anjali Devi B. 1562 (YVUH); Shilathoranam, on rock, 08.02.12, Anjali Devi B., 1602 (YVUH).

32. Parmotrema praesorediosum (Nyl.) Hale-Phytologia 28: 338. 1974. Plate 6.6B

It is a foliose lichen, found growing on bark, mineral grey, yellowish grey to dark grey in colour, up to 8 cm across, lobes rotund, up to 8 mm wide, sorediate, lower surface black with shiny, brown to mottled, erhizinate margin, contains atranorin, fatty acids.

Distribution in India: Andhra Pradesh, Assam, Himachal Pradesh, Jammu and Kashmir, Karnataka, Kerala, Madhya Pradesh, Maharashtra, Manipur, Nagalnd, Orissa, Rajasthan, Tamil Nadu, Uttarakhand, Uttar Pradesh and West Bengal

Specimens examined: Andhra Pradesh, Chittoor district, Horsley Hills, on bark, 04.03.2010, A Madhusudhana Reddy 0807C (YVUH); Talakona, near water fall, on bark, 07.12.11, Anjali Devi B. 1251, 1254, 1392,1397, 1481 (YVUH); on bark, 09.02.12, Anjali Devi B. 1520, 1534 (YVUH); Tirumala hills, down the way from Tirumala hills, on bark, 15.11.11, Anjali Devi B. 1194, 1195, 1213 (YVUH); on rock, 09.02.12, Anjali Devi B. 1512, 1513, 1514, 1518, 1519,1522, 1524, 1532 (YVUH); on bark, 09.02.12, Anjali Devi B. 1520 (YVUH); Tirumala hills, Japali Anjaneya Swamy Temple, on bark, 16.11.11, Anjali Devi B. 1220, 1224, 1226, 1240, 1244 (YVUH); 05.12.11, Anjali Devi B. 1362, 1363 (YVUH); Papavinasanam, on bark, 15.11.11, Anjali Devi. B. 1185 (YVUH); Shilathoranam, on rock, 06.12.11, Anjali Devi B. 1301 (YVUH); 08.02.12, Anjali Devi B. 1578, 1587, 1590, 1594, (YVUH); on bark, 08.02.12, Anjali Devi B. 1584, 1585 (YVUH); on rock, 06.12.11, Anjali Devi B. 1275, 1304, 1307,1337, 1351, 1351,

1359, 1352, (YVUH); Srivari Mettu, on bark, 05.12.11, Anjali Devi B. 1327 (YVUH); Vedapatashala, on rock, 04.01.12, Anjali Devi B. 1443, 1445, 1452, 1455,1463, (YVUH); 06.12.11, Anjali Devi B. 1270, 1273 (YVUH).

33. Parmotrema reticulatum (Taylor) Choisy-Bull. Mens. Soc. Linn. Soc. Bot. Lyon. 21: 175. 1952.

It is a foliose lichen, found growing on bark and rock, grey to grey green in colour, loosely attached, up to 13 cm in diam., sorediate, upper surface reticulately maculate or cracked, lobes subrotund to laciniate, imbricate, up to 12 mm wide, margin ciliate, lower surface black with narrow, brown, erhizinate margin, contains atranorin and salazinic acid.

Distribution in India: Andhra Pradesh, Arunachal Pradesh, Assam, Himachal Pradesh, Jammu and Kashmir, Karnataka, Kerala, Madhya Pradesh, Manipur, Maharashtra, Meghalaya, Nagaland, Sikkim, Tamil Nadu, Uttarakhand and West Bengal.

Specimens examined: Andhra Pradesh, Chittoor district, Horsley Hills, on bark, 04.03.2010, A. Madhusudhana Reddy and B. Ravi Prasad Rao 0805 (YVUH); on rock, 04.03.2010 A. Madhusudhana Reddy 0801 (YVUH); Tirumala hills, down the way from Tirumala hills, on bark, 15.11.11, Anjali Devi B. 1511 (YVUH); on rock, 09.02.12, Anjali Devi B. 1533 (YVUH); Shilathoranam, on rock, 06.12.11, ADB. 1355 (YVUH); Tirumala hills, Dharmagiri, on bark, 03.01.12, ADB. 1415, 1411 (YVUH); Srivari padalu, on rock, 03.01.12, Anjali Devi B. 1426 (YVUH); on bark, 03.01.12, Anjali Devi B. 1420, 1472 (YVUH); Srivari padalu, 08.02.12, ADB. 1551 (YVUH); Vedapatashala, on bark, 04.01.12, Anjali Devi B. 1460 (YVUH); on rock, 1283 (YVUH); 08.02.12, Anjali Devi B. 1542, 1546 (YVUH); on bark, 06.12.11, Anjali Devi B. 1288 (YVUH); on rock, 03.01.12, Anjali Devi B. 1427 (YVUH).

34. Parmotrema saccatilobum (Taylor) Hale-Phytologia 28: 339, 1974

It is a foliose lichen, found growing on bark, mineral grey to grey, closely adnate to the substratum, 4.5 - 7.5 cm across, emaculate, reticulately cracked in centre, isidiate, isidia laminal, simple, granular to filiform, rarely branched, often black tiped, lobes rotund 4.5 - 6.5 mm wide, margin eciliate, contains atranorin and protocetraric acid.

Distribution in India: Andaman and Nicobar Islands, Assam, Goa, Kerala, Maharashtra, Nagaland, Sikkim, Tamil Nadu, Uttarakhand and West Bengal plains. It is a new record for Andhra Pradesh and new addition for Chittoor district.

Specimens examined: Andhra Pradesh, Chittoor district, Tirumala hills, down the way from Tirumala hills, on bark, 09.02.12, Anjali Devi B. 1529 (YVUH); 15.11.11, Anjali Devi B. 1196 (YVUH).

35. Parmotrema stuppeum (Taylor) Hale-Phytologia 28: 339, 1974. Plate 6.6C

It is a foliose lichen, found growing on bark of trees, mineral grey to grey in colour, loosely adnate to the substratum, 9.0 -14.0 cm across, smooth, emaculate, older part reticulately cracked, sorediate, soredia farinose, on margin or submargins, lobes rotund 10 - 15 mm wide, margins subascending, irregularly crenate-dentate, ciliate, cilia sparse to dense, simple, 1.5 to 2.5 mm long, contains atranorin, salazinic acid and consalazinic acid.

Distribution in India: Andhra Pradesh, Kerala, Nagaland and Tamil Nadu.

Specimens examined: Andhra Pradesh, Chittoor district, Horsley Hills, on bark, 24.09.11, Anjali Devi B. 1117 (YVUH); Tirumala hills, Vedapatashala, on bark, 08.02.12, Anjali Devi B. 1544 (YVUH).

36. Parmotrema tinctorum (Despr. ex Nyl.) Hale-Phytologia 28: 339. 1974. Plate 6.6D

It is a foliose lichen, found growing loosely on bark, whitish to mineral grey in colour, shining, loosely attached to the substratum, up to 10 cm across, isidiate, isidia branched and brown tipped, lobes rotund, up to 15 mm wide, lower surface black with wide bare zone, brown, erhizinate margin, contains atranorin and lecanoric acid.

Distribution in India: Andhra Pradesh, Arunachal Pradesh, Assam, Chhattisgarh, Himachal Pradesh, Jammu and Kashmir, Jharkhand, Karnataka, Kerala, Madhya Pradesh, Maharashtra, Manipur, Meghalaya, Nagaland, Orissa, Rajasthan, Sikkim, Tamil Nadu, Uttarakhand and West Bengal

Specimens examined: Andhra Pradesh, Chittoor district, Horsley Hills, on bark, 04.03.2010, A. Madhusudhana Reddy 0803 (YVUH); 24.09.11, Anjali Devi B. 1118 (YVUH); Japali Anjaneya Swamy Temple, on bark, 16.11.11, Anjali Devi B. 1246, 1247 (YVUH); on rock, 16.11.11, Anjali Devi B. 1225 (YVUH); Talakona, near water fall, on bark, 07.12.11, Anjali Devi B. 1394, 1482 (YVUH); Tirumala hills, down the way from Tirumala hills, on bark, 09.02.12, Anjali Devi B. 1527 (YVUH); 15.11.11, Anjali Devi B. 1195, 1197, 1211 (YVUH); Tirumala hills, Dharmagiri, on rock, 03.01.12, Anjali Devi B. 1410 (YVUH); on bark, 03.01.12, Anjali Devi B. 1412, 1413, 1416 (YVUH); Srivari Padalu, on rock, 08.02.12, Anjali Devi B. 1559 (YVUH); on bark, 08.02.12, Anjali Devi B. 1561, 1563 (YVUH); Vedapatashala, on bark, 06.12.11, Anjali Devi B. 1271, 1277 (YVUH); 04.01.12, Anjali Devi B. 1448, 1458, 1459, 1462 (YVUH); on rock, 04.01.12, Anjali Devi B. 1457 (YVUH); on bark, 08.02.12, Anjali Devi B. 1541, 1543 (YVUH).

Usnea Dill. ex Adans.

37. Usnea baileyi (Stirt.) Zahlbr.-Denkschr. Kaiserl. Akad. Wiss., Wien. Math. Naturwiss. Kl. 83: 182. 1909. Plate 6.8E

It is a fruticose lichen, found growing on bark, greenish grey to brown in colour, sub- erect to pendulous, dichotomously to subsympodially branched, main branches up to 2 mm wide, pseudocyphellate and isidiate, isidia dense on cortex or along margin of pseudocyphellae, periaxial part of medulla yellowish red, centrally hollow, contains norstictic and salazinic acid.

Distribution in India: Arunachal Pradesh, Assam, Kerala, Manipur, Meghalaya, Nagaland, Sikkim, Tamil Nadu and West Bengal –hills. It is a new record for Andhra Pradesh and new addition for Chitoor district.

Specimens examined: Andhra Pradesh, Chittoor district, Tirumala hills, Shilathoranam, on bark, 07.02.2013 Anjali Devi B. 1604 (YVUH).

Xanthoparmelia (Vain.) Hale

38. Xanthoparmelia pseudocongensis Hale-Mycotaxon 30: 327. 1987. Plate 6.8F

It is a foliose lichen, found growing on exposed and shady rocks, yellowish green to yellowish grey in colour, closely attached to the substratum, 3.5 - 7.5 cm across, isidiate, isidia cylindrical, simple, black tipped, sometimes branched, lobes sublinear, lower side black, rhizinate, thallus contains usnic, constictic and stictic acids.

Distribution in India: Andhra Pradesh, Madhya Pradesh and Rajasthan.

Specimens examined: Andhra Pradesh; Chittoor district, Japali Anjaneya Swamy temple, on rock, 13.06.12, A. Madhusudhana Reddy and Sanjeeva Nayaka 1801(YVUH).

PELTULACEAE

Peltula Nyl.

1.   Thallus grey, circular to irregular .... *P. farinosa*

1a. Thallus brownish to olive green, areolate to placodioid .... *P. placodizans*

39. Peltula farinosa Büdel-Cryptogamic Bot., 4: 262–269. 1994.

It is a squamulose lichen, found growing on exposed rock, circular to irregular in shape, 1.5 - 10.0 mm in size, strikingly grey in colour and pruinose thallus, undulating and sorediate margins, and orange to orange–brown lower surface.

Distribution in India: It is known only from Andhra Pradesh.

Specimens examined: Andhra Pradesh, Chittoor district, Horsley Hills, on rock, 12.06.12, Sanjeeva Nayaka and A. Madhusudhana Reddy 1702 (YVUH).

40. Peltula placodizans (Zahlbr.) Wetmore-Ann. Mo. Bot. Gdn. 57: 196, 1971.

It is a squamulose lichen, found growing on exposed rocks, areolate to placodioid, lobate at margins, brownish to olive green in colour, sorediate, soredia farinose, black, lower surface paler than upper surface, apothecia one per squamule, restricted in central squamules, immersed, disc punctiform, yellowish brown to brownish black, up to 0.3 mm in diam., ascus multispored, ascospores globose to ellipsoid.

Distribution in India: Andhra Pradesh, Madhya Pradesh and Uttar Pradesh. It is new addition for Chittoor district.

Specimens examined: Andhra Pradesh, Chittoor district, Horsley Hills, on rock, on rock, N 13°.39.123′ E 078° 34.117′, alt. 1265 m, 08.03.2014, Anjali Devi B. and Satish Mohabe 3973 (YVUH).

PERTUSARIACEAE

Pertusaria DC.

1.   Thallus saxicolous, sorediate .... *P. leucosora*

1a.   Thallus corticolous, esorediate 2

2.   Ascus 6 - 8 spored, lacking lichen substances .... *P. melastomella*

2a.   Ascus 2 - 4 spored, containing lichen substances 3

3.   Fertile verrucae conical to subhemispherical .... *P. pustulata*

3a.   Fertile verrucae tuberculed-verrucose .... *P. quassiae*

41. Pertusaria leucosora Nyl.-Flora 60: 223. 1877.

It is a crustose lichen, found growing on rock, grey to brownish grey in colour, rimose areolate, sorediate, soredia granular, lacking apothecia, contains atranorin.

Distribution in India: Andhra Pradesh, N. W. Himalaya, Karnataka, Kerala, Madhya Pradesh, Sikkim and Tamil Nadu.

Specimens examined: Andhra Pradesh, Chittoor district, Horsley Hills, on rock, 24.09.11, Anjali Devi B. 1132 (YVUH); Shilathoranam, on the back side of the arch, 10 ft away, alt. ca. 938.4 m, on rock, 13.06.12, A. Madhusudhana Reddy and Sanjeeva Nayaka 2012 (YVUH); 07.02.2013, Anjali Devi B. and Satish Mohabe, 3975 (YVUH).

42. Pertusaria melastomella Nyl.-Acta Soc. Sci. Fenn. 26(10): 16. 1900. Plate 6.6E

It is a crustose lichen, found growing on bark of tree, whitish-grey or greenish-grey, smooth to verrucose, apothecia perithecioid, 1 – 2 per verrucae, varrucae not constricted at base, ostioles 1 – 2 per verrucae, ascus 6 – 8 spored, ascospores large, elongate to ellipsoidal, double walled, smooth, non-costulate, lacking lichen compounds.

Distribution in India: Andhra Pradesh, Himachal Pradesh, Madhya Pradesh and Tamil Nadu.

Specimens examined: Andhra Pradesh, Chittoor district, Horsley Hills, on bark, 12.06.2012, A. Madhusudhana Reddy, Sanjeeva Nayaka and Anjali Devi B 2009 (YVUH).

43. Pertusaria pustulata (Ach.) Duby, Bot. Gall. 2(2): 673. 1830.

It is a crustose lichen, found growing on rock, greenish grey to greyish brown in colour, verruculose to verrucose, fertile verrucae conical to subhemispherical, ostiloles 1 – 4 per verruca, ascus 2 – 4 spored, ascospores simple, colourless, large, double walled, sometimes inner wall radially costulate, contains norstictic, stictic, constictic acids.

Distribution in India: Madhya Pradesh, Manipur, Karnataka, Sikkim and Tamil Nadu. It is a new record for Andhra Pradesh and new addition for Chittoor district.

Specimens examined: Andhra Pradesh, Chittoor district, Tirumala hills, Vedapatashala, on bark, 04.01.12, Anjali Devi B. 1453 (YVUH).

44. Pertusaria quassiae (Fée) Nyl.-Ann. Sci. Nat., Bot., ser. 4, 15: 45. 1861. Plate 6.6F

It is a crustose lichen, found growing on bark, whitish or greenish grey in colour, verruculose to verrucose, fertile verrucae and ostiolar region tuberculed, ascus 2 – 4 spored, ascospores simple, large, both the walls or only the inner wall radially costulate, contains norstictic, stictic and constictic acids.

Distribution in India: Andhra Pradesh, Arunachal Pradesh, Andaman and Nicobar Islands, Andhra Pradesh, Himachal Pradesh, Jammu and Kashmir, Karnataka, Madhya Pradesh, Maharashtra, Nagaland and Sikkim. It is a new addition for Chittoor district.

Specimens examined: Andhra Pradesh, Chittoor district, Horsley Hills, on bark, 12.06.2012, A. Madhusudhana Reddy, Sanjeeva Nayaka and Anjali Devi B 2012 (YVUH).

PHYSCIACEAE

Buellia De Not.

45. Buellia hemispherica S.R. Singh & D.D. Awasthi-Biol. Mem. 6(2): 186. 1981. Plate 6.2A

It is a crustose lichen, found growing on siliceous rock, whitish to yellowish or yellowish-grey in colour, rimose-areolate, UV+ yellow, apothecia adnate to sessile, less than 0.5 mm, disc black, convex, lecidine, internal stipe brown, K+ red crystals, ascus 8 spored, ascospores brown, 2 celled, spore wall smooth, contains norstictic acid.

Distribution in India: Madhya Pradesh, Orissa and Tamil Nadu. It is a new record for Andhra Pradesh, new addition for Chittoor distrcit and endemic to India.

Specimens examined: Andhra Pradesh, Chittoor district, Tirumala Hills, Shilathoranam, on Rock, N 13°41.282′ E079°20.418′ alt. 937 m, 07.02.2013, Anjali Devi B. and Satish Mohabe 3361 (YVUH).

Dirinaria (Tuck.) Clem.

1.   Soredia on isidioid verrucae ....*D. aegialita*

1a. Soredia directly on lamina .... *D. applanata*

46. Dirinaria aegialita (Afz. in Ach.) Moore-Bryologist 71: 248. 1968.

It is a foliose lichen, found growing closely attached to the bark, grey in colour, up to 5 cm diam., sorediate, soredia on isidioid verrucae, lobes small, up to 1.5 mm broad, lower surface erhizinate, contains divaricatic acid.

Distribution in India: Andhra Pradesh, Andaman and Nicobar Islands, Arunachal Pradesh, Chhattisgarh, Kerala, Madhya Pradesh, Odisha, Sikkim, Tamil Nadu and West Bengal.

Specimens examined: Andhra Pradesh, Chittoor district, Horsley Hills, on bark, 04.03.2010, A. Madhusudhana Reddy 0811A (YVUH).

47. Dirinaria applanata (Fee) D. D. Awasthi in D.D. Awasthi & M.R. Agarwal-J. Indian Bot. Soc. 49: 135. 1970. Plate 6.3E

It is a foliose lichen, found growing closely attached to the bark, greyish white in colour, up to 3.0 - 6.0 cm diam., sorediate, soredia capitate to granular; lobes large flabellate, 1.3 - 1.8 mm broad, contains divaricatic acid, triterpenoids.

Distribution in India: Andhra Pradesh, Andaman and Nicobar Islands, Karnataka, Madhya Pradesh, Maharashtra, Nagaland, Sikkim, Tamil Nadu, Uttar Pradesh, Uttarakhand and West Bengal -plains

Specimens examined: Andhra Pradesh, Chittoor district, Talakona, before jungle thrills, N 14°28.698' E 078°42.749' alt. 539 m, on bark of *Mangifera indica*, 16.03.2013, Anjali Devi B. and Satish Mohabe 3642, 3677, 3680, 3737, 3759 (YVUH).

Hafellia Kalb and al.

48. Hafellia curatellae (Malme) Marbach-Biblioth. Lichenol. 74: 255. 2000.

It is a crustose lichen, found growing on bark, grey in colour, apothecia rounded, disc black, hypothecium brown, hymenium inspersed with oil globules, ascus 8 spored, ascospores brown, 2-celled, wall uniformly thickened, ornamented on surface, contains norstictic acid.

Distribution in India: Andhra Pradesh and Tamil Nadu.

Specimens examined: Andhra Pradesh, Chittoor district, Horsley Hills, on bark, 04.03.2010, A. Madhusudhana Reddy 0811C (YVUH). Talakona, before jungle thrills, N 14°28.698' E 078°42.749' alt. 539 m, on bark of *Mangifera indica*, 16.03.2013, Anjali Devi B. and Satish Mohabe 3681 (YVUH).

Heterodermia Trevis.

1.   Thallus corticated on both sides 2

1a. Thallus corticated only on upper side .... *H. obscurata*

2.   Thallus isidiate or sorediate 3

2a. Thallus lacking isidia and soredia .... *H. diademata*

3.   Thallus isidiate, lacking soredia 4

3a. Thallus sorediate, lacking isidia 5

4.   Thallus with cylindrical to coralloid isidia .... *H. isidiophora*

4a. Thallus with marginal subsidial lobules .... *H. dissecta*

5.   Soralia continuous along margin of lobes, lobe tips esorediate .... *H. albicans*

5a. Soralia capitate to labriform on main and lateral lobes .... *H. speciosa*

49. Heterodermia albicans (Pers.) Swinsc. & Krog-Lichenologist 8: 113. 1976.

It is a foliose lichen, found growing on bark, greyish white in colour, up to 10 cm in diam., lobes up to 4 mm wide, stretched, corticated on both the surfaces, sorediate, soredia only in lateral margins, contains zeorin and salazinic acid.

Distribution in India: Andhra Pradesh, N. W. Himalayas, Maharashtra and Tamil Nadu

Specimens examined: Andhra Pradesh, Chittoor district, Horsley Hills, on bark, 04.03.2010, A. Madhusudhana Reddy 0810 (YVUH).

50. Heterodermia diademata (Taylor) D.D. Awasthi-Geophytology 3: 113. 1973.

It is a foliose lichen, found growing on rock, whitish-grey to grey in colour, up to 12 cm in diam., corticated on both side, lobes linear, stretched, up to 2.5 mm wide, lacks isidia and soredia, contains atranorin and zeorin.

Distribution in India: Andhra Pradesh, Arunachal Pradesh, Assam, Himachal Pradesh, Jammu and Kashmir, Karnataka, Kerala, Madhya Pradesh, Maharashtra, Manipur, Meghalaya, Nagaland, Rajasthan, Sikkim, Tamil Nadu, Uttarakhand and West Bengal hills.

Specimens examined: Andhra Pradesh, Chittoor district, Horsley Hills, on rock, 04.03.2010, A. Madhusudhana Reddy and B. Ravi Prasad Rao 0802 (YVUH); on bark, 24.09.11, Anjali Devi B. 1140 (YVUH); Tirumala hills, Dharmagiri, on bark, 03.01.12, A. Madhusudhana Reddy 1403 (YVUH).

51. Heterodermia dissecta (Kurok.) D. D. Awasthi-Geophytology 3: 113. 1973.

It is a foliose lichen, found growing on rock, up to 8 cm diam., lobes linear, 1.5 – 1.8mm wide, corticated on both the surface, margins microphyllous with subisidial lobules, contains zeorin, norstictic, salazinic acids and unknown pigments.

Distribution in India: Andhra Pradesh, Himachal Pradesh, Karnataka, Kerala, Madhya Pradesh, Manipur, Nagaland, Sikkim, Tamil Nadu, Uttarakhand and West Bengal–hills.

Specimens examined: Andhra Pradesh, Chittoor district, Tirumala hills, Vedapatashala, on rock, 06.12.11, A. Madhusudhana Reddy 1272 (YVUH).

52. Heterodermia isidiophora (Vain.) D.D. Awasthi-Geophytology 3: 114. 1973.

It is a foliose lichen, found growing on bark and rock, greyish white in colour, closely attached to the substratum, up to 8 cm in diam., corticated on both the surface, lobes linear, stretched, up to 2.5 mm wide, densely isidiate, contains atranorin and zeorin.

Distribution in India: Andhra Pradesh, Kerala, Maharashtra, Manipur, Nagaland, Tamil Nadu and West Bengal–hills.

Specimens examined: Andhra Pradesh, Chittoor district, Horsley Hills, on bark, 04.03.2010, A. Madhusudhana Reddy and S. Raja Gopal Reddy 0813 (YVUH); Tirumala hills, down the way from Tirumala hills, on bark, 15.11.11, Anjali Devi B.

1204 (YVUH); on bark, 09.02.12, Anjali Devi B. 1537 (YVUH); on rock, 09.02.12, Anjali Devi B. 1537 (YVUH); Tirumala hills, Dharmagiri, on rock, 03.01.12, Anjali Devi B. 1408 (YVUH); on bark, 03.01.12, Anjali Devi B. 1402, 1406, 1414 (YVUH); Japali Anjaneya Swamy Temple, 05.12.11, Anjali Devi B. 1360 (YVUH); Shilathoranam, on bark, 06.12.11, Anjali Devi B. 1348 (YVUH); Srivari Padalu, on rock, 03.01.12, Anjali Devi B. 1421 (YVUH); Vedapatashala, on bark, 06.12.11, Anjali Devi B. 1285 (YVUH).

53. Heterodermia obscurata (Nyl.) Trevis.-Nouvo Giorn Bot. Ital 1: 114. 1869.

It is a foliose lichen, found growing on bark, greyish to dark grey in colour, closely attached to the substratum, up to 12 cm diam., lobe apices curved, labriform sorediate, lobes 1.0 - 1.8 mm wide, corticated only on upper side, lower side deep yellow, contains zeorin and unknown pigments.

Distribution in India: Andhra Pradesh, Arunachal Pradesh, Himachal Pradesh, Jammu and Kashmir, Kerala, Maharashtra, Manipur, Nagaland, Sikkim, Tamil Nadu, Uttarakhand and West Bengal – hills. It is a new record for Andhra Pradesh and new addition for Chittoor district.

Specimens examined: Andhra Pradesh, Chittoor district, Tirumala hills, Srivari padalu, on bark, 08.02.12, Anjali Devi B. 1552 (YVUH).

54. Heterodermia pseudospeciosa (Kurok.) W.L. Culb.-Bryologist 69: 484. 1966. Plate 6.3F

It is a foliose lichen, found growing on bark and rock, greyish white in colour, up to 6 - 12 cm diam., lobes plane, not ascending, 1.3 mm wide, corticated on both the surface, sorediate, soralia capitate to labriform, contains zeorin.

Distribution in India: Andhra Pradesh, West Bengal-hills, Maharashtra.

Specimens examined: Andhra Pradesh, Chittoor district, Tirumala hills, Srivari mettu, on bark, 05.12.11, Anjali Devi B. 1318, 1326, 1330 (YVUH); Vedapatashala, on rock, 06.12.11, Anjali Devi B. 1289, 1291 (YVUH).

Hyperphyscia Müll. Arg.

1.   Medulla white, K- (traces of skyrin) .... *H. adglutinata* var. *adglutinata*

1a.   Medulla orange-red, K+ violet (erythrin) .... *H. adglutinata* var. *pyrithrocardia*

55. Hyperphyscia adglutinata var. **adglutinata** (Flörke) H. Mayrhofer & Poelt in Hafellner & al.- Herzogia 5: 62 1979. Plate 6.4A

It is a foliose lichen, found growing closely adnate on bark and rocks, greenish grey to greenish brown in colour, smooth on surface, sorediate, sorelia maculiform, capitate or globose, medulla white, contains traces of skyrin.

Distribution in India: Himachal Pradesh, Jammu and Kashmir, Madhya Pradesh and Uttarakhand. It is a new record for Andhra Pradesh and new addition for Chittoor district.

Specimens examined: Andhra Pradesh, Chittoor district, Horsley Hills, on the way, near 4th curve, alt. ca. 882 m, on rock, 12.06.12, A. Madhusudhana Reddy, Sanjeeva Nayaka and Anjali Devi B. 2041 (YVUH); Talakona, before jungle thrills, N 14°28.698' E 078°42.749' alt. 539 m, on bark of *Mangifera indica,* 16.03.2013, Anjali Devi B. and Satish Mohabe 3717, 3756, 3759 (YVUH).

56. Hyperphyscia adglutinata var. **pyrithrocardia** (Müll. Arg.) D.D. Awasthi-Comp. Macrolich. India, Nepal & Sri Lanka : 197. 2007.

It is a foliose lichen, found growing closely adnate on tree, greenish grey to greenish brown in colour, smooth on surface, sorediate, soralia maculiform, capitate or globose, medulla white and intermittently orange-red, K+ violet, contains skyrin and erythrin.

Distribution in India: Madhya Pradesh and Tamil Nadu. It is a new record for Andhra Pradesh and new addition for Chittoor district.

Specimens examined: Andhra Pradesh, Chittoor district, Talakona, before jungle thrills, N 14°28.698′ E 078°42.749′ alt. 539 m, on bark of *Mangifera indica,* 16.03.2013, Anjali Devi B. and Satish Mohabe 6362 (YVUH).

Phaeophyscia Moberg
57. Phaeophyscia hispidula (Ach.) Moberg-Bot. Not. 131: 260. 1978.

It is a foliose lichen, found growing adnate to the bark, grey to darker in colour, sorediate, soredia laminal, capitate, sometimes extending up to margin, lower side black, rhizinate, rhizines black, long, projecting beyond the lobes, medulla white, lacking lichen substances.

Distribution in India: Andhra Pradesh Arunachal Pradesh, Himachal Pradesh, Jammu and Kashmir, Madhya Pradesh, Maharashtra, Manipur, Nagaland, Rajasthan, Sikkim, Tamil Nadu and Uttarakhand.

Specimens examined: Andhra Pradesh, Chittoor district, Horsley Hills, on bark, 24.09.11, Anjali Devi B. 1120 (YVUH).

Physcia (Schreb.) Michx.
1.	Thallus sorediate, soredia marginal, expanding	2
1a.	Thallus lacking soredia and isidia	*Physcia* sp.
2.	Lower cortex prosoplectenchymatous	*P. abuensis*
2a.	Lower cortex paraplectenchymatous	*P. tribacoides*

58. Physcia sp.
It is a foliose lichen, found growing on bark, whitish grey in colour, loosely attached, delicate, lobes linear, rotund, up to 5 mm wide, pruinose, lacking isidia and soredia, containg atranorin.

Distribution in India: Andhra Pradesh.

Specimens examined: Andhra Pradesh, Chittoor district, Horsley Hills, on bark, 04.03.2010, A. Madhusudhana Reddy 0804 (YVUH).

59. Physcia abuensis D.D. Awasthi & S.R. Singh-Norweg. J. Bot. 26(2): 93. 1979. Plate 6.7A

It is a foliose lichen, found growing closely adnate to the rock and bark, whitish-grey to greenish grey in colour, densely pruinose, marginally sorediate, soredia expending to laminal region, cortex prosoplectenchymatous, lower side greyish, contains zeorin and unknown substances.

Distribution in India: Andhra Pradesh, Madhya Pradesh and Rajasthan. It is a new addition for Chittoor district and endemic to India.

Specimens examined: Andhra Pradesh, Chittoor district, Tirumala hills, Shilathoranam, on rock, N 13°41.282′ E 079°20.418′, 07.02.2013, Anjali Devi B. and Satish Mohabe 3301 (YVUH).

60. Physcia tribacoides Nyl.-Flora 57: 307. 1874. Plate 6.7B

It is a foliose lichen, found growing on bark, grey in colour; up to 5 cm across, lobes rotund, crenate at tips, pruinose, sorediate, soralia marginal, lower side grey, ascospores brown, 2 celled, contains atranorin and zeorin.

Distribution in India: Andhra Pradesh, Arunachal Pradesh, Himachal Pradesh, Madhya Pradesh, Maharashtra, Manipur, Nagaland, Sikkim and Tamil Nadu.

Specimens examined: Andhra Pradesh, Chittoor district, Horsley Hills, on bark, 04.03.2010, A. Madhusudhana Reddy and B. Ravi Prasad Rao 0808 (YVUH).

Pyxine Fr.

1. Thallus isidiate, sorediate or pseudocyphellate 2
1a. Thallus lacking isidia, soredia or pseudocyphellate 3
2. Thallus isidiate, isidia nodular to dactylate .... *P. punensis*
2a. Thallus pseudocyphellate-sorediate .... *P. subscinerea*
3. Medulla P+ yellow, lobes smaller ....*P. minuta*
3a. Medulla P- 4
3. Internal stipe brown, K+ violet .... *P. petricola* var. *petricola*
4a. Internal stipe colourless, K- ....*P. petricola* var. *pallida*

61. Pyxine minuta Vain.-Acta Soc. Fauna Fl. Fenn. 7(1): 156. 1890

It is a foliose lichen, found growing closely on rock, yellow to brownish grey in colour, apices olivaceous brown, subareolate to eventually evanescent in central part, lacking isidia and soredia, rarely maculate, lobes up to 1 mm wide, medulla white, ascospores 2 celled, contains lichenoxanthon and triterpenes.

Distribution in India: Karnataka, Tamil Nadu and Uttarakhand. It is a new record for Andhra Pradesh and new addition for Chittoor district.

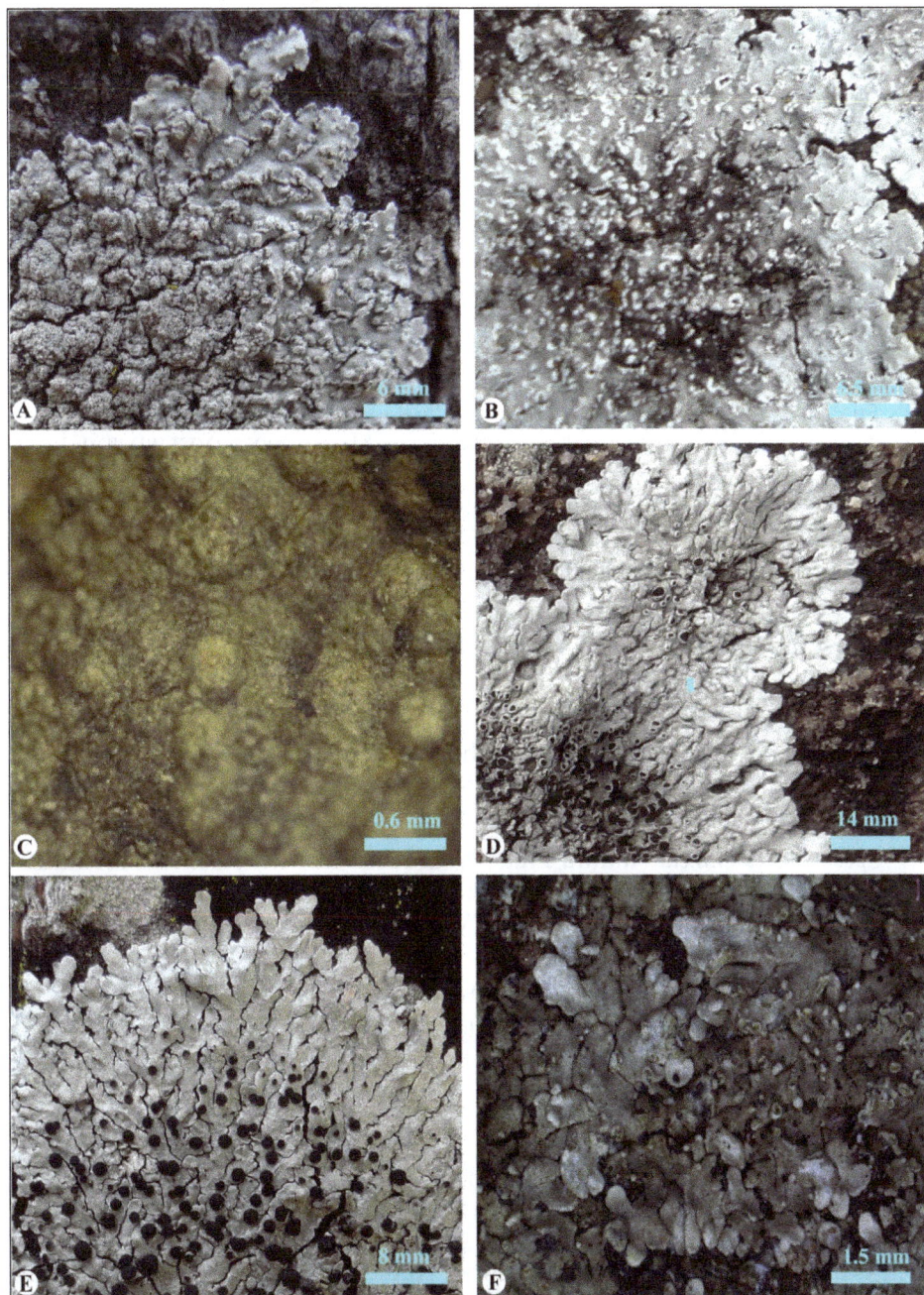

Plate 6.7: Some Prominent Lichens of Chittoor District.

A. *Physcia abuensis,* **B.** *P. tribacoides,* **C.** *Porina tetracerae,* **D.** *Pyxine petricola* **var.** *pallida,* **E.** *P. petricola* **var.** *petricola,* **F.** *P. punensis.*

Specimens examined: Andhra Pradesh, Chittoor district, Horsley Hills, near checkpost 4th curve, alt. ca. 882 m, on rock, 12.06.12, A. Madhusudhana Reddy, Sanjeeva Nayaka and Anjali Devi B. 2013 (YVUH).

62. Pyxine petricola var. **pallida** Swinscrow & Krog-Norweg. J. Bot. 22: 62. 1975. Plate 6.7D

It is a foliose lichen, found growing closely on bark and rock, the var. is similar to *Pyxine petricola* var. *petricola* but differs by narrower lobes with laminal to marginal sub-reticulate maculae, colourless to brown internal stipe with K-, contains lichenoxanthon.

Distribution in India: Andhra Pradesh and Karnataka. It is a new addition for Chittoor district.

Specimens examined: Andhra Pradesh, Chittoor district, Horsley Hills, on rock, 12.06.12, A. Madhusudhana Reddy and Sanjeeva Nayaka 2052 (YVUH); Talakona, before jungle thrills, N 14°28.698′ E 078°42.749′ alt. 539 m, on bark of *Mangifera indica*, 16.03.2013, Anjali Devi B. and Satish Mohabe 3642, 3720, 3758, 3760, (YVUH).

63. Pyxine petricola var. **petricola** Nyl. in Cromb.-J. Bot. London 14: 263. 1876. Plate 6.7E

It is a foliose lichen, found growing closely, on bark and rock, pale to greenish grey in colour, glistening plaques of pruina adglutinated in younger parts, medulla white, internal stipe red brown, K+ red violet, contains lichenxanthone.

Distribution in India: Andhra Pradesh, Assam, Himachal Pradesh, Jammu and Kashmir, Jharkhand, Karnataka, Kerala, Madhya Pradesh, Maharashtra and Tamil Nadu. It is a new addition for Chittoor district.

Specimens examined: Andhra Pradesh, Chittoor district, Talakona, on bark of *Mangifera indica*, before jungle thrills, N 14°28.698′ E 078°42.749′ alt. 539 m, 16.03.12, Anjali devi B. and Satish Mohabe 3772 (YVUH).

64. *Pyxine punensis* Nayaka & Upreti- Lichenologist, 45: 3–8. 2013. Plate 6.7F

It is a foliose lichen, found growing closely on rock, yellowish grey to yellowish brown in colour, isidiate, isidia nodular to dactylate and do not produce soredia, medulla yellow, contains lichenxantone.

Distribution in India: Andhra Pradesh and Maharashtra and Endemic to India.

Specimens examined: Andhra Pradesh, Chittoor district, Horsley Hills, on rock, Sanjeeva Nayaka and A. Madhusudhana Reddy 2201 (YVUH).

65. Pyxine subcinerea Stirt.-Trans. & Proc., New Zealand Inst. 30: 397. 1898. Plate 6.8A

It is foliose lichen, found growing closely on bark, greenish to greyish in colour, margins intermittently pseudocyphellate, developing into soralia and spreading on to lamina, soredia white to stramineous, medulla yellow, contains lichenxanthone and triterpenes.

Plate 6.8: Some Prominent Lichens of Chittoor District.

A. *Pyxine subcinerea,* **B.** *Ramalina conduplicans,* **C.** *Rinodina badiella,* **D.** *Roccella montagnei,* **E.** *Usnea baileyi,* **F.** *Xanthoparmelia pseudocongensis.*

Distribution in India: Andhra Pradesh, Himachal Pradesh, Jammu and Kashmir, Madhya Pradesh, Nagaland, Sikkim, Tamil Nadu, Uttarakhand and West Bengal-hills.

Specimens examined: Andhra Pradesh, Chittoor district, Talakona, before jungle thrills, N 14°28.698' E 078°42.749' alt. 539 m, on bark of *Mangifera indica*, 16.03.2013, Anjali Devi B. and Satish Mohabe 3737, 3756 (YVUH).

Rinodina (Ach.) Gray

66. Rinodina badiella (Nyl.) Th. Fr.-Lichenogr. Scand. 1: 197. 1871. Plate 6.8C

It is a crustose lichen, found growing on rock, greenish brown to brown in colour, verruculose to areolate, apothecia sunken to sessile, biatorine to lecanorine, up to 1 mm in diam, epihymenium red brown, ascus 8-spored, ascospores brown, 2-celled, *Pachysporina*-type, contains zeorin.

Distribution in India: Arunachal Pradesh, Jammu and Kashmir and Tamil Nadu. It is a new record to Andhra Pradesh and new addition for Chittoor district.

Specimens examined: Andhra Pradesh, Chittoor district, Tirumala Hills, Shilathoranam, on Rock, N 13°41.282' E079°20.418' alt. 937 m, 07.02.2013, Anjali Devi B. and Satish Mohabe 3359 (YVUH).

PORINACEAE

Porina Ach.

67. Porina tetracerae (Afz.) Müll. Arg.-Bot. Jahrb. Syst. 6: 401. 1885. Plate 6.7C

It is a crustose lichen, found growing on bark of tree, greenish brown to brown in colour, smooth to verruculose, ecorticated, perithecia semiglobose, ostiolar region K+ red, ascus 6 spored, ascospores colourless, transversely 1 – 7 septate.

Distribution in India: Andaman and Nicobar Islands, Arunachal Pradesh, Goa, Karnataka, Madhya Pradesh, Nagaland, Odisha, Sikkim, Tamil Nadu and West Bengal.

Specimens examined: Andhra Pradesh, Chittoor district, Japali Anjaneya Swamy temple, alt. 746.5 m, on bark, 13.06.2012, A. Madhusudhana Reddy and Sanjeeva Nayaka 1826 (YVUH).

RAMALINACEAE

Ramalina Ach.

68. Ramalina conduplicans Vain.-Ann. Soc. Zool. Bot. Fenn 1(3): 35. 1921. Plate 6.8B

It is a fruticose lichen, found growing on bark, greenish grey to yellowish brown, branches uniformly up to 4.5 mm wide, upper side smooth, pseudocyphellate, lower side rugose, with raised round to oblong prominent pseudocyphellae, chondroid tissue uneven in thickness, distinctly cracked into hyphal bundles, medulla solid, contains usnic, sekikaic, and salazinic acid.

Distribution in India: Andaman and Nicobar Islands, Andhra Pradesh, Arunachal Pradesh, Himachal Pradesh, Jammu and Kashmir, Kerala, Meghalaya, Sikkim, Tamil Nadu, Uttarakhand and West Bengal hills. It is new addition for Chittoor district.

Specimens examined: Andhra Pradesh, Chittoor district, Talakona, before jungle thrills, N 14°28.698′ E 078°42.749′ alt. 539 m, on bark of *Mangifera indica*, 16.03.2013, Anjali Devi B. and Satish Mohabe 3734, 3760, 3772 (YVUH); Tirumala Hills, Shilathornam, on bark, N 13°41.282′ E079°20.418′ alt. 937 m, 07.02.2013, Anjali Devi B. and Satish Mohabe 3356 (YVUH).

ROCCELLACEAE

Roccella DC.

69. Roccella montagnei Bél-Voy. Ind. Or. 2: 117. 1838. Plate 6.8D

It is a fruticose lichen, found growing on rock, greenish grey to brownish in colour, erect or pendulous, attached by basal hold fast, branched, strap shaped, irregularly widened, up to 5.0 mm long, tapering, isidia and soredia absent, apothecia pedicellate, up to 1 mm, contains erythrin, traces of lecanoric acid, roccellic acid.

Distribution in India: Andhra Pradesh, Gujarat, Karnataka, Kerala, Odisha, Pondicherry and Tamil Nadu.

Specimens examined: Andhra Pradesh, Chittoor district, Horsley Hills, on rock, N 13°.39.123′ E 078° 34.117′, alt. 1265 m, 08.03.2014, Anjali Devi B. and Satish Mohabe 3969 (YVUH).

STEREOCAULACEAE

Lepraria Ach.

70. Lepraria coriensis (Hue) Sipman-Herzogia 17: 28. 2004. Plate 6.5A

It is a leprose lichen, found growing on rocks, powdery to membranous, margin delimited, lobes present, obscure or more often well-developed up to 1.5 mm wide, with raised marginal rim, medulla usually present, thin to medium, white, sorediate, soredia fine to coarse, projecting hyphae usually absent, contains usnic acid, zeorin.

Distribution in India: Andhra Pradesh, Madhya Pradesh, Karnataka. It is a new addition for Chittoor district.

Specimens examined: Andhra Pradesh, Chittoor district, Horsley Hills, near checkpost 4th curve, alt. ca. 882 m, on rock, 12.06.12, A. Madhusudha Reddy, Sanjeeva Nayaka and Anjali Devi B. 2056 (YVUH); Tirumala hills, Papavinasanam, down to Govardhan Dam side, alt. ca. 660 m, on rock, 13.06.12, Sanjeeva Nayaka and A. Madhusudhana Reddy 1864 (YVUH).

TELOSCHISTACEAE

Caloplaca Th. Fr.

1.   Thallus corticolous, yellow, isidiate .... *C. bassiae*

1a. Thallus saxicolous 2

2.   Thallus crustose, whitish grey or greyish brown, apothecia lecidine 3

2a.   Thallus crustose-effigurate, yellow orange to dark orange,
      apothecia lecanorine .... *C. cinnabarina*

3.   Thallus distinct, with dark brown to black prothallus .... *C. poliotera*

3a.   Thallus indistinct, prothallus absent .... *C. tropica*

71. Caloplaca bassiae (Willd. ex Ach.) Zahlbr.-Cat. Lich. Univ. 7: 78. 1930. Plate 6.2C

It is a crustose lichen, found growing on bark, greenish yellow to yellowish orange, thin, smooth, continuous to areolate, isidiate, isidia numerous, yellowish orange, simple to coralloid branched, apothecia rare, scattered, rounded, sessile, 0.3 – 0.8 mm in diam., disc orange to brownish orange, plane to subconvex, margin thin, paler than disc, sometimes isidiate, biatorine, ascus 8 spored, ascospores polaribilocular, hyaline, elongate to ellipsoidal, contains parietin.

Distribution in India: Andhra Pradesh, Andaman and Nicobar Islands, Arunachal Pradesh, Assam, Himachal Pradesh, Jammu and Kashmir, Madhya Pradesh, Odisha, Rajasthan, Sikkim, Tamil Nadu and Uttar Pradesh. It is a new addition for Chittoor district.

Specimens examined: Andhra Pradesh, Chittoor district, Horsley hills, on bark, 12.06.2012, A. Madhusudhana Reddy, Anjali Devi B and Sanjeeva Nayaka 2009 (YVUH); Talakona, before jungle thrills, N 14°28.698′ E 078°42.749′ alt. 539 m, on bark of *Mangifera indica*, 16.03.2013, Anjali Devi B. and Satish Mohabe 3756 (YVUH).

72. Caloplaca cinnabarina (Ach.) Zahlbr. in Engl. & Prantl-Nat. Pflanzenfam.1(1): 228. 1908. Plate 6.2D

It is a crustose-effugurate lichen, found growing on rock, reddish orange to darker orange in colour, areole regular, separated by deep cracks, margins of areole thickened, flat or convex, apothecia restricted in central part, numerous, immersed in central areole or finally raised, disc little darker than thallus, flat to subconvex, ascus 8 spored, ascospores polaribilocular, hyaline, elongate to ellipsoidal, contains parietin.

Distribution in India: Andhra Pradesh, Himachal Pradesh, Karnataka, Madhya Pradesh, Meghalaya Odisha, Rajasthan Tamil Nadu and Uttarakhand.

Specimens examined: Andhra Pradesh, Chittoor district, Horsley Hills, on rock, 12.06.2012, A. Madhusudhana Reddy, Sanjeeva Nayaka and Anjali Devi B. 1715 (YVUH); Tirumala hills, Shilathoranam, N 13°41.282′ E 079°20.418′, alt. 937 m, on rock, 07.02.2013, Anjali Devi B. and Satish Mohabe 3377, 3976, 3977 (YVUH).

73. Caloplaca poliotera (Nyl.) Stein-Sitzungsber Kaiserl Akad. Wiss., Wien, Math.-Naturwiss. Cl. Abt. 1, 106: 219. 1897.

It is a crustose lichen, found growing on rock, greenish-grey to grey in colour; rimose-areolate, prothallus black, apothecia rounded, sessile, 0.2 – 0.5 mm in diam., disc yellowish to reddish brown, flat to convex, epruinose, margin brownish to black, biatorine to lecidine, ascus 8 spored, ascospores colourless, polaribilocular, elongate to ellipsoidal, contains anthraquinons.

Distribution in India: Andhra Pradesh, Madhya Pradesh and West Bengal.

Specimens examined: Andhra Pradesh, Chittoor district, Tirumala hills, Japali, Anjaneya Swamy temple, alt. 746.5 m, on Bark, 13.06.2012, A. Madhusudhana Reddy and Sanjeeva Nayaka 1850/A, (YVUH); Shilathoranam, on rock, 07.02.2013, Anjali Devi B. and Satish Mohabe 3978 (YVUH).

74. Caloplaca tropica Y. Joshi & Upreti-Lichenologist 39(6): 505. 2007. Plate 6.2E

It is a crustose lichen, found growing on rock, grey in colour, thin, indistinct or almost absent, apothecia scattered, sessile, round to irregular, lecidine, ascus 8 spored, ascospores polaribilocular, hyaline, elongate to ellipsoidal, contains parietin.

Distribution in India: Andhra Pradesh, Madhya Pradesh and Uttar Pradesh. It is a new addition for Chittoor district and endemic to India.

Specimens examined: Andhra Pradesh, Chittoor district, Horsley Hills, on rock, 12.06.2012, A. Madhusudhana Reddy, Sanjeeva Nayaka and Anjali Devi B 1715 (YVUH); Thamballapalli, Mallaiah Konda hills, temple premises, alt. 992 m, on bark, 05.01.2013, Satish Mohabe and A. Madhusudhana Reddy 2869 (YVUH).

THELOTREMATACEAE

Diploschistes Norman

75. Diploschistes actinostomus (Pers. ex Ach.) Zahlbr.-Hedwigia 31: 34. 1892 Plate 6.3D

It is a crustose lichen, found growing on exposed rock, whitish grey in colour, scarcely to densely pruinose, rimose cracked, sometimes maculate, apothecia urceolate, immersed in the areolae, ascus 8 spored, ascospores muriform, contains lecanoric acid.

Distribution in India: Himachal Pradesh, Jammu and Kashmir, Meghalaya, Tamil Nadu and Uttarakhand. It is a new record for Andhra Pradesh and new addition for Chittoor district.

Specimens examined: Andhra Pradesh, Chittoor district, Horsley Hills, on rock, 12.06.12, A. Madhusudhana Reddy and Sanjeeva Nayaka 2039, 2039 dup. (YVUH); Tirumala hills, Shilathoranam, backside of the arch, alt. 938.4 m, on rock, 13.06.12, A. Madhusudhana Reddy and Sanjeeva Nayaka, 1802, 1804, 1811 (YVUH).

Acknowledgements

Authors are thankful to Council of Scientific and Industrial Research and Department of Science and Technology, New Delhi for financial assistance; the Director and Dr. D.K. Upreti, Chief Scientist, CSIR-National Botanical Research Institute, Lucknow for their kind permission to utilize the infrastructure facilities of Lichenology Laboratory for identification of lichens; to the Vice-Chancellor of Yogi Vemana University, Kadapa for his support to carry out the research; and to the Forest Official of Chittoor district for the permission.

References

Awasthi, D.D. 1991. *A Key to the Microlichens of India, Nepal and Sri Lanka*. Bibliotheca Lichenologica, J Cramer, Berlin, Stuttgart.

Awasthi, D.D. 2007. *A Compendium of the Macrolichens from India, Nepal and Sri Lanka*. Bishen Singh Mahendra Pal Singh, Dehra Dun.

Chetty, K.M, K.Sivaji and K.Thulasi Rao. 2008. *Flowering plants of Chittoor district, Andhra Pradesh, India*. Student offset printers, Tirupati.

Devi, B.A., Mohabe, S., Reddy, M.A., Nayaka, S. And Shankar, P.C. 2013. Diversity and distribution of Lichens in YSR district, Andhra Pradesh with several new additions. *Indian J. Plant Sci.* **2(4):** 1-9.

Divakar, P.K. and Upreti, D.K. 2005. *Parmelioid Lichens in India (A Revisionary Study)*. Bishen Singh Mahendra Pal Singh, Dehra Dun.

Joshi, Y. 2008. *Morphotaxonomic studies on lichen family Teloschistaceae from India*. Ph.D. Thesis, University of Kumaun, Nainital.

Laundon, J.R. 1981. The species of *Chrysothrix*. *Lichenologist* **13(2):** 101-121.

Lumbsch, H.T. and Huhndorf, S.M. 2007. Outline of Ascomycota–2007. *Myconet* **13:** 1-58.

Mayrhofer, H., Matzer, M., Wippel, A. And Elix, J.A. 1996. Genus *Dimelaena* (Lichenized Ascomycetes, Physciaceae) in southern hemisphere. *Mycotaxon* 58: 293-311.

Miria, A., Khan, A.B. and Rao, B. R.P. 2012. Orchids of Talakona Sacred Grove, Andhra Pradesh, India. *American-Eurasian J. Agric. and Environ. Sci.*, **12 (4):** 469 - 471.

Mohabe, S., Reddy, M.A., Devi, B.A., Nayaka, S. and Chandramati, P.S. 2014. Further new additions to the lichen mycota of Andhra Pradesh, India. *Journal of Threatened Taxa,* **6(8):** 6122-6126.

Nayaka, S. 2005. *Revisionary studies on lichen genus Lecanora sensu lato in India*. Ph.D. Thesis, Dr. Ram Manohar Lohia Avadh University, Faizabad, India

Nayaka, S., Reddy, A.M., Ponmurugan, P., Devi, B.A., Ayyappadasan, G. and Upreti, D.K. 2013. Eastern Ghats, biodiversity reserves with unexplored lichen wealth. *Curr. Sci.* **104(7):** 821-825.

Nayaka, S., Upreti, D. K., Ponmurugan, P. and Ayyappadasan, G. 2013. Two new species of Saxicolous *Pyxine* with yellow medulla from southern India. *Lichenologist* 45(1): 3–8.

Orange, A., James, P.W. and White, F.J. 2001. *Microchemical methods for the identification of lichens*. British Lichen Society, U.K.

Reddy, M.A., Nayaka, A., Shankar, P.C., Reddy, S.R. and Rao, B.R.P. 2011. New distributional records and check list of lichens for Andhra Pradesh, India. *Indian Forester* **137:** 1371-1376.

Singh, K.P. and Sinha, G.P. 2010. *Indian Lichens: Annotated Checklist*. Botanical Survey of India, Kolkata.

Upreti, D.K., Joshi, Y. and Bajpai, R. 2010. New records of lichens growing on monuments in central India. *Geophytology* **38 (1-2):** 37-40.

Vedavathy, S., Sudhakar, A. and Mrdula, V. 1997. Tribal medicinal plants of Chittoor. *Ancient Science of Life* **4:** 307-331.

White, F.J. and James, P.W. 1985. A new guide to the microchemical technique for the identification of lichen substances. *British Lichen Society Bulletin* **57(suppl.):** 1-41.

Chapter 7

Ethnic Plant-Based Nutraceutical Values in Kodagu Region of the Western Ghats

A.A. Greeshma and K.R. Sridhar*

Department of Biosciences, Mangalore University,
Mangalagangotri, Mangalore – 574 199, Karnataka

ABSTRACT

The present contribution highlights traditional plant-derived nutritional and health-promoting practices and products in Kodagu region of the Western Ghats of India. An attempt has been made to document the cultural and traditional heritage on utilization of major wild plant species as food security mainly by the Kodava community of Western Ghats (10 greens, 9 fruits and 4 tubers). According to recent reports, leaves are most valuable ethnomedicinally followed by fruits and roots/tubers. Specific plant materials utilized include greens (leaves and tender shoots), fruits (whole fruits, rind, pulp, juice and seeds) and tubers. Although these practices mainly followed in view of nutritional than medicinal perspective, several products derived from native or wild plant species possesses pharmaceutical advantages. Many plants used for nutritional purpose possess a variety of bioactive compounds of nutraceutical value deserves further insight especially for utilization as feedstock. Recent developments on bioactive potential or principles of plant species traditionally employed are also inventoried.

Keywords: Ethnobotany, Traditional knowledge, Western Ghats, Kodava, Greens, Fruits, Tubers.

---

* Corresponding Author: E-mail: kandikere@gmail.com

Introduction

Indigenous non-conventional folk or ethnic practises pertaining to nutrition and medicine or health-promotion has a long historical background. Indian subcontinent being multi-cultural, multi-ethnic and multi-geographic landscape endowed with treasure of ethnic resources, practices and products across the length and breadth of the country. Diverse, region- and tribe-dependent wild plant species serve as a major resource of ethnic practices to fulfil the food and health requirements. Such practices are followed mainly due to intimate association of local people or tribes with forest ecosystem and utilizing non-timber forest products. Several records document about the richness of wild plant species of the Western Ghats and their importance. For example: 171 edible plant species (67 families) were documented based on the experience of Palliyar tribe in Tamil Nadu (Arinathan *et al.*, 2007); based on traditional knowledge, 70 plant species (42 families) were documented in the southern Western Ghats of Tamil Nadu (Revathi and Parimelazhagan, 2010); 45-50 wild edible fruit-yielding plant species have been identified in Kodagu region of Karnataka (Uthaiah, 1994; Karun *et al.*, 2014); 35 tuberous edible wild plant species (26 genera and 17 families) have been documented from the Western Ghats of Tamil Nadu (Balakrishna, 2014); about 126 plant species (60 families) used traditionally for treating several human ailments (using latex, bark, whole plant, root, flower/fruit and leaves) in the Western Ghats of Karnataka have been documented recently (Lingaraju *et al.*, 2014).

Kodagu, a tiny district in southern India has a geographical area up to 4100 km² situated in the Western Ghats of Karnataka (11°56-12°52N, 75°22-76°11E) with mountain forests at elevation between 300 and 2200 m asl (Pascal and Meher-Homji 1986). The moist rainy monsoon climate is prevalent during June-October with annual rainfall over 500 cm and the temperature generally fluctuates between 15°C and 25°C. A wide range of vegetation exists at different altitudes (*e.g.* grasslands and scrub jungles; shola, moist deciduous, moist-dry deciduous and evergreen forests). This tiny strip of the Western Ghats, Kodagu is a major hub of endemic angiosperms, macrofungi, invertebrates and vertebrates (Myers *et al.*, 2000; Mohana *et al.*, 2011). Over 1300 flowering plants in Kodagu account for 8 per cent and 35 per cent of diversity in India and Karnataka, respectively (Keshavamurthy and Yoganarasimhan, 1990).

The Kodava is a patrilineal ethnolingual community of Kodagu ethnically and culturally differs from the other tribes in southern India. Kodava represented by rich tribal communities for many centuries include Adias, Airis, Betta-Kurubas, Binepadas, Holeyas, Jenu-Kurubas, Kaplas, Kavatis, Koyuvas, Kudiyas, Kurubas, Madivalas, Maleyas, Medas, Nainda, Pales and Yeravas. Among these tribes, some have been migrated to high stream civilised life. Dependence on non-timber forest products for livelihood reveals how important diversity of plants in the Western Ghats. It is not surprising that these tribes have their own style of utilizing diverse plant species in their daily life as source of nutrition in spite of modernization in food practices. Although innumerable plant species serve as nutraceutical source, their documentation and precise methods of utility are less known (*e.g.* Lingaraju *et al.*, 2014; Karun *et al.*, 2014). Moreover, the land use pattern is changing over a period to time towards commercial crops and many wild plant species may not reach kitchen due to modern food practices. Hence, an attempt has been made here to identify selected major exotic plant resources as greens, fruits and tubers routinely used as source of nutrition and comment on their nutraceutical benefits.

Greens

Greens derived mainly from herbs (climbers and subshrubs) constitute a major nutritional source in daily diets of Kodava (Table 7.1). Tender leaves, tender shoots, succulent petioles, fiddle heads and inflorescence are commonly used. Methods of processing, cooking and preparation of dishes of each herb differ. Greens serve as main stuff with addition of other ingredients especially coconut, spices and edible oils. The dishes prepared will be used as fresh (starters or sides) and some can be preserved for some time (*e.g.* sauce, custard and cake). Most of these wild herbs are not cultivated deliberately and they constitute common in and around domestic habitats, forests and plantations. In this section, 10 wild herbs commonly used by Kodava as nutritional source with their possible pharmaceutical value are given.

Alternanthera sessilis (L.) R. Br. ex DC. (Amaranthaceae)

As leafy seasonal wild vegetable, it has a wide pantropical distribution (Africa, Southeast Asia, China, Indonesia, Malaysia and Philippines). Cleaned tender leaves and shoots were pan fried to remove the raw odour followed by seasoning with oil, chilli, onion and spices. This product serves as a starter dish, which can be preserved up to two days and reserved on heating.

Several therapeutic benefits of *A. sessilis* have been recognized, which include anti-inflammatory potential, nootropic activity, cytotoxic effects on pancreatic cancer cell lines and antioxidant property (George *et al.*, 2010; Subhashini *et al.*, 2010; Borah *et al.*, 2011; Kumar *et al.*, 2011). In addition to several ethnomedicinal potential, ethyl acetate fraction of aerial parts of *A. sessilis* serve as an anti-diabetic agent helpful in management of type 2 diabetes (Tan and Kim, 2013).

Cassia tora L. (Fabaceae)

It is a wild seasonal herb distributed in India, Sri Lanka, West China and other tropics (Jain and Patil, 2010). The tender leaves, shoots and flowers cleaned followed by frying until reduce the volume in half and to eliminate raw odour. It was then seasoned with oil, garlic, onion and spices. As starter dish, it has a shelf life of about 3 days.

Aqueous extract of seeds of *C. tora* possess inhibitory effect on germination of *Parthenium* leading to weed control (Vitonde *et al.*, 2014). It is a well known for anthraquinone and all parts (leaves, seeds and roots) are extensively used in Ayurvedic and Chinese medicines (Jain and Patil, 2010).

Centella asiatica (L.) Urb. (Apiaceae) (Figures 7.1A, B)

This tiny plant species has a wide distribution especially in tropical and subtropical regions of Asia and Africa (Singh *et al.*, 2010). Cleaned tender leaves were chopped into pieces and dry fried in a pan for 2 min, ground with coconut and spices to serve as sauce for 2 days.

Table 7.1: Plant Species, Parts and Products Traditionally Used for Nutritional Source in Kodagu Region of the Western Ghats (*Name: En, English; Ka, Kannada; Ko, Kodava)

Plant Species	Name*	Habit	Part Used	Preparation	Product
		Greens			
Alternanthera sessilis (L.) R. Br. ex DC. (Amaranthaceae)	Sessile joyweed (En); Honganne (Ka); Kolike thoppe (Ko)	Herb (creeper)	Tender leaves and shoots	Fried and seasoned with spices and coconut	Starter dish
Cassia tora Linn. (Fabaceae)	Feotida cassia (En); Tagache (Ka); Thathe thoppe (Ko)	Herb	Tender leaves, shoots and pods	Fried by addition of spices and coconut	Starter dish
Centella asiatica (L.) Urb. (Apiaceae) (Figures 7.1A, B)	Indian pennywort (En); Ondelaga (Ka); Onti yele thoppe (Ko)	Herb (creeper)	Tender leaves and shoots	Dry frying, ground with spices and coconut	Sauce
Colocasia esculenta (L.) Schott (Araceae) (Figures 7.1C-G)	Green Taro (En); Kesavina gadde (Ka); Kembu kande (Ko)	Herb	Tender leaves and succulent petioles	Baked with water and mix with coconut and spices	Starter dish and curry
Cucumis dipsaceus Ehrenb. ex Spach (Curcurbitaceae)	Teasle gourd (En); Pavakke (Ko)	Herb (creeper)	Leaves and fruits	Leaves are pan fried and seasoned with spices; Fruits are baked with milk and spices	Starter dish and soup
Diplazium esculentum (Retz.) Sw. (Athyriaceae) (Figures 7.1I-K)	Vegetable fern (En); Therme thoppe (Ko)	Riparian Fern (subshrub)	Fiddle heads	Fried and seasoned with spices and coconut	Starter dish
Drymaria cordata (L.) Willd. ex Schult (Caryophyllaceae)	Tropical chickweed (En); Pana thoppe (Ko)	Herb	Tender leaves	Dry fried, ground with spices and coconut	Sauce
Justicia wynaadensis (Nees) Heyne ex T. Anders. (Acanthaceae) (Figure 7.1L)	Wayanad justicia (En); Aati soppu (Ka); Madd thoppe (Ko)	Subshrub	Tender leaves and shoot extract	Prolonged cooking in low flame and ground with soaked rice	Custard

Contd...

Table 7.1–*Contd...*

Plant Species	Name*	Habit	Part Used	Preparation	Product
Remusatia vivipara (Roxb.) Scott (Araceae)	Hitchhiker elephant ear (En); Marakesu (Ka); Marakembu (Ko)	Herb	Tender leaves and petiole	Leaves/petioles with ground rice, coconut and spices, rolled and steamed; crumbled cake shallow fried with oil	Starter dish
Solanum nigrum L. (Solanaceae) (Figures 7.1M, N)	Blackberry night shade (En); Ganike soppu (Ka); Kake thoppe (Ko)	Herb	Tender leaves and inflorescence	Fried and seasoned with spices and coconut	Starter dish
Fruits					
Carissa inermis Vahl (Apocyanaceae)	Bush plum (En); Waka (Ka); Karmanji pann (Ko)	Thorny shrub	Endocarp	–	Edible as raw and ripened
Citrus aurantium L. (Rutaceae)	Bitter orange (En); Kaipuli (Ko)	Thorny small tree	Pulp	Ground with coconut; baked, roasted and ground	Sauce and juice
Cucumis dipsaceus Ehrenb. ex Spach (Curcurbitaceae)	Teasle gourd (En); Pavakke (Ko)	Herb (creeper)	Leaves and fruits	Leaves pan fried and seasoned with spices; fruits baked with milk and spices	Starter dish and soup
Garcinia gummi-gutta (L.) Roxb. (Clusiaceae)	Malabar Tamarind (En); Mantulli (Ka); Panapuli (Ko)	Tree	Fruits and seeds	Pulp extract heated in low flame to get condensed sour sauce; oil extraction from seeds	Sour sauce and edible fat
Garcinia indica (Thouars) Choisy (Clusiaceae)	Kokam butter (En); Kake mara (Ka); Punarpuli (Ko)	Tree	Fruits	Sun-dried and preserved	Beverage and souring agent

Contd...

Table 7.1-*Contd...*

Plant Species	Name*	Habit	Part Used	Preparation	Product
Mangifera indica L. (Anacardiaceae)	Wild mango (En); Kaad mavu (Ka); Kaad mange (Ko)	Tree	Endocarp	Tender fruits salted/pickled; ripened fruits used in curries	Pickle and curry
Physalis peruviana L. (Solanaceae)	Cape gooseberry (En); Budde hannu (Ka); Goomatte pann (Ko)	Perennial herb	Endocarp	Gently boiled, sugar and lime juice added; heated until become cream	Sauce and cream
Syzygium jambos (L.) Alston (Myrtaceae)	Rose apple (En); Pannerale (Ka); Jammu nerale (Ko)	Tree	Endocarp	–	Edible as raw and ripened
Ziziphus rugosa Lam. (Rhamnaceae)	Wild jujube (En); Mulluhannu (Ka); Kotte pan (Ko)	Thorny shrub	Endocarp	–	Edible as raw and ripened
Tubers					
Amorphophallus paeoniifolius (Dennst.) Nicolson (Araceae)	Elephant foot yam (En); Suvarna gadde (Ka); Chenaekande (Ko)	Herb	Tubers	Deskinned tuber soaked in water, sliced into cubes and baked with salt	Curry
Colocasia esculenta (L.) Schott (Araceae) (Figure 7.1H)	Taro (En); Kesavina gadde (Ka); Kembu kande (Ko)	Herb	Tubers	Deskinned tuber soaked in water, sliced into cubes and baked with salt	Curry
Dioscorea esculenta (Lour.) Burkill (Dioscoreaceae)	Lesser yam (En); Mullu genasu (Ka); Puthari kalanji (Ko)	Herb (creeper)	Tubers	Deskinned tubers baked with limited water and salted	Cake
Manihot esculenta Crantz (Euphorbiaceae)	Cassava/Tapioca (En); Maragenasu (Ka); Marakalanji (Ko)	Herb	Tubers	Deskinned tuber soaked in water, sliced into cubes and baked with salt	Curry

Centella asiatica is one of the most important medicinal plants in the international market of medicinal plant trade (Singh *et al.*, 2010). It is used for enhancing the tranquillizing activity (Aithal and Sirsi, 1961). It has the property to improve learning and memory *in vivo* (Nalini *et al.*, 1992). The leaf juices are commonly used to feed young babies as an attempt to improve intelligence. Leaves also possess curative properties against hypertension and jaundice (Lingaraju *et al.*, 2014). A variety of health-promoting products based on *C. asiatica* have been launched in the market (Singh *et al.*, 2010). The annual demand for *C. asiatica* in 1990 was 12,700 tonnes valued up to Rupees 1.5 billion.

Colocasia esculenta (L.) Schott (Araceae) (Figures 7.1C-G)

This tuberous herbaceous plant distributed in Asia, Southern Africa, Southeast United States and South Western Australia (García-de-Lomas *et al.*, 2012). Some possess green leaves with green petioles (Figures 7.1C, F), some posses reddish tinge on the leaves with reddish-purple petiole (Figures 7.1E, G).

Besides leaves, tender tubers are also delicacy in Kodagu region (Figures 7.1H). Leaves, petioles and tubers of one of the varieties growing on the tree canopies are

Figure 7.1: Leaves of *Centella asciatica* (A, B); Leaves and petioles of *Colocasia esculenta* (C, D), Red petiole of *C. esculenta* (E), White and red exudates from petioles of *C. esculenta* (F, G) and main tuber with baby tubers of *C. esculanta* (H); Leaves of riparian fern *Diplazium esculentum* (I, J) and Edible tender terminal part with fiddle heads (K); Leaves of *Justicia wynaadensis* (L); Leaves and fruits of *Solanum nigrum* (M) and Harvested leaves of *S. nigrum* for dish preparation (N).

most delicacy in Kodagu. Cleaned tender leaves and pealed petioles cut into small pieces and washed in saline water (to prevent throat itching). Leaf and petiole pieces are baked in pressure up to 15 min with onion, spices, ground coconut and tamarind extract. This dish serves as a starter and also used to prepare curry.

Cucumis dipsaceus Ehrenb. ex Spach (Curcurbitaceae)

This annual creeping herb has distribution in India and Ethiopia (Nivedhini *et al.*, 2014). The tender leaves are fried and seasoned with spice and coconut milk followed by baking. Sliced tender fruits are blanched for 5 min followed by pan frying, seasoning with spices and grated coconut or milk are also used to prepare soup. Besides leaves and fruits as vegetable, they possess potent antioxidant property justifying its nutraceutical value (Bussman and Glen, 2010; Chandran *et al.*, 2013).

Diplazium esculentum (Retz.) Sw. (Athyriaceae) (Figures 7.1I-K)

Besides India, this riparian fern has distribution in Cambodia, China, Laos, Malaysia, Philippines, Thailand and Vietnam. The fiddle heads of *D. esculentum* are delicacy in Kodagu. Fiddle heads are pan fried followed by seasoning with a small quantity of edible oil, spice and grated coconut to serve as starter dish. It is the most commonly consumed fern in hilly tribes of North Eastern India and Philippines (Copeland and Collado, 1936).

It is believed by native tribes of India found that this fern is useful in constipation and also as an appetizer (Kala, 2005; Das *et al.*, 2008). It is reported that the edible fronds are rich in iron, phosphorus, potassium and protein (Seal *et al.*, 2012). Decoction of dried rhizomes has the capacity to cure haemoptysis and cough, besides laxative, anti-inflammatory, antioxidant, anthelmintic, antimicrobial, cytotoxic activities are also reported (Akter *et al.*, 2014).

Drymaria cordata (L.) Willd. ex Schult (Caryophyllaceae)

It is distributed in tropical regions of Asia, Africa, Central/South America, and tropical and sub tropical India extending into the Himalayas up to an elevation of 2100 m (Kashyap *et al.*, 2014). Cleaned tender leaves are chopped following by dry frying for 2 min to eliminate mild flavor. Fried leaves were ground with coconut and spices to serve as sauce.

A variety of medicinal properties of leaves of *D. cordata* has been documented by Kashyap *et al.* (2014). Nono *et al.* (2014) reported anti-inflammatory, anti-tussive, anti-bacterial, cytotoxic, anxiolytic activity, analgesic, anti-nociceptive and anti-pyretic properties of *D. cordata* extract. As a vigorously fast growing herb it has been included in the Global Compendium of Weeds (Randall, 2012).

Justicia wynaadensis (Nees) Heyne ex T. And. (Acanthaceae) (Figure 7.1L)

It has distribution mainly in the Western Ghats of India (Ponnamma and Manjunath, 2012). It is believed that the juice of *J. wynaadensis* consists of 18 medicinal properties, hence designated as 'Madd thoppe' meaning 'medicinal leaves'. There is a notion that medicinal principals are present in leaves during mid of July-mid of

August (called 'Ashada Masa' in Hindu calendar), hence it is appropriate time to use this plant as dish. Cleaned leaves and tender shoots are shredded, followed by immersion in cold water, cooked up to 2 hr in low flame to obtain purple extract. The deep coloured extract after eliminating residue ground with soaked rice to make a fine paste, placed on moulds and baked for about 15 min to form a custard, which has shelf life up to 3-4 days.

It is one the most used traditional plants in Kodagu and believed as blood purifier, improves immunity and serve as anthelminthic agent. This plant is known for polyphenols, flavonoids, catalase activity, peroxidise activity and anti-inflammatory ability (Nigudkar *et al.*, 2014). Pounded whole plant, boiled in water to prepare decoction and mixing with honey is useful to treat Asthma (Lingaraju *et al.*, 2014). Whole plant decoction can also be used to cook rice, Its consumption is known to improve immunity (Lingaraju *et al.*, 2014). Leaf and stem possess good antioxidant activity (Medapa *et al.*, 2011). Aqueous extract has bluish purple pigment and serve as potential natural dye in foods as well as in cosmetics (Nigudkar *et al.*, 2014).

Remusatia vivipara (Roxb.) Scott (Araceae)

Besides India, it has a wide distribution in South West China, Indonesia, Nepal, Myanmar, Sri Lanka, Thailand, Vietnam and West Africa (Asha, 2013). This plant also grows luxuriantly in the crevices on tree tops during monsoon season (June-September). Cleaned fresh tender leaves are chopped, ground (with rice, coconut and spices) moulded in rolled leaves and steam-baked for about 20 min. The baked cakes were crumbled and seasoned with oil and mustard to serve as starter dish.

The dish made out of tender leaves helps fighting cold and maintaining body temperature especially during rainy season (Asha *et al.*, 2013). Presence of high quantity of total phenolics in leaves of *R. vivipara* resulted in high antioxidant potential (Asha *et al.*, 2013).

Solanum nigrum L. (Solanaceae) (Figures 7.1M, N)

Distributed in Africa, Indonesia and North America (Saleem *et al.*, 2009). The tender leaves, shoots and inflorescence are washed in cold water, finely chopped followed by pan fry, seasoned with spices and grated coconut, which serve as starter dish up to 2 days.

This plant has been extensively used in traditional medicine in India and other parts of world to cure liver disorders, chronic skin ailments (psoriasis and ringworm), inflammatory conditions, painful periods, fevers, diarrhoea, eye diseases and hydrophobia (Kritikar and Basu, 1935). Decoction prepared out of leaves and unripe fruits prevent diabetes (Lingaraju *et al.*, 2014). The crude extract of leaves of *S. nigrum* showed significant inhibitory activity against filamentous fungi (*Aspergillus flavus* and *A. niger*) (Prakash and Jain, 2011).

Fruits

A wide variety of wild fruits are available in Kodagu region and their edibility has been identified by the tribals. They are rich source of minerals, vitamins, carbohydrates, proteins, fats and fiber (Karun *et al.*, 2014). Besides richness in nutrients, wild fruits help overcome health disorders (Deshmukh and Waghmode, 2011). Many wild fruits also serve as products like pickles, wine, jams, juice and dry fruits. Uthaiah (1994) documented over 50 species of edible fruits in Kodagu region. Valvi *et al.* (2011) documented 30 fruit yielding wild plant species from the Western Ghats of Maharashtra. Recently, up to 45 species of edible wild fruit-yielding plant species have been identified in coffee-based agroforests of Kodagu region entirely on traditional knowledge (Karun *et al.,* 2014). Exotic fruits are known for aroma, texture and pigments, moreover they also serve to produce excellent low calorie dietetic products (Ramadan, 2011). We have repeated eight fruits commonly used in Kodagu (Table 7.1).

Carissa inermis Vahl (Apocyanaceae)

Distributed mainly in the Eastern and Western Ghats (Jayasuriya, 1996). Ripened fruits are sweet, flavoured, eaten raw and also used to prepare pickle (Valvi *et al.,* 2011).

Citrus aurantium L. (Rutaceae)

Besides India, it has widely distributed in China and Indonesia. Fruit have orange-shape, slightly rougher and darker than the sweet orange. Its fruit is also served as vegetable. Sauce can be prepared from the peeled and deseeded pulp mixed with ground coconut. Pulp can also be baked, roasted to decrease moisture content and ground to prepare sauce. The juice extracted from the pulp could be used as squash with addition of sugar.

Decoction made out of shade-dried powder of fruit rind with a small quantity of ginger boiled in water helps to treat diabetes (Lingaraju *et al.,* 2014).

Cucumis dipsaceus Ehrenb. ex Spach (Curcurbitaceae)

Besides India, it has distribution in Ethiopia (Nivedhini *et al.,* 2014). Fruits are good source of proteins, essential aminoacids (including sulphur aminoacids) and minerals (especially calcium) (Nivedhini *et al.,* 2014). It is also useful as a leafy vegetable as described above in Section Greens.

Fruits are very good source of antioxidants and serve as nutraceutical supplement and useful in preparation of health-promoting diets (Nivedhini *et al.,* 2014). Fruit juice also serves in topical applications to prevent hair loss (Bussman and Glen, 2010).

Garcinia gummi-gutta (L.) Roxb. (Clusiaceae)

It is mainly distributed in tropical Asia and Africa (Naveen and Krishnakumar, 2013). Fully ripened fruits are loaded into woven basket and hanged to release fruit extract, which will be collected in earthen pots. Initially, the juice has straw-yellow colour, upon heating it turns into dark red and known as 'Kachampuli'. Usually, the juice will be heated on charcoal or wood fire till it thickens and turns into red. This thick paste will serve as a source of sour alternative to tamarind in dish preparation, which can be stored in glass jars where the colour further deepens into black. This paste has shelf life for several decades. Seeds of *G. gummi-gutta* are also a potential source of fat. Dehusked seeds dried in sunlight up to 15 days, coarsely powdered and heated in low flame or fire in a vessel. The oozed out oil will be strained and stored, which solidifies and serve as edible fat like butter of a considerable period.

It is a common medicinal plant historically known to treat respiratory infections especially sore throat and cough (Oluyemi *et al.*, 2007). Fruit juice serves in topical applications to treat eczema and also possesses anti-cholesterolemic activity, prevent cholesterol accumulation and obesity (Lingaraju *et al.*, 2014). Fruit extract has many ethnomedicinal properties (antioxidant, astringent, anthelmintic and prevents rheumatism, moreover other parts like leaf, stem, bark and roots are also known for a number of disease preventing properties (see Madappa and Bopaiah, 2012).

Garcinia indica (Thouars) Choisy (Clusiaceae)

It is distributed mainly in the Western Ghats of Karnataka and Kerala states. It is a slender, pyramid-shaped, evergreen tree with drooping branches. Rind portion cut from the fresh fruits are sundried and it is used in gravies as alternative to tamarind. Ripe fruit yield juice, which can serve as squash and its shelf life can be extended by addition of salt.

It is a traditional home remedy especially for flatulence, heat strokes and infections (Dushyantha *et al.*, 2010). It is also used in as an appetizer, an anti-inflammatory agent, a liver tonic and to relieve muscle tremor. Methanol extract of fruits of *G. indica* showed the capacity of neuroprotective capacity against Parkinson's disease (Antala *et al.*, 2012).

Mangifera indica L. (Anacardiaceae)

Besides India, it has a wide distribution in Bangladesh, Burma, China, Myanmar and Sri Lanka. It is consumed afresh in tender and ripened stage. Tender fruits are salted and preserved or pickled, which can be preserved for several years. Ripened fruits are used routinely to prepare gravies.

The ripened fruit extracts of *M. indica* serve as a potential protective agent against mild cognitive impairment, increases cholinergic function and decreases oxidative stress leading to enhanced memory in Wistar rats (Wattanathorn *et al.*, 2014). Certainly, further research is needed on the usefulness of fruits of *M. indica* in pharmaceutical products.

Physalis peruviana L. (Solanaceae)

It is mainly distributed in India and tropical part of South America. It is a perennial, soft-wooded, vining plant and purplish branches with hairs. The fruit is berry, smooth, waxy, orange yellow having juicy pulp. Fruits were boiled with a small quantity of water to soften the skin, followed by addition of sugar, lime juice and cooked up to 10 min. The resulting cream can be stored in jars, which has shelf life of 2-3 days. This fruit is highly valued due to its unique flavour (aroma), texture and pigment, hence serve in food industry as functional food especially as beverages, yoghurts and jams (Ramdan, 2011).

Fruits possess high content of vitamin K_1, which has unique health-promoting property. This vitamin functions as a coenzyme and involve in the synthesis of a number of proteins participating in blood clotting and bone metabolism (Damon *et al.*, 2005; Shearer, 1992). Ramdan (2011) reviewed nutritional, bioactive principles and health benefits of fruits of *P. peruviana*. Fruits also endowed with high quantities of phosphorus and vitamin C, the latter has high antioxidant potential. Fruit juice consists of oleic, palmitic, palmitoleic, linoleic and γ-linolenic acids. Fruits are also known for health benefits like anti-hepatotoxic and anti-hepatoma. Owing to elegant nutraceutical potential, fruits of *P. peruviana* are in high demand for its product promotion in industrial scale.

Syzygium jambos (L.) Alston (Myrtaceae)

It is very common in Asia, Central America and Sub-Saharan Africa (Sharma *et al.*, 2013). It is a large shrub to a small tree with wide spread branches. Fruits are almost round, hallow centre with hard brown seeds in the cavity, loosen from the inner wall when ripe seeds rattle indicating its maturity and edibility.

Infusion of the fruit acts as a diuretic as well as sweetening agent, besides seeds are useful to prevent diarrhea and dysentery (Murugan *et al.*, 2011). Acetone extract of bark has potential antibiotic activity against Gram-positive as well as Gram-negative pathogenic bacteria (Murugan *et al.*, 2011).

Ziziphus rugosa Lam. (Rhamnaceae)

Besides India, it has distribution in Bangladesh, Myanmar and Sri Lanka. It is a small tree or straggling thorny shrub. Fruits are globose and become white on ripening and locally marketed. Besides edibility, fruit are useful in treating rheumatism and the decoction of the bark has wound healing property and also prevents diarrhoea (Prashith *et al.*, 2011).

Tubers

Plant species yielding tubers are of immense value as food and pharmaceuticals. Up to 35 tuberous edible wild plant species (26 genera and 17 families) have been reported from the Western Ghat region of Tamil Nadu (Balakrishna, 2014). Their ecological status (*e.g.* rare, endangered and threatened species) was documented and conservation strategies have been suggested. A wide variety of tuberous plant species exists in Kodagu region and the tribals are knowledgeable in identification, available regions and season. On reaping sufficient amount of tubers, tribals sell them as part of their livelihood. This section deals with four mainly used tubers as nutritional source with their possible pharmaceutical value.

Amorphophallus paeoniifolius (Dennst.) Nicolson (Araceae)

Its massive starchy tuber is usually called 'elephant foot yam'. It has distribution in Pacific Islands and Sri Lanka besides India. It is usually found on terrestrial or humus accumulated rocky and shady regions. The corm has characteristic coarse furrows. Tubers are highly acrid and irritate mouth and throat on ingestion due to the presence of calcium oxalate crystals. It has a prominent spike inflorescence with a bulbous knob. Leaf is solitary and emerges after flowering. On peeling skin of corm, soaked in water for half an hour, cut into cubes and baked with a little quantity of salt and water, which can be added to gravies or starters along with other vegetable.

Besides, its stem and petioles are used as thickening agent in fermentation of rice to prepare dish like 'dosa' in Southwest India (Bhagya *et al.*, 2013).

This tuber endowed with a variety of phytochemicals of pharmocological and insecticidal potential (Singh and Wadhwa, 2014). In Ayurvedic system of medicine, tubers are highly valued in treatment of piles, haemophilic conditions, skin diseases, intestinal warms, obesity, restorative in dyspepsia and debility. The tubers also serve as appetizer, tonic and prevent stomach ache (Nadkarni and Nadkarni, 2000; Prajapati *et al.*, 2004).

Colocasia esculenta (L.) Schott (Araceae) (Figure 7.1H)

Besides its usefulness as leafy vegetable (see above; Table 7.1), its baby tubers are commonly edible in Kodagu. Deskinned baby tubers are soaked in water for 30 min, cut into cubes and baked with a small quantity of salt, which is useful as ingredient in preparation of gravies or serve as starters as in *A. paeoniifolius*.

Tubers contain globulins up to 80 per cent of the total proteins. The total amino acids in the tubers is in the range of 1.4-2.4 mg/100 g with relatively low lysine content (Khare, 2007). Up to 70-80 per cent of starch in bulbils of *C. esculenta* called taro starch possesses small granules and highly digestible (Ahmed and Khan, 2013). Compared to conventional starches (*e.g.* maize, potato and wheat), taro starch is ideal in cosmetic formulations and aerosol dispersant. Taro has been used by Hawaiians to treat illness ranging from constipation to tuberculosis. However, more research is warranted on taro starch especially its usefulness in nutrition and pharmaceuticals.

Dioscorea esculenta (Lour.) Burkill (Dioscoreaceae)

It has a wide distribution in Central/South America, Southeast Asia and West Africa (Horrocks and Nunn, 2007). In Kodagu, these tubers are consumed on the occasion of Puthari festival. Cleaned deskinned tubers are baked in steam up to 20 min with a small quantity of salt and consumed as cakes with sugar syrup similar to sweet potato.

This tuber has several ethnomedicinal properties like anti-fatigue, anti-inflammatory, anti-stress, anti-spasmodic and prevention of immune deficiency. The peel of the tuber is also known for anti-cancer and anti-fungal properties (Olayemi and Ajaiyeoba, 2007).

Manihot esculenta Crantz (Euphorbiaceae)

It has a wide distribution in tropical and subtropical regions in Africa, Asia and Latin America (Raji *et al.*, 2009). Roots being a main storage organ, the radical of the germinating roots grow vertically downwards and develop into taproot, from which adventitious roots originate and become storage roots (Alves, 2002). Cassava tubers after peeling outer dark skin are soaked in water for 30 min. It was then cut into cubes and baked with a little amount of salt and water. This dish can be added to gravies or used as starters as in *A. paeoniifolius* and *C. esculenta*.

Cassava has several folk remedies especially to treat cancer, abscesses, boils, conjunctivitis, diarrhoea, dysentery, fever, flu, haemorrhoids, headache, hernia, inflammation, marasmus, prostatitis, rheumatism, snake bite and sore (Awe *et al.*, 2012).

Discussion

The present study embodies commonly used edible greens, fruits and tubers of wild plant species in Kodagu region of the Western Ghats. There are several plant species of nutraceutical interest needs proper exploration, documentation and utilization. Some examples of bioresources serve as delicacy in Kodagu region include bamboo shoots, sweet potato and mushrooms. Documentation of traditional knowledge is the first step to identify a plant species useful nutritionally or pharmaceutically or both. Climatic conditions of the Kodagu region is utmost important to support innumerable number of wild plant species of nutraceutical value. It is a major challenge to document distribution, current status and vulnerability of nutraceutically valued plant species in the Western Ghats. Agroforestry should support existing wild plant species or develop strategies of cultivation to reap benefits besides their main plantation crops. Properly identified and domesticated plant species of nutraceutical value may result in several industrial setup to serve value-added products leading to improve economic status of the tribes, farmers and village dwellers. Besides, greens, fruits and tubers derived from wild plant species serve as suitable feedstock for product generation in industrial scale. For instance, wild fruits are valuable feedstock for production of wines with different aroma and nutraceutical potential (Ramdan, 2011; Karun *et al.*, 2014). Similarly, starches derived from the tubers are potential value in food as well as pharmaceutical industries (*e.g.* taro starch derived from tubers of *Colocacia esculenta*) (Ahmed and Khan, 2013).

Ecosystem and geographical location in the Western Ghats embody useful wild plant species need to be identified to safeguard for future (*e.g.* Sacred groves). According Lingaraju *et al.* (2014), leaves of wild plants of the Western Ghats have foremost importance in ethnomedicinals (39 per cent) followed by flowers/fruits (18 per cent) and root/tubers (16 per cent). Studies on the ethnomedicinal properties of wild plant species used by tribals in Tamil Nadu also showed that leaves of highest number of plant species possess medicinal properties (45 per cent) compared to root, fruit, bark, twig and seeds (Revathi and Parimelazhagan, 2010). The trend of usefulness of aerial part of wild plant species (*e.g.* foliage, twig and fruits) possessing nutritional value is always safer than roots and bark in preventing massive extraction of roots/bark, which may lead to eradication of desired plant species. Besides, some of the major threats for the wild plant resource of Kodagu region include roads, electrical lines, underground gas pipelines, railway tracks and agricultural activities. Wild fire is another major threat, which can be circumvented by cultivating wild plant species in the agroforests. Use of agricultural chemicals like pesticides also prevents growth of wild plant species in agroforests.

Mere documentation of nutritional or medicinal property of a wild plant is insufficient, because so called active ingredient may not elicit function in isolation and the consortia may be beneficial. Decision should be made to select pure compound or consortia in deriving maximum benefit. After assessment of active principles of wild plants, their structural elucidation is valuable to follow availability of similar/

relative natural or synthetic compounds, which leads to venture synthesis through approaches like combinatorial chemistry. Several questions need to be addressed to sustainably utilize the wild plant resources as nutraceuticals: i) Which are the plant species that play a major role in imparting nutritional and health benefits? ii) How much wild resource can be tapped for production of a product in large scale? iii) How to conserve the natural habitats possessing these wild plant species? iv)What are the possibilities to cultivate wild plant species of interest along with agroforestry to improve the economic incentive of natives or tribals?

Acknowledgements

Authors are grateful to Mangalore University for permission to carry out this study in the Department of Biosciences. One of us (GAA) greatly acknowledges the award of INSPIRE Fellowship, Department of Science and Technology, New Delhi, Government of India (Fellowship # IF140953). KRS is grateful to the University Grants Commission, New Delhi, India for the award of UGC-BSR Faculty Fellowship. We thank Karun Chinnappa, N. (Department of Biosciences) and Keshava Chandra (Department of Applied Botany) for timely help and stimulating discussion.

References

Ahmed, A. and Khan, F. 2013. Extraction of starch from taro (*Colocasia esculenta*) and evaluating it and further using taro starch as disintegrating agent in tablet formulation with overall Evaluation. *Inventi Rapid: Novel Excipients,* **2013:** 1-5.

Aithal, H.N. and Sirsi, M. 1961. Preliminary pharmacological studies on *C. asiatica* Linn. (N.O. Umbelliferae). *J. Ethnopharmacol.* **62:** 183-193.

Akter,S., Hossain, M.M., Ara1, I. and Akhtar, P. 2014. Investigation of *in vitro* antioxidant, antimicrobial and cytotoxic activity of *Diplazium esculentum* (Retz.) Sw. *Intern. J. Advances in Pharmacy, Biology and Chemistry,* **3:** 723-733.

Alves, A.A.C. 2002. Cassava botany and physiology, In: Cas*sava: Biology, Production and Utilization* (Eds.) Hillocks, R.J., Thresh, J.M. and Bellotti, A.C. CAB International Publishers, UK, pp. 67-90.

Antala, B.V., Patel, M.S., Bjiva, S.V., Gupta, S., Rabadiya, S. and Lahkar, M. 2012. Protective effect of methanolic extract of *Garcinia indica* fruits in 6-OHDA rat model of Parkinson's disease. *Indian J. Pharmacol.* **44:** 683-687.

Arinathan, V., Mohan, V.R., Britto, A.J. and Murugan, C. 2007. Wild edibles used by Palliyars of the Western Ghats, Tamil Nadu. *Indian J. Traditional Knowl.* 6: 163-168.

Asha, D., Nalini, M.S. and Shylaja, M.D. 2013. Evaluation of Phytochemicals and Antioxidant Activities of *Remusatia vivipara* (Roxb.) Schott., an edible genus of Araceae. *Scholars Research Library,* **5:** 120-128.

Awe, E.O., Kolawole, T.O. and Olaniran, O.B. 2012. Synergistic hepatotoxic potential of *Manihot esculenta* Crantz leaf extract on paracetamol-induced liver damage in rats. *Intern. J. Pharmacology and Therapeutics,* **2:** 38-46.

Balakrishna, S.B. 2014. Ethnomedicinal importance of wild edible tuber plants from tribal areas of Western Ghats of Coimbatore, Tamil Nadu, India. *Lifesciences Leaflets*, **58:** http://dx.doi.org/10.1234/lsl.v58i0.154

Bhagya, B., Ramakrishna, A. and Sridhar, K.R. 2013. Traditional seasonal health food practices in Southwest India: Nutritional and medicinal perspectives. *Nitte University J. Health Sci.* **3:** 30-34.

Borah, A., Yadav, R.N.S. and Unn, B.G. 2011. *In vitro* antioxidant and free radical scavenging activity of *Alternanthera sessilis*. *Intern. J. Pharmaceutical Sciences and Research*, **2:** 1502–1506.

Bussmann, R.W. and Glenn, A. 2010. Medicinal Plants used in Northern Peru for reproductive problems and female health. *J. Ethnobiology and Ethnomedicine*, **6:** 1-12.

Chandran, R., Nivedhini, V. and Parimelazhagan, T. 2013. Nutritional composition and antioxidant properties of *Cucumis dipsaceus* Ehreng. ex Spach leaf. *The Scientific World Journal*: http://dx.doi.org/10.1155/2013/890451

Copeland, E.B. and Collado, T.G. 1936. Crop ferns. *The Philippine J. Agricultural Scientist*, **7:** 367–377.

Damon, M., Zhang, N.Z., Haytowitz, D.B. and Booth, S.L. 2005. Phylloquinone (vitamin K_1) content of vegetables. *J. Food Composition and Analysis*, **8:** 751–758.

Das, A.K., Dutta, B.K. and Sharma, G.D. 2008. Medicinal plants used by different tribes of Cachar District Assam. *Indian J. Trad. Knowl.* **7:** 446–454.

Deshmukh, B.S. and Waghmode, A. 2011. Role of wild edible fruits as a food resource: Traditional knowledge. *Intern. J. Pharmacy and Life Sciences*, **2:** 919–924.

Dushyantha, D.K., Girish, D.N., Suvarna, V.C. and Dushyantha, D.K. 2010. Native Lactic acid bacterial isolates of Kokum for preparation of fermented beverage. *European J. Biol. Sci.*, **2:** 21-24.

García-de-Lomas, J. Dana, E.D. and Ceballos, G. 2012. First report of an invading population of *Colocasia esculenta* (L.) Schott in the Iberian Peninsula. *REABIC BioInvasions Records* **1:** 139–143.

George, S., Bhalerao, S.V. and Lidstone, E.A. 2010. Cytotoxicity screening of Bangladeshi medicinal plant extracts on pancreatic cancer cells. *BMC Complementary and Alternative Medicine*, **10:** 1-11.

Horrocks, M. and Nunn, P.D. 2007. Evidence for introduced taro (*Colocasia esculenta*) and lesser yam (*Dioscorea esculenta*) in Lapita-era (c. 3050–2500 cal. yr BP) deposits from Bourewa, southwest Viti Levu Island, Fiji. *J. Archaeological Sci.*, **34:** 739-748.

Jain, S. and Patil, U.K. 2010. Phytochemical and pharmacological profile of *Cassia tora* Linn – An overview. *Indian J. Natural Products and Resources*, **1:** 430-437.

Jayasuriya, A.H.M. 1996. Two new plant species records from Sri Lanka. *J. South Asian Natural History*, **2:** 43-48.

Kala, C.P. 2005. Ethnomedicinal botany of the Apatani in the Eastern Himalayan region of India. *J. Ethnobiology and Ethnomedicine*, **1**: 1–8.

Karun, N.C., Vaast, P. and Kushalappa, C.G. 2014. Bioinventory and documentation of traditional ecological knowledge of wild edible fruits of Kodagu - Western Ghats, India. *J. Forestry Res.* **25**: 717-721.

Kashyap, K., Sarkar, P., Kalita, M.C. and Banu, S. 2014. A Review on the widespread therapeutic application of the traditional herb *Drymaria cordata*. *Intern. J. Pharmacology and Bio Sciences*, **5**: 696–705.

Keshavamurthy, K.R. and Yoganarasimhan, S.N. 1990. *Flora of Coorg*. Vimsat Publishers, Bangalore.

Khare, C.P. 2007. *Indian medicinal plants: An illustrated dictionary.* Springer, New Delhi.

Kirtikar,K.R. and Basu, B.D. 1935. *Indian Medicinal Plants*, Second Edition, Volume 2. Lalit Mohan Basu, Allahabad.

Kumar, S.M., Rani, S., Kumer, S.L.V.V.S.N.K. and Astalakshmi, N. 2011. Screening of aqueous and ethanolic extracts of aerial parts of *Alternanthera sessilis* Linn. R. Br. ex. DC. for nootropic activity. *J. Pharmaceutical Sciences and Research*, **3**: 1294–1297.

Lingaraju, D.P., Sudarshana, M.S. and Rajashekar, N. 2014. Ethnopharmacological survey of traditional medicinal plants in tribal areas of Kodagu district, Karnataka, India. *J. Pharmacy Research*, **6**: 284-297.

Madappa, M.B. and Bopaiah, A.K. 2012. Preliminary phytochemical analysis of leaf of *Garcinia gummi-gutta* from Western Ghats. *IOSR J. Pharmacy and Biol. Sci.*, **4**: 17-27.

Medapa, S., Singh, G.R.J. and Rivikumar, V. 2011. The phytochemical and antioxidant screening of *Justicia wynaadensis*. *African J. Plant Sci.* **5**: 489-492.

Mohana, G.S., Chittiappa, S., Sinclair, F.L., Kushalappa, C.G., Raghuramulu, Y. and Vaast, P. 2011. *Essence of farmers knowledge on coffee agroforestry systems in Kodagu, Karnataka, India*. College of Forestry, Ponnampet, Kodagu, Karnataka, India.

Murugan, S., Devi, P.U., Parameswari, N.K. and Mani, K.R. 2011. Antimicrobial activity of *Syzygium jambos* against selected human pathogens. *Intern. J. Pharmacy and Pharmaceutical Sciences*, **3**: 44-47.

Myers, N., Mittermeier, R.A., Mittermeier, C.G., da Fonseca, G.A.B. and Kent. J. 2000. Biodiversity hotspots for conservation priorities. *Nature*, 403: 853–858.

Nadkarni, K.M. and Nadkarni, A.K. 2000. *Indian Materia Medica-2*, Third Edition. Popular Prakasan Publication, Bombay.

Nalini, K., Aroor, A.R., Karanth, K.S. and Rao, A. 1992. Effect of *Centella asiatica* fresh leaf aqueous extract on learning and memory and biogenic amine turnover in albino rats. *Fitoterapia*, **63**: 232-237

Naveen, G.P.A.N. and Krishnakumar, G. 2013. Traditional and medicinal uses of *Garcinia gummi-gutta* fruit - A review. *Discovery*, 4: 2-5.

Nigudkar, M., Nital, P., Ramesh, S. and Datar, A. 2014. Preliminary phytochemical screening and finger printing analysis of *Justicia wynaadensis* (Nees). *Intern. J. Pharma Sci*. **4**: 601-605.

Nivedhini, V., Chandran, R. and Parimelazhagan, T. 2014. Chemical composition and antioxidant activity of *Cucumis dipsaceus* Ehrenb. ex Spach fruit. *International Food Research Journal*, **2**: 1465-1472.

Nono, N.R., Nzowa, K.L., Barboni, L. and Tapondjou, A.L. 2014. *Drymaria cordata* (Linn.) Wild. (Caryophyllaceae): Ethnobotany, pharmacology and phytochemistry. *Advances in Biological Chemistry*, **4**: 160-167.

Olayemi, J.O. and Ajaiyeoba, E.O. 2007. Anti-inflammatory studies of yam (*Dioscorea esculenta*) extract on wistar rats. *African journal of Biotechnology*, **6**: 1913-1915.

Oluyemi, K.A., Omotuyi, I.O., Jimoh, O.R., Adesanya, O.A., Saalu, C.L. and Josiah, S.J. 2007. Erythropoietic and anti-obesity effects of *Garcinia cambogia* (bitter kola) in Wistar rats. *Biotechnology and Applied Biochemistry*, **46**: 69-72.

Pascal, J.P. and Meher-Homji, V.M. 1986. Phytochorology of Kodagu (Coorg) District, Karnataka. *J. Bombay Nat. Hist. Soc*. **83**: 43–56.

Ponnamma, S.U. and Manjunath, K, 2012. GC-MS analysis of phytocomponents in the methanolic extract of *Justicia wynaadensis* (Nees) T. Anders. *Intern. J. Pharma and Biosciences*, **3**: 570 – 576.

Prajapati, B.N.D., Purohit, S.S., Sharma, A.K. and Kumar, T. 2004. *A Hand Book of Medicinal Plants – A Complete Source Book*. Agrobios, Jodhpur, India.

Prakash, S. and Jain, A. 2011. Antifungal activity and preliminary phytochemical studies of leaf extract of *Solanum nigrum* Linn. *Intern. J. Pharmacy and Pharmaceutical Sci*. **3**: 352-355.

Prashith, K.T.R., Raghavendra, H.L. and Vinayaka, K.S. 2011. Evaluation of pericarp and seed extract of *Zizyphus rugosa* Lam. for cytotoxic activity. *Intern. J. Pharmaceutical and Biological Archives*, **3**: 887-890.

Raji, A.A.J., Anderson, J.V., Kolade, O.A., Ugwu, C.D., Dixon, A.G.O. and Ingelbrecht, I.L. 2009. Gene-based Microsatellites for cassava (*Manihot esculenta* Crantz): prevalence, polymorphisms, and cross-taxa utility. *BMC Plant Biology*, **9**, 118: 10.1186/1471-2229-9-118.

Ramadan, M.F. 2011. Bioactive phytochemicals, nutritional value, and functional properties of cape gooseberry (*Physalis peruviana*): An overview. *Food Research International*, **44**: 1830-1836.

Randall, R.P. 2012. *A Global Compendium of Weeds*. Second edition. Department of Agriculture and Food, Western Australia.

Revathi, P. and Parimelazhagan, T. 2010. Traditional knowledge on medicinal plants used by the Irula tribe of Hasanur Hills, Erode Distinct, Tamil Nadu, India. *Ethnobotanical Leaflets*, **14**: 136-160.

Saleem, T.S.M., Chetty, C.M., Ramkanth, S., Alagusundaram, M., Gnanaprakash, K., Rajan, V.S.T. and Angalaparamesware, S. 2009. *Solanum nigrum* Linn. - A review. *Pharmocognosy Review*, **3:** 342-345.

Seal, T. 2012. Evaluation of nutritional potential of wild edible plants, traditionally used by the tribal people of Meghalaya state in India. *American J. Plant Nutrition and Fertilization Technology*, **2:** 19–26.

Sharma, R., Kishore, N., Hussein, A. and Lall, N. 2013. Antibacterial and anti-inflammatory effects of *Syzygium jambos* L. (Alston) and isolated compounds on acne vulgaris. *BMC Complementary and Alternative Medicine*, **13,** 292: http://www.biomedcentral.com/1472-6882/13/292

Shearer, M.J. 1992. Vitamin K metabolism and nutriture. *Blood*, **6:** 92–104.

Singh, A. and Wadhwa, N. 2014. A review on multiple potential of aroid: *Amarphophallus paeoniifolius*. *Intern. J. Pharmaceutical Sciences Review and Research*, **24:** 55-60.

Singh, S., Gautam, A., Sharma, A. and Batra, A. 2010. *Centella asiatica* (L.): A plant with immense medicinal potential but threatened. *International J. Pharmaceutical Sciences Review and Research*, **4:** 9-17.

Subhashini, T., Krishnaveni, B. and Reddy, C.S. 2010. Antiinflammatory activity of the leaf extract of *Alternanthera sessilis*. *HYGEIA J. for Drugs and Medicines*, **2:** 54–57.

Tan, K.K. and Kim, K.H. 2013. *Alternanthera sessilis* Red ethyl acetate fraction exhibits antidiabetic potential on obese type 2 diabetic rats. *Evidence-Based Complementary and Alternative Medicine*, 2013: http://dx.doi.org/10.1155/2013/845172

Uthaiah, B.C. 1994. Wild edible fruits of Western Ghats – A survey, In: *Higher Plants of Indian Subcontinent, Additional series of Indian J. Forestry*, Volume 3. (Singh, B.S.M.P. Ed.), Dehra Dun, Uttar Pradesh, India, pp. 87–98.

Valvi, S.R., Deshmukh, S.R. and Rathod, V.S. 2011. Ethnobotanical survey of wild edible fruits in Kolhapur District. *International J. Applied Biology and Pharmaceutical Technology*, **2:** 194-197.

Vitonde, S., Thengane, R.J., Ghole, V.S. 2014. Allelopathic effects of *Casia tora* and *Cassia unifolra* on *Parthenium hysterophorous*. *J. Medicinal Plants Reseach*, **8:** 194-196.

Wattanathorn, J., Muchimapura, S., Thukham-Mee, W., Ingkaninan, K. and Wattya-Areekul, S. 2014. *Mangifera indica* fruit extract improves memory impairment, cholinergic dysfunction, and oxidative stress damage in animal model of mild cognitive impairment. *Oxidative Medicine and Cellular Longevity*: http://dx.doi.org/10.1155/2014/132097

Chapter 8

Conservation of Bryophytes in the Sacred Groves

K.B. Aruna and M. Krishnappa*

*Department of PG Studies and Research in Applied Botany,
Kuvempu University, Jnana Sahyadri,
Shankaraghatta – 577 451. Shimoga District, Karnataka, India*

ABSTRACT

Bryophytes are less concentrated and neglected group of plants compared to other fauna and flora. They play important role in the ecosystem dynamics. The present study is mainly concentrated on documentation and distribution of bryophytes in the sacred groves of Sringeri taluks of Chikmagalur Dist., Karnataka during December 2012 to October 2013. The present study revealed that 37 species of bryophytes belonging to 29 genera and 20 families from six small sacred groves (banas). Naga bana has highest species richness compared to other banas. Because, less human disturbance and scare and believes about God and snakes. In sacred groves, plant and animal communities's were conserved in the name of God.

Keywords: Bryophytes, Sacred groves, Banas, Conservation.

Introduction

Sacred groves are territories of virgin forest with rich diversity, which have been protected by the local people for centuries for their cultural and religious beliefs and

---

* Corresponding Author: E-mail: krishnappam4281@gmail.com

prohibitions that the divinities reside in them and protect the villagers from different calamities (Khan *et al.,* 2008). Sacred groves are forest patches conserved by the local people intertwined with their socio-cultural and religious practices. These groves harbour rich biodiversity and play a significant role in the conservation of biodiversity. Sacred groves occur in many parts of India *viz.,* Western Ghats, Central India, northeast India, particularly where the indigenous communities live. The Western Ghats region of India is well known for its biological diversity and has always been a "Botanist paradise". It's awesome, diversified land forms and environmental conditions support a wide range of forest types. This region has been studied with a highlighting on flowering plants, pteridophytes and few bryophytes.

Bryophytes (Liverworts, Hornworts and Mosses) are diverse and a distinct group of primitive plants (Mishler, 2001; Buck and Goffinet, 2000; Crum, 2001) with about 25,000 species distributed over the world, making in the second largest group of land plants (Shaw and Renzaglia, 2004). In the present scenario, studies of bryophytes are less and neglected compared to other fauna and flora. In the sense of bryophytes conservation, it's really left to God. Because, forests are disturbed and transformed into agricultural land each year and the majority of remaining forests undergo frequent disturbance by human activities, such as timber extraction and agriculture (Achard *et al.,* 2002; Ariyanti, 2008).

As a bryologists, we are happy to know that conservation of bryophytes is going on in the name God's worship. As bryophytes are substrate specific, their ecological niche is nothing but a world within a world. They grow on a wide range of substrates. The present investigation has been undertaken for a detailed enumeration of bryophyte communities and micro habitats interaction in sacred groves of Chikmagalur District, Central Western Ghats, Karnataka.

Materials and Methods

Study Area

The study was conducted in Kigga (13°25′11″ N, 75°11′21″ E, elevation 2212ft), Sringeri taluk of Chikmagalur District, Karntaka. Most of the area can be regarded as 'forest' such as evergreen and semi-evergreen forests which governed by Karnataka state forest department. We have selected six banas (small sacred groves) for our study (Table 8.1).

The season of the annual rain is during June to October. Thus humidity varies from 55 per cent during dry months to 99 per cent during monsoon months. Annual rainfall measured during last 10 years is in the range 4000 – 6000mm. The mean daily maximum temperature is between 22.8°C (July) and 29.1°C (April) and mean daily minimum temperature between 13.2°C (Jan) and 19.8°C (May).

Methods

Frequent field visits were undertaken from December 2012 to October 2013. We have selected six banas (small sacred groves). In each banas, we laid quadrants of size 20 m × 20 m. In each quadrant, presence of bryophytes, their substrate trees, their substrates like tree, main trunk, dead logs, leaf surface, soil and rock surface were

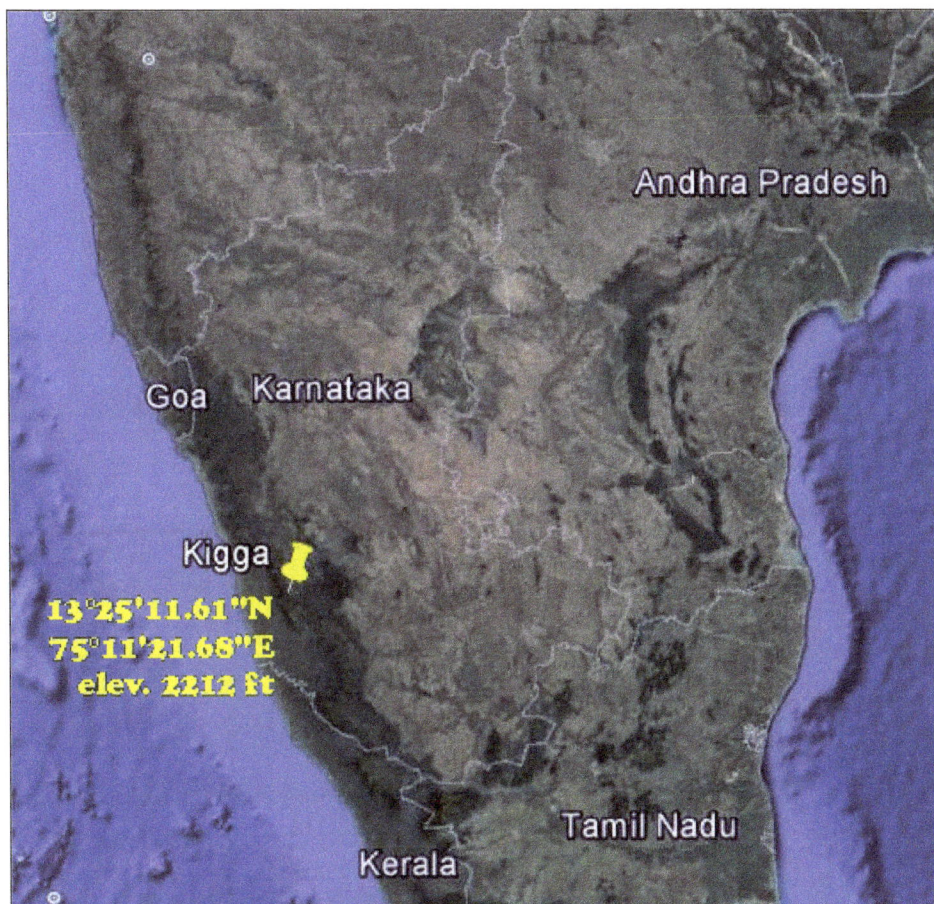

**Figure 8.1: Map Showing Location of the Study Area in
Sringeri Taluk of Chikmagalur Dist., Karnataka.**

documented. The herbarium was prepared by air-drying the collected specimens and stored in paper pockets of 5½ × 4½. The collected materials were also kept in 4 per cent formalin for detailed studies. The external features of the specimens were studied using a stereo trinocular microscope and internal features using the compound microscope. Identification of the specimens was based on the gametophytic and sporophytic characters using the standard manuals and literatures (Kashyap, 1929-1932; Chopra, 1975; Gangulee, 1985; Nair *et al.*, 2005 and Sathisha, 2007). The evaluation of the taxonomical status was done by using TROPICOS. A critical judgment of possible misidentification was not executed and will be subject to future studies. Voucher herbarium specimens of all the bryophyte were preserved in the Department of Applied Botany, Kuvempu University, Shankaraghatta, Shivamogga Dist. Karnataka.

Table 8.1: Status of Banas in the Study Sites in Sringeri Taluks of Chikmagalur Dist., Karnataka

Sl.No.	Name of Banas and Name of Gods and Spirits	Ecology of the Banas	Disturbance Status
1.	**Naga Bana:** God: Naga: Naga statues in stones.13°25'2"N, 75°11'12"E.	Termites mounts in the small patch of forest, surrounded by trees such as *Hopea* spp., *Aporosa lindleyana*, *Memecylon* spp. *Artocarpus hirsutus*, *A. heterophyllus*, Lianas *Gnetum* sp. Little closed canopy, moderate temperature and humidity, totally cool environment.	Disturbance only when Nagarapanchami festival. Here, little scare about Cobra snakes. So human interference is very low.
2.	**Choudi Bana:** Spirits: Choudi. Unstructured two stones treated as the spirit Choudi.13°25'13.46"N, 75°11'25.91"E.	Stones are kept under the big tree of *Ficus* sp., surrounded by trees such as *Hopea* spp., *Memecylon* spp. lianas and climbers. Open canopy, moderate temperature and humidity, cool environment.	Disturbance only when Mahalaya amavaase (Worship of Spirits). But here, litter and fallen twigs collection are going on regularly.
3.	**Arali Katte:** God: Naga.13°25'4.44"N, 75°11'32.29"E.	Naga statues in stones are kept under the tree *Ficus religiosa*. Stage built artificially to the *Ficus* tree by using rocks with cement.	Human interference is too much.
4.	**Raktha Panjrulli Bana:** Spirit: Raktha Panjrulli. Unstructured two stones treated as the spirit Panjrulli. 13°25'12.60"N,75°11'23.29"E.	Stones are kept under the big tree of *Hopea parviflora*, surrounded by trees such as other *Hopea* spp., *Memecylon* spp. lianas and climbers. Closed canopy, moderate temperature and humidity, cool environment.	Disturbance only when Mahalaya amavaase (Worship of Spirits). But here, litter and fallen twigs (for fuel) collection are going on regularly. But no one can touch particular tree *Hopea parviflora*, and the spirit stones, because scare about Raktha panjrulli (Raktha= Blood).
5.	**Varaha Katte:** Spirit: Varaha panjrulli: Statue (Metal) of Varaha (Pig) with panjrulli.13°25'5.89"N,75°11'18.69"E.	Varaha statues of metal are kept under of tree *Artocarpus heterophyllus*. Stage built artificially to the tree by using rocks with cement.	Disturbance is limited. Only in the time of Worship of Spirits.
6.	**Brahma panjrulli Bana:** Spirit: Brahma panjrulli. Unstructured a stone as treated as the spirit Brahma Panjrulli. 13°25'6.51"N,75°11'19.20"E	Stone is kept under the big tree of *Ficus* sp. surrounded by no trees.	Disturbance is very high. Human interference, animal interference etc., are too much.

Results and Discussion

In the present study, we have collected 163 specimens of bryophyte species. Our study revealed that 37 species of bryophytes belonging to 29 genera and 20 families. Of these, mosses comprise 30 species belonging to 22 genera and 13 families, liverworts comprise six species, six genera and six families and a hornworts species (Table 8.2).

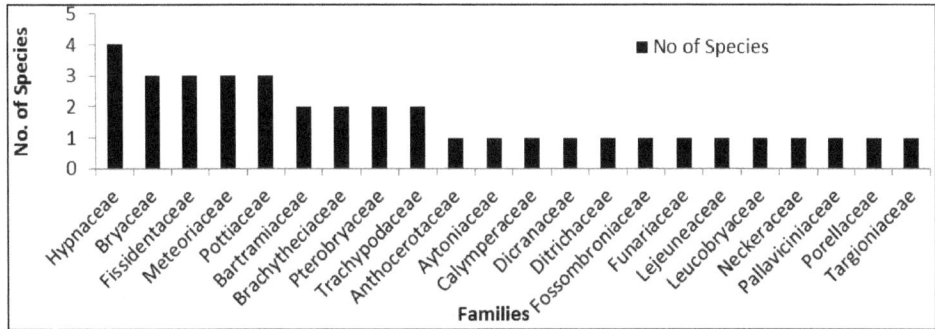

Figure 8.2: Number of Species in the Families.

Banas and Bryophytes Distribution

Naga Bana

In this small sacred grove, we have documented 24 bryophytes species. A notable feature in this bana is the occurrence of termite mounds, which anchorage a variety of mosses such as *Fissidens ceylonensis*, *F. zollingeri* and *Cyathodium cavernarum* and are very commonly distributed. *Brachythecium buchananii*, *Bryum wightii*, *Campylopus flexuosus*, *Floribundaria walkeri*, *Garckea flexuosa* and *Meteoriopsis squarrosa* are well distributed on bark of the trees. There is no human and other animal interference. Disturbance by human only when Nagarapanchami festival. Cobra and other snakes are found commonly in this bana. So, little scare about snakes in human being. Hence, rate of disturbance is low. A detailed list of species occurring in this bana is given in Table 8.2.

Choudi Bana

This bana is the belief of Spirit (Daiva) Choudi. In this bana, 21 species of bryophytes were recorded. *Bryum wightii*, *Campylopus flexuosus*, *Floribundaria walkeri*, *Garckea flexuosa* and *Meteoriopsis squarrosa*, *Cyathodium cavernarum*, *Fissidens crenulatus* and *Trachypus bicolor* are common. Disturbance is little, only when the celebration of Mahalaya amavaase (Worship of Spirits). Bryophytes on soil are very low, compared to bark and rock substrates. Because, maximum of litter collected in the soil surface; decomposition of litter, checks the development of bryophytes in soil surface. But here, local people have been collecting litter and fallen twigs. A detailed list of species occurring in this bana is given in Table 8.2.

Arali Katte

Arali Katte is a small bana of God Naga. It is kept under the tree *Ficus religiosa*. Here, we documented only 12 species. Mainly terricolous species such as *Anthoceros*

Table 8.2: Distribution of Bryophytes in Microhabitat of Banas of the Study Area

Sl.No.	Species Bames	Family	Naga Bana	Choudi Bana	Arali Katte	Raktha Panjrulli Bana	Varaha Katte	Brahma Panjirulli Bana
1.	Aerobryopsis longissima	Meteoriaceae	cc: bt	–	–	cc: bt	–	–
2.	Anthoceros sp.	Anthocerotaceae	–	–	tc: sc	–	tc	tc
3.	Asterella khasiana	Aytoniaceae	–	–	rc: sr	–	tc: sc	tc: sc
4.	Barbula indica	Pottiaceae	–	–	rc: sr	–	rc: sr	tc: sc
5.	Brachythecium buchananii	Brachytheciaceae	cc: bt; rc	cc: bt,br; rc	–	cc: bt	–	cc: bt
6.	Bryum coronatum	Bryaceae	rc: sr	–	cw, sr	–	cw	–
7.	Bryum pseudotriquetrum	Bryaceae	rc: sr	cc: bt	–	–	–	–
8.	Bryum wightii	Bryaceae	cc: ms	cc: ms	cc: bt, ms	cc: bt, ms	cc: ms	cc: bt
9.	Calymperes afzelii	Calymperaceae	–	–	–	cc: ms	cc: ms	–
10.	Calyptothecium sp.	Neckeraceae	rc	rc	–	cc: bt	–	cc: bt
11.	Campylopus flexuosus	Dicranaceae	cc: ms	cc: ms	–	cc: ms	–	–
12.	Cyathodium cavernarum	Targioniaceae	cc: bt; tc	rc	tc: sc	cc: bt; tc	tc: sc	–
13.	Fissidens ceylonensis	Fissidentaceae	tm, rc	–	–	rc	–	–
14.	Fissidens crenulatus	Fissidentaceae	cc: bt, tc	tc	–	–	–	–
15.	Fissidens zollingeri	Fissidentaceae	tm, tc	–	tc	–	tc	–
16.	Floribundaria walkeri	Meteoriaceae	cc: bt; tc	cc: bt	cc: bt	cc :bt; rc	cc: ms	–
17.	Fossombronia indica	Fossombroniaceae	tc, rc	–	tc	–	–	tc
18.	Funaria hygrometrica	Funariaceae	–	–	cw; rc: sr	–	cw, rc	rc
19.	Garckea flexuosa	Ditrichaceae	cc: ms	cc: ms	–	cc: ms	–	–
20.	Hyophila involuta	Pottiaceae	–	–	cw; rc: sr	–	cw; rc: sr	–
21.	Hyophila sp.	Pottiaceae	–	–	cw; rc: sr	–	cw	–

Contd...

Table 8.2–Contd...

Sl.No.	Species Names	Family	Naga Bana	Choudi Bana	Arali Katte	Raktha Panjrulli Bana	Varaha Katte	Brahma Panjrulli Bana
22.	Isopterygium albescens	Hypnaceae	cc: bt	lc	–	cc:ms; lc	–	–
23.	Isopterygium sp.	Hypnaceae	cc: bt	lc	–	lc	–	–
24.	Lopholejeunea subfusca	Lejeuneaceae	cc: bt; fc	fc	–	cc: bt; fc	–	–
25.	Meteoriopsis squarrosa	Meteoriaceae	cc: bt, ms	cc: bt, ms	–	cc: bt, ms	rc	–
26.	Octoblepharum albidum	Leucobryaceae	cc: ms, bt	ef	cc: ms	–	ef	ef
27.	Pallavicinia lyellii	Pallaviciniaceae	–	tc	–	–	tc: sc	tc:sc
28.	Philonotis fontana	Bartramiaceae	tc	tc	tc	tc: sc	tc	tc
29.	Philonotis hastata	Bartramiaceae	–	tc	–	–	–	–
30.	Porella campylophylla	Porellaceae	cc: bt, ms	cc: bt, ms	–	cc: bt, ms	–	–
31.	Pterobryopsis orientalis	Pterobryaceae	–	–	–	cc: bt, ms	–	–
32.	Pterobryopsis sp.	Pterobryaceae	cc: bt, ms	–	–	–	–	–
33.	Rhynchostegium herbaceum	Brachytheciaceae	–	–	–	cc: bt, ms	–	–
34.	Taxiphyllum taxirameum	Hypnaceae	lc	lc	–	lc	–	–
35.	Trachypodopsis serrulata	Trachypodaceae	cc: bt, ms	–	–	–	–	–
36.	Trachypus bicolor	Trachypodaceae	–	cc: bt, ms	–	cc: bt, ms	–	–
37.	Vesicularia reticulata	Hypnaceae	–	rc	–	–	–	–

cc: Corticolous; ms: Main stem; bt: Base of the trees; br: Branches of trees; tc: Terricolous; sc: Soil cuttings; rc: Rupicolous; sr: Submerged rocks; tm: Termite mounds; lc: Lignicolous; cw: Cement walls; fc: Foliicolous; ep: On epiphytic ferns.

sp., *Cyathodium cavernarum, Fissidens zollingeri, Floribundaria walkeri, Fossombronia indica, Funaria hygrometrica* and *Hyophila* spp. are found commonly. Here disturbance is more. So, bryophytes species are comparably low. A detailed list of species occurring in this bana is given in Table 8.2.

Raktha Panjrulli Bana

In this bana, 21 species of bryophytes are recorded. *Aerobryopsis longissima, Brachythecium buchananii, Bryum wightii, Campylopus flexuosus, Floribundaria walkeri, Garckea flexuosa* and *Meteoriopsis squarrosa* are well distributed. People in this bana also believe in the Spirit (Daiva) Raktha panjrulli. Raktha means blood, people believe that this spirit wants blood and punish them very cruelly for their faults. So, here the limit of disturbance is very low. Litter and fallen logs collection is going on far away from the divine tree *Hopea parvifolia*. A detailed list of species occurring in this bana is given in Table 8.2.

Varaha Katte

In this bana, 16 species of bryophytes are documented. Varaha statues are kept under of tree *Artocarpus heterophyllus*. The tree is substrate for some bryophytes such as *Bryum wightii, Calymperes afzelii, Floribundaria walkeri* etc. Disturbance is limited, only in the time of Worship of Spirits. A detailed list of species occurring in this bana is given in Table 8.2.

Brahma Panjrulli Bana

In this bana, we documented only 11 species. This bana is facing some disturbance from human beings and animals. Stone is kept under the big tree of *Ficus* sp. surrounded by no trees. Bark of this *Ficus* tree is used in the preparation of dye for Areca nut, making of tags from *Ficus* barks. A detailed list of species occurring in this bana is given in Table 8.2.

Hypnaceae, Bryaceae, Fissidentaceae, Meteoriaceae and Pottiaceae are dominant families in the study area (Plates 8.1–8.4). *Bryum wightii, B. coronatum, B. pseudotriquetrum, Brachythecium buchananii, Campylopus flexuosus, Cyathodium cavernarum, Floribundaria walkeri, Hyophila involuta, Fissidens* sps., *Porella campylophylla, Barbula indica, Pterobryopsis* sps., *Isopterygium* sps., *Asterella khasiana* and *Fossombronia indica* are commonly distributed in study area. Corticolous bryophytes are well distributed compared to other microhabitat in all banas, it is followed by rupicolous, terricolous, lignicolous and cement wall.

The microhabitat classification is mainly depends on types classifications of Pocs (1982) and Nair *et al.* (2005) with some modifications to deal with distribution of bryophytes in the study area. Though the bryophytes are cosmopolitan in distribution, most of them exhibit strong preference to specific microhabitats. Usually they do not exist as isolated populations but are found growing mixed with other species of bryophytes, ferns, herbs and grasses etc. as biological associations of communities.

Generally, the conservation strategies and special care programmes are meant for the higher group of organisms only. But lower groups of plants especially bryophytes play key role in the ecosystem dynamics (Nair *et al.*, 2005). In the study region, there is no idea about bryophytes. People treated them as 'weed' and also they

Fig 1. *Aerobryopsis longissima*, 2. *Anthoceros* sp., 3. *Asterella khasiana*, 4. *Barbula indica*, 5. *Brachythecium buchananii*, 6. *Bryum coronatum*, 7. *B. pseudotriquetrum*, 8. *B. wightii*.

Plate 8.1

Fig 9. *Calymperes afzelii*, 10. *Calyptothecium* sp., 11. *Campylopus flexuosus*, 12. *Cyathodium cavernarum*, 13. *Fissidens ceylonensis*, 14. *F. crenulatus*, 15. *F. zollingeri*, 16. *Floribundaria walkeri*.

Plate 8.2

Fig 17. *Fossombronia indica*, 18. *Funaria hygrometrica*, 19. *Garckea flexuosa*, 20. *Hyophila involuta*, 21. *Hyophila* sp., 22. *Isopterygium albescens*, 23. *Lopholejeunea subfusca*, 24. *Meteoriopsis squarrosa*.

Plate 8.3

Fig 25. *Octoblepharum albidum*, 26. *Pallavicinia lyellii*, 27. *Philonotis fontana*, 28. *Philonotis hastata*, 29. *Porella campylophylla*, 30. *Pterobryopsis orientalis*, 31. *Pterobryopsis* sp., 32. *Rhynchostegium herbaceum.*

Plate 8.4

believed that bryophytes are 'algae'. They are commonly called by names for bryophytes as 'paachi', 'haasumbe' and 'maragaja'.

Although, there has been no inclusive study on the sacred groves of the entire country, experts estimate the total number of sacred groves in India could be in the range of 100,000 – 150,000 (Malhotra *et al.*, 1998). The forested districts in the Western Ghats namely Uttara Kannada, Shimoga, Udupi, Mangalore, Dakshina Kannada and Kodagu harbour 1424 Sacred Groves (Kalam, 1996; Gokhale, 2000). In Karanataka 1531 sacred groves are documented Studies of sacred groves in Chikmagalur district has not been studied comprehensively. Sacred groves play important role in the ecosystem. They aid to conserve flora and fauna. No systematic data is available on bryophytes of sacred groves of Chikmagalur district. So this preliminary baseline data gives their documentation, distribution and species richness.

Conclusion

Bryophytes are substrate specific and weak competitors and as colonists mostly occur only in temporarily available small microhabitats like the soil, barks of trees, leaves, dead logs, rocks and stones. For these small microhabitat must need strong and healthy macrohabitats like sacred groves. People are scared about God and spirits; but, actually there is no need of scare and believes. There is need of proper knowledge about ecosystem and its application. Today, if we conserve one small sacred grove's like banas, devarakaadu then we can conserve many of species in the ecosystem, directly and/or indirectly.

References

Achard, F., Eva, H.D., Stibig, H.J., Mayaux, P., Gallego, J., Richards, T. and Malingreau, J.P. 2002. Determination of deforestation rate of the world's humid tropical rain forest. *Science*, **297**: 999-1002.

Ariyanti, N.S., Bos, M.M., Kartawinata, K., Tjitrosoedirdjo, S.S., Guhardja, E. and Gradstein, S.R. 2008. Bryophytes on tree trunks in natural forests, selectively logged forests and cacao agroforests in Central Sulawesi, Indonesia. *Biological Conservation*, doi:10.1016/j.biocon.2008.07.012.

Buck, W.R. and Goffinet, B. 2000. Morphology and classification of mosses. In: *Bryophyte Biology* (Shaw A.J. and Goffinet B., eds.), Cambridge University Press. pp. 71–123.

Chopra, R.S. 1975.*Taxonomy of Indian Mosses*. Botanical Monograph 10, CSIR, New Delhi.

Crum, H. 2001. *Structural diversity of bryophytes*. University of Michigan Herbarium, Ann Arbor.

Gangulee, H.C. 1985. *Handbook of Indian Mosses*. Amerind Publishing Co. New Delhi.

Gokhale, Y. 2000. Sacred conservation tradition in India: An overview. *Abstract National Workshop on Community Strategies on the Management of Natural Resources*. Bhopal.

Kalam, M.A. 1996. *Sacred Groves in Kodagu district of Karnataka (South India): A socio historical study*. Institute Francais de Pondichery, Pondichery.

Kashyap, S.R. 1929-1932. *Liverworts of the Western Himalayas and the Punjab Plain, Part I & II (Reprints 1972)*. Research Co. Publications, Trinagar. Delhi.

Khan, M.L., Bongmayum, A. D. K. and. Tripathi, R.S. 2008. The Sacred Groves and their significance in conserving biodiversity: An Overview. *International J. Ecol. Environ. Sci.* **34**: 277-291.

Malhotra, K.C., Gokhale, Y., Chatterjee, S. and Srivastava, S. 1998. *Cultural and ecological dimensions of Sacred Groves in India.* Indian National Science Academy, New Delhi & Indira Gandhi Rashtriya Manav Sangrahalaya, Bhopal.

Mishler, B.D. 2001. The biology of bryophytes – Bryophytes aren't just small tracheophytes. *American J. Bot.,* **88**: 2129–2131.

Nair, M. C., Rajesh, K. P. and Madhusoodanan, P. V. 2005. *Bryophytes of Wayanad in Western Ghats. Malabar Natural History Society*, Calicut, India, pp. 284.

Pocs, T. 1982. Tropical Forest Bryophytes. In : A.J.E. Smith (ed.), *Bryophyte Ecology*, pp. 59-104, Chapman and Hall, Landon.

Sathisha, A.M. 2007. *Survey and Documentation of Bryophytes in Bhadra Wildlife sanctuary, Karnataka. Ph.D. Thesis*, Kuvempu University.

Shaw, A.J. and Renzaglia, K.S. 2004. Phylogeny and diversification of bryophytes. *American J. Bot.* **91**: 1557–1581.

Chapter 9

Some New Additions to Flora of Nagpur District (Maharashtra, India)

M.T. Thakre* and T. Srinivasu**

PGTD of Botany, Rashtrasant Tukadoji Maharaj Nagpur University,
Amravati Road, Nagpur – 440 033, Maharashtra, India

ABSTRACT

Nagpur is one of the districts in the Vidarbha of Maharashtra with rich biodiversity of plants. In 1986, Ugemuge studied 'The Flora of Nagpur district'. Recent exploration of Nagpur flora after urbanization and industrialization, some alterations have been noted in the existing flora. In view of it, it was very necessary to update and revise the existing floristic structure of Nagpur district. During this study, eight plant species belonging to 7 different families were collected, identified and recorded as new additions to Flora of Nagpur district. These species are *Nasturtium officinale* R. Br. (Brassicaceae), *Mimosa rubicaulis* Lam. (Mimosaceae), *Cestrum diurnum* L. (Solanaceae), *Torenia indica* Sald., *Torenia fournieri* Lind. (Scrophulariaceae), *Ocimum gratissimum* L. (Lamiaceae), *Amaranthus dubius* Mart. ex Thells. (Amaranthaceae) and *Phyllanthus tenellus* Roxb. (Euphorbiaceae). All these species have medicinal value and some are rare which needs conservation.

Keywords: Species, Nagpur district, Rare plants.

Corresponding Author: E-mail: *madhu_swt10@rediffmail.com, **dr_srinivasu_t@hotmail.com

Introduction

Nagpur is the most popular district of eastern Maharashtra. It is the largest city in central India and third largest city in Maharashtra after Mumbai and Pune. It is also winter capital of Maharashtra. Nagpur is one of the districts in the Vidarbha (Maharashtra) with rich biodiversity of plants, many of them are economically and medicinally important plants and some of them are rare and endangered plants which need immediate attention for conservation. Nagpur district lies between the latitudes 20°35′ and 21°44′ North and longitudes 78°15′ and 79°40′ East and has an area of 9930 sq km. The average rainfall is 1205 mm and average humidity is 45 per cent. (www.maharashtraonline.in). The climate follows a typical seasonal weather pattern. The peak temperatures are usually reached in May-June and can be as high as 48-50°C. The onset of monsoon is usually from July to September, with monsoon peak during July to August. After monsoon the average temperature varies between 27°C and approx 6-7°C through December and January (http://www.indianngos.com). The district is divided into 14 talukas: Ramtek, Umrer, Kalmeshwar, Katol, Kamthi, Kuhi, Narkhed, Nagpur, Nagpur (Rural), Parseoni, Bhiwapur, Mauda, Saoner and Hingna. The flora of Nagpur District was earlier studied by Ugemuge in 1986. It includes total 1136 plants comprising 841 dicotyledons and 295 moncotyledons. Later on some attempts have been made to upgrade the flora of Nagpur District. Those are *Saccopetalum tomentosum* Hook. (Annonaceae), *Anogeissus pendula* Edgew. (Combretaceae), *Luffa acutangula* (L.) Roxb., *Luffa tuberosa* Roxb., *Momordica charantia* L. var. *muricata* (Willd.) Chakravarty of Cucurbitaceae, *Ceropegia bulbosa* Roxb. (Asclepiadaceae), *Cuscuta chinensis* Lam. and *Cuscuta hyalina* Roth. of Cuscutaceae, *Pedalium murex* L. (Pedaliaceae), *Rhinacanthus nastus* (L.) Kurz. (Acanthaceae), *Boerhavia erecta* L. and *B. repens* L. of Nyctaginaceae, *Aristolochia indica* L. (Aristolochiaceae), *Borasus flabellifer* L. (Arecaceae) and *Eriocaulon duthiei* Hook. (Eriocaulaceae) by Bhuskute (1989); *Nigella sativa* L. (Ranunculaceae), *Coronopus didymus* (L.) Smith. (Brassicaceae), *Indigofera glabra* L. (Fabaceae), *Terminalia paniculata* Roth. (Combretaceae), *Canscora heteroclite* (L.) Gilg. (Gentianaceae), *Ipomoea pes-tigridis* L. (Convolvulaceae), *Striga gesneriodes* (Willd.) Vatke (Scrophulariaceae), *Ocimum canum* Sims. (Lamiaceae), *Aerva javanica* (Burm. f.) Juss. and *Alternanthera tenella* Colla var. *versicolor* (Lem.) Veldk. (Amaranthaceae), *Acalypha lanceolata* Willd. and *Phyllanthus debilis* Klein ex Willd. (Euphorbiaceae), *Curcuma pseudomontana* Grah. (Zingiberaceae), *Dioscorea hispida* Dennstedt, *D. pentaphylla* L., *D. oppositifolia* L. (Dioscoreaceae), *Dichanthium maccanii* Blatt. (Poaceae) by Bhuskute (1990); *Desmodium scorpiurus* (Sw.) Desv., *Indigofera caerulea* Roxb., *Medicago polymorpha* L., *Stylosanthes fruticosa* (Retz.) Alst (Fabaceae) by Thakre and Srinivasu (2012a); *Urena lobata* L. subsp. *sinuata* (L.) Borss. var. *glauca* (Bl.) Borss. (Malvaceae), *Cassia alata* L. (Caesalpiniaceae), *Neptunia oleracea* Lour. (Mimosaceae), *Cuscuta campestris* Yancker. (Cuscutaceae), *Morinda pubescens* J.E. Sm. (Rubiaceae), *Synedrella vialis* (Less.) A. Gray (Asteraceae), *Ipomoea sinensis* (Desv.) Choisy (Convolvulaceae), *Polygonum plebeium* R. Br. var. *brevifolia* Hook. (Polygonaceae) by Thakre and Srinivasu (2012b); *Polycarpon prostratum* (Forsk.) Asch. (Caryophyllaceae), *Acacia ferruginea* DC. (Mimosaceae), *Mollugo nudicaulis* Lam. (Molluginaceae), *Holoptelea integrifolia* (Roxb.) Planch. (Ulmaceae) by Thakre and Srinivasu (2013); *Spigelia anthelmia* L. (Loganiaceae), *Vitex trifolia* L. (Verbenaceae),

Cassia uniflora Mill. (Caesalpiniaceae), *Solanum sisymbrifolium* Lam. (Solanaceae), *Oxalis dehradunensis* Raiz. (Oxalidaceae), *Rauvolfia tetraphylla* L. (Apocynaceae) by Kamble *et al.* (2013a); *Chirita hamosa* R. Br. (Gesneriaceae), *Murdannia semiteres* (Dalz.) Sant. (Commelinaceae), *Habenaria roxburghii* Nicols. (Orchidaceae), *Costus speciosus* (Koen) J.E.Sm. (Zingiberaceae) by Kamble *et al.* (2013b); *Byttneria herbacea* Roxb. (Byttneriaceae), *Alysicarpus ovalifolius* (Schum.), *Millettia peguensis* Ali, *Vigna trilobata* (L.) Verdc. var. *trilobata* (Fabaceae), *Ceiba pentandra* (L.) Gaertn. (Bombacaceae), *Exacum petiolare* Griseb. (Gentianaceae), *Leucas longifolia* Bth. (Lamiaceae), *Justicia betonica* L. (Acanthaceae), *Persicaria barbata* (L.) Hara.var. *gracilis* (Danser) (Polygonaceae), *Cyanotis axillaris* (L.) D. Don (Commelinaceae) by Kamble *et al.* (2013c); *Mutingia calabura* L. (Mutingiaceae), *Mikania micrantha* Kunth and *Centratherum punctatum* Cass. (Asteraceae), *Clerodendrum splendens* G. Don (Verbenaceae) by Kamble and Chaturvedi (2014); *Cardiospermum microcarpa* (Sapindaceae), *Ficus amplissima* and *F. carica* (Moraceae), 3 species from dicotyledons and *Geodorum densiflorum* (Orchidaceae) from monocotyledons reported by Gadpayale *et al.* (2014). However, it was thought worthwhile to undertake current study in the preparation of digital database and electronic herbarium of dicot biodiversity of Nagpur District (Srinivasu, 2003). This work was done by using software, DELTA (Descriptive Language for Taxonomy) (Dalwitz *et al.*, 2000), which provide taxonomic description, identification and information retrieval package system and stores data with interactive key facility. During this floristic study, some plants which are new to the flora of Nagpur district were collected and identified.

Materials and Methods

The plant exploration tours were conducted during various seasons such as pre and post monsoon, winter and summer and visits to forests, crop fields, road sides, vacant places, lakes, tanks and rivers for vegetation and collected the digital photos of plants in their natural habitat and plant specimens for observation, identification and data preparation in the laboratory during the study period. Plant specimens identified with Flora of Maharashtra State: Dicotyledons Volume 1 & 2 (Singh *et al.*, 2000, 2001); Flora of Marathwada Volume I & II (Naik, 1998); The Flora of The Presidency of Bombay (Cooke, 1958) and Flora of Maharashtra (Almeida, 1998, 2001, 2003). The digital images were attached (after processing them) to the respective plant descriptions in the database and the medicinal uses of these plants were noted from different literature.

Result and Discussion

During the preparation of electronic herbarium and digital database of plants of Nagpur district, 8 plant species belonging to 7 different families were recorded as new additions to flora of Nagpur district. The brief descriptions of new records along with their medicinal uses are given below:

Family: Brassicaceae

Nasturtium officinale R. Br. in Ait. Hort. Kew ed. 2, 4: 111. 1812; Hook. f. & T. And. in Hook. f., Fl. Brit. India 1: 133. 1872; Cooke, Fl. Pres. Bombay 1: 31. 1958 (Repr.); Hajra & Chowdhary in Sharma *et al.*, Fl. India 2: 125. 1993.

Much branched herbs, stem rooting at the nodes, sparsely hairy, slightly striate. Leaves alternate, pinnatifid, 2-12 cm long, petiolate. Petiole grooved on upper side, sparsely hairy. Segments 3-9, oblong-orbicular, margin crenate- shallowly lobed, glabrous on both sides. Inflorescence in corymbose, later turns to racemose at the fruiting stage. Flowers white, pedicillate. Calyx divided up to the base, lobes 4, linear-lanceolate. Petals 4, obovate, narrow at the base. Stamens 6, tetradynamous. Ovary elongated with capitate stigma. Fruits linear, cylindrical, 2-3 cm long, with up to 1 cm long stalk, dehisce longitudinally. Seeds ovoid, muricate, biseriate.

Flowering & Fruiting: July-November.

Habitat: Common in wet places.

Distribution in Nagpur district: Rare. University campus, Hajaripahad (Figure 9.1).

Uses: Used for garnishing the salads.

Family: Mimosaceae

Mimosa rubicaulis Lam., Encycl. 1:20. 1783; Baker in Hook. f., Fl. Brit. India **2**: 291, 1878; Cooke, Fl. Pres. Bombay 1: 470, 1958 (Repr.); Sanj. Legumes of India 69, 1991.

Straggling shrubs; branches grooved, furnished with numerous straw colored hooked prickles. Leaves 2-pinnate, up to 18 cm long; pinnae 5-12 pairs; leaflets 8-15 pairs, oblong, obtuse, mucronate, petiolules minute. Flowers in globose heads, pink fading to white, tetramerous, peduncle up to 5 cm long. Calyx minute, ciliolate. Stamens 8. Pods up to 10 cm long and up to 1.5 cm broad, flat, falcate, sutures not prickly.

Flowering & Fruiting: June-October.

Habitat: Common in hilly regions and also in wastelands around fields

Distribution in Nagpur district: Seminary Hills, Bajargaon, Satnavari (Figure 9.2).

Medicinal uses: The leaves are prescribed as an infusion for piles; the bruised leaves are applied to burns. The powdered root is given when from weakness the patients vomits his food (Kirtikar and Basu, 1975).

Family: Solanaceae

Cestrum diurnum L., Sp. Pl. 191. 1753; Bailey, Man. Cult. Pl. ed. 2, 874. 1949; Deb. in J. Econ. Tax. Bot. 1: 37. 1980.

Shrubs; leaves elliptic-oblong, membranous, thinly hairy beneath. Flowers ivory white in terminal panicles. Berries deep purple or nearly black.

Flowering & Fruiting: Almost throughout the year.

Habitat: Rarely planted in gardens, surrounding areas of temples.

Distribution in Nagpur district: Khaperkheda, Kelwad (Figure 9.3).

Uses: The plant is used as ornamental and grown in gardens for its beautiful fragrant flowers.

Figure 9.1

Figure 9.2

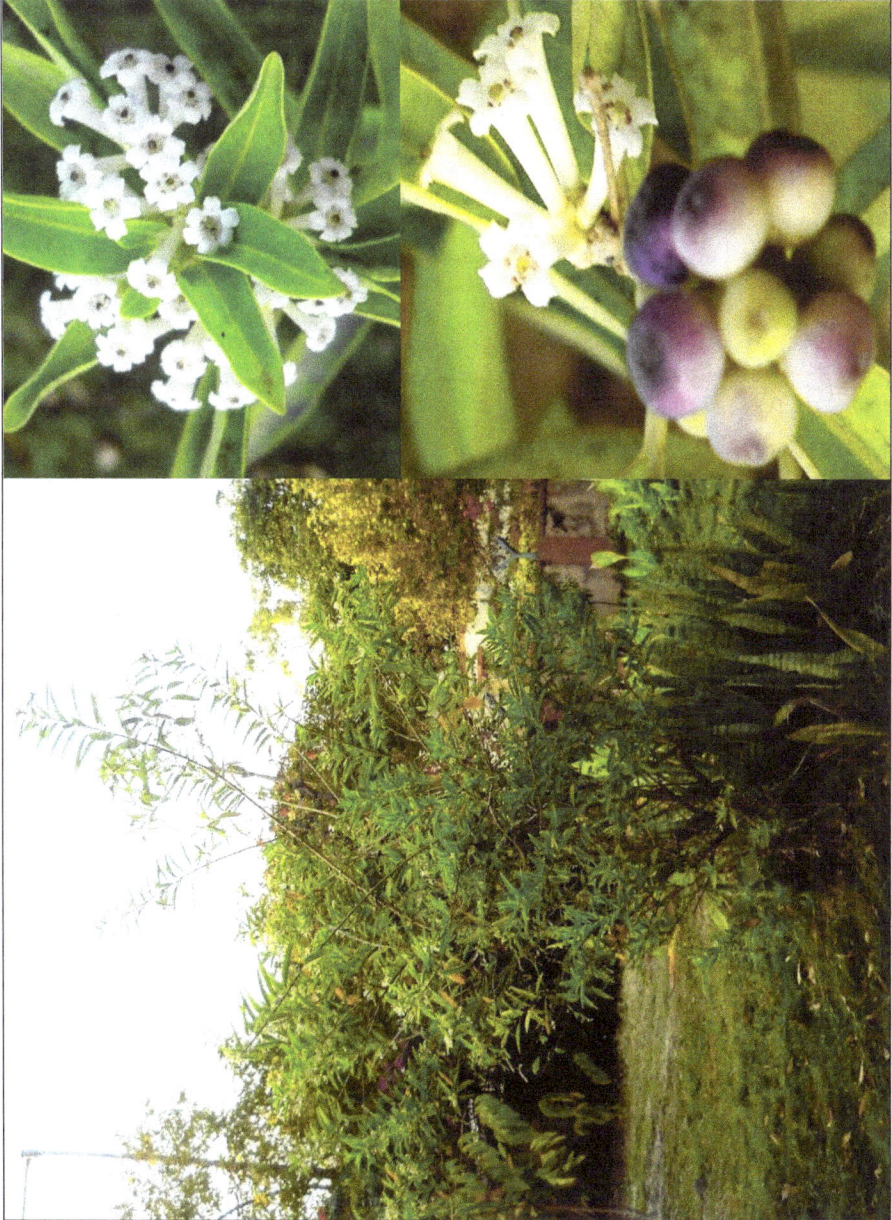

Figure 9.3

Family: Scrophulariaceae

Torenia indica Sald. in Bull. Bot. Surv. India 8: 126, f. 1. 1967. *T. cordifolia* Hook. in Bot. Reg. t. 3715. 1839 non Roxb. 1802; Hook. f. Fl. Brit. India 4: 276. 1884; Cooke, Fl. Pres. Bombay 2: 364. 1958 (Repr.).

Small annual herb, 10-15 cm long. Stem quadrangular, 4-winged, sparsely hairy. Leaves opposite-decussate, 2-4 cm long, 1-2.5 cm broad, ovate, apex obtuse, base cuneate, margin serrate, glabrous on both sides. Petiole 0.5-1.2 cm long, glabrous. Inflorescence solitary cyme or sometimes subumbellate cyme. Flowers axillary, towards the end of branches, pedicillate. Pedicel 1-2 cm long. Calyx tubular, winged, hairy at the margin of the wings. Corolla tubular-bilipped, bluish purple, 1.5-2.5 cm long, upper lip emarginated at apex, lower lip equally 3-lobed, corolla tube pubescent outside. Stamens 4, epipetalous, didynamous, filaments and anthers white or light blue. Capsules oblong, enclosed in the persistent calyx, glabrous. Seeds many, truncate at both ends.

Flowering & Fruiting: August-November.

Habitat: In moist places, in forest undergrowth.

Distribution in Nagpur district: Rare. University campus, Pench forest (Figure 9.4).

Medicinal uses: The juice of leaves used to cure gonorrhea (Kirtikar & Basu, 1975).

Torenia fournieri Lind. ex Fourn. in Illustri.Hortic.23:129, t. 249. 1876.

Annual herb. Stem angular or tetragonal, pubescent. Leaves opposite decussate, petiolate, ovate-lanceolate, apex acute, base truncate-cuneate, margin serrate. Inflorescence solitary or subumbellate cymes at the top of the branches. Flowers axillary or terminal, tubular-bilabiate, light blue with dark purple and dark yellow patches, pedicillate. Calyx broadly winged, pubescent on the wings, lobes 5, ovate-oblong. Corolla bilabiate, white bulbous at the base, golden yellow in the middle and lilac or light blue colour in the upper portion. Stamens 4, epipetalous, didynamous. Carpels 2, syncarpous. Capsule oblong, enclosed in the persistent calyx. Seeds numerous, minute.

Flowering & Fruiting: August-October.

Habitat: In moist places.

Distribution in Nagpur district: Sometimes ornamentally cultivated, also found as an escape. Seminary hills, Ravinagar (Figure 9.5).

Uses: Used as a seasonal ornamental plant in gardens.

Family: Lamiaceae

Ocimum gratissimum L., Sp. Pl. 1197. 1753; Hook. f. Fl. Brit. India 4: 608. 1885; Cooke, Fl. Pres. Bombay 2: 522. 1958 (Repr.); Mukerjee in Rec. Bot. Surv. India 14(1): 20. 1940; Cramer in Dassan. & Fosb. Rev. Handb. Fl. Ceylon 3: 112. 1981.

Herb or undershrub; much branched; young parts pubescent. Leaves 4-10 cm long and 3-5 cm broad, ovate or ovate-elliptic, membranous, pubescent and gland dotted on both sides, margins crenate-serrate. Flowers greenish-yellow in softly hairy

Figure 9.4

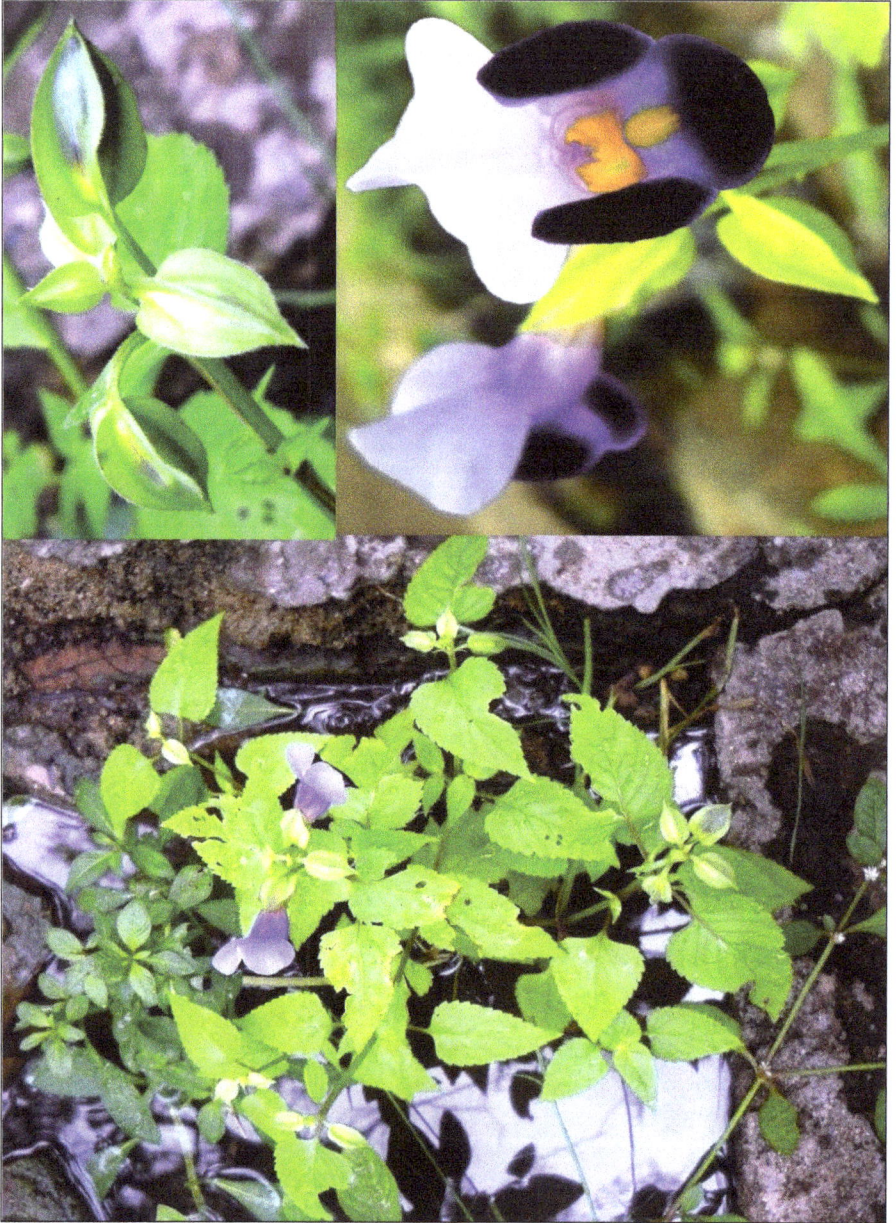

Figure 9.5

whorls in simple or branched racemes; bracts sessile, decussate, longer than the calyx, broadly ovate-lanceolate; calyx campanulate, tube sparsely hispidulous or strigose and dotted with oil globules; corolla up to 0.7 cm long, upper lip up to 0.4 cm long and lower lip up to 0.3 cm long. Stamens exerted. Nutlets subglobose, up to 0.2 cm across, brown.

Flowering & Fruiting: August – November.

Habitat: In open situations in deciduous forests.

Distribution in Nagpur district: Rare. Pench forest, Maharajbag, Rajbhavan (Figure 9.6).

Medicinal uses: The plant is useful in vomiting, fits, "vata", skin diseases, inflammation. The decoction of leaves is of value in cases of seminal weakness, and remedy in gonorrhoea. The seeds are given in headaches and neuralgia (Kirtikar and Basu, 1975).

Family: Amaranthaceae

Amaranthus dubius Mart. ex Thells. in Aschers & Graebn. Syn. 5, 1: 265. 1914; Backer in Steenis, Fl. Males. 1, 4: 79. 1949; Naik, Fl. Marathwada 2: 746. 1998.

Erect annual herb, 30-90 cm tall, much branched, stems often greenish pink, striate, thinly hairy. Leaves 4-15 cm long and 3-7 cm broad, margin entire or undulate, ovate or broadly lanceolate, apex acute or emarginated and minutely mucronate, base cuneate. Flowers in dense terminal simple or branched spikes; bract and bracteoles shorter than tepals, tepals 5, ovate, green. Stamens 5. Ovary ovoid, with 3 short styles. Utricles smooth, irregularly dehiscent.

Flowering & Fruiting: July-November.

Habitat: Common in wasteland around fields and on old walls.

Distribution in Nagpur district: Khaperkheda, University campus (Figure 9.7).

Medicinal uses: The plant is astringent, and highly recommended in menorrhagia, diarrhea, dysentery (Kirtikar and Basu, 1975). The plant is also used as vegetable mostly in rural areas.

Family: Euphorbiaceae

Phyllanthus tenellus Roxb., Fl. India 3:668. 1832; Webster in J. Arnold Arbor. 37:257, t. 1, f. 3. 1956 & 38:52,f.6.1957; R. L. Mitra in Bull. Bot. Surv. India 27:154, f. 1(1985)1987.

Annual erect herb. Stem obtusely subquadrangular, green but at the nodes purplish red. Leaves stipulate, linear-acuminate, reddish. Petiole 0.1 cm long. Leaves alternate-distichous, elliptic, apex obtuse or subacute, base rounded, margin entire, 0.5–1.8 cm long, 0.4–0.9 cm broad, glabrous on both surfaces. Flowers axillary, unisexual monoecious, pentamerous, shallow bowl-like, 0.15–0.2 cm. long, 0.15–2 cm. across. Male flowers sub sessile 1–2 present at the axil, caducous, saucer shaped, white with green midvein, the tepal lobes overlapping at their bases, stamens 5 in the centre of the flower, pistillode absent. Female flower pedicillate solitary from the

Figure 9.6

Figure 9.7

Figure 9.8

same axil of the male flower, sometimes 2, pedicil 0.5 cm long, tepals green, united near the base, staminodes absent, style spreading, connate at base, carpels 3, syncarpus, stigma bifid, seeds 2 in each cocci. Tepals 5, ovate. Fruit a loculicidal capsule, globose, brown, glabrous, 0.15–0.2 cm. long. Seeds trigonous, rounded on the back, pale brown.

Flowering & Fruiting: September-November.

Habitat: Wet places.

Distribution in Nagpur district: Rare, University Campus, Maharajbag (Figure 9.8).

Conclusion

The present study indicates that Nagpur district is one of the biodiversity rich regions for medicinal and economically important plants and some rare plants. This study will give insight for rare plants which need conservation. Electronic herbarium and digital database forms an important centre with data for faster dispersal of information to any corner of the world *i.e.* bio-informatics of plant species in information technology era.

Acknowledgement

One of the authors (M. T. Thakre) is thankful to UGC SAP DRS I for providing financial assistance and Department of Botany, Rashtrasant Tukadoji Maharaj Nagpur University, Nagpur for providing us necessary facilities during the study period.

References

Almeida, M. R. 1998. *Flora of Maharashtra. Vol. II*, Orient Press, Mumbai.

Almeida, M. R. 2001. *Flora of Maharashtra. Vol. III A -B*, Orient Press, Mumbai.

Almeida, M. R. 2003. *Flora of Maharashtra. Vol. IV*, Orient Press, Mumbai.

Bhuskute, S.M. 1989. New Records for Nagpur District (Maharashtra). *Ind. Bot. Reptr.* **8(1)**: 39-42.

Bhuskute, S.M. 1990. New Records for Nagpur District (Maharashtra)-II. *Ind. Bot. Reptr.* **9(2)**: 61-65.

Cooke, T. 1901-1908. *The Flora of the Presidency of Bombay. Vol. I-II.* Taylor Francis, London; reprinted edition, 1958, BSI, Calcutta.

Dallwitz, M. J., Paine, T. A. and Zurcher, E. J. 2000. *User's Guide to the DELTA System: A general system for processing taxonomic descriptions.* 4th ed. http:// biodiversity.uno.edu/delta

Gadpayale, J.V., Somkuwar, S.R. and Chaturvedi, A. 2014. Some noteworthy addition to the Flora of Nagpur district (M.S.), India. *Int. J. of Life Science,* Special issue, **A2**: 35-38.

Kamble, R.B., Hate, S. and Chaturvedi, A. 2013a. New additions to the Flora of Nagpur District, Maharashtra. *J. New Biol. Rep.*, **2(1)**: 09-13.

Kamble, R.B., Hate, S., Mungole, A. and Chaturvedi, A. 2013b. New Record of Some Rare Plants to the Flora of Nagpur District, Maharashtra. *J. New Biol. Rep.,* **2(2)**: 103-107.

Kamble, R.B., Hate, S. and Chaturvedi, A. 2013c. Some new plant reports to the Flora of Nagpur District, Maharashtra - III. *Sci. Res. Rept.,* **3(2)**: 124-128.

Kamble, R. B. and Chaturvedi, A. 2014. New additions to the Flora of Nagpur District, Maharashtra – IV. *Bioscience Discovery,* **5(2)**:160-162.

Kirtikar, K. R. and Basu, B. D. 1975. *Indian Medicinal Plants*. 2nd edition, Vol. I-III Jayyed Press, Delhi.

Thakre, M.T. and Srinivasu, T. 2012a. New (Fabaceae Members) Records to Nagpur District. MFP News, A Quarterly Newsletter of the Centre of Minor Forest Products (COMFORPTS) for Rural development & Environment Conservation, Dehradun (India), **22**: 4-5.

Thakre, M.T. and Srinivasu, T. 2012b. New Plant Species Records to Flora of Nagpur District. *MFP News*, A Quarterly Newsletter of the Centre of Minor Forest Products (COMFORPTS) for Rural development & Environment Conservation, Dehradun (India), **22**: 6-10.

Thakre, M.T. and Srinivasu, T. 2013. New Plant Species Records to Flora of Nagpur District (Maharashtra). *J. Global Biosciences,* **2(6)**: 202-205.

Naik, V. N. 1998. *Flora of Marathwada. Vols. I & II*, Amrut Prakashan, Aurangabad.

Singh, N. P. and Karthikeyan, S. 2000. *Flora of Maharashtra State: Dicotyledons. Vol. 1.* BSI, Calcutta.

Singh, N. P., Lakshminarasimhan, P., Karthikeyan, S. and Prasanna, P. V. 2001. *Flora of Maharashtra State: Dicotyledons. Vol. 2*, BSI, Calcutta.

Srinivasu, T. 2003. Proc. First National Training Workshop on Electronic herbarium and digital database preparation held at Institute of Science, Mumbai, 17-20th Feb. 2003, pp. 1-4.

Ugemuge, N.R. 1986. *Flora of Nagpur District*. Shree Prakashan, Nagpur.

Index to Species and Families

www.ingramcontent.com/pod-product-compliance
Lightning Source LLC
Chambersburg PA
CBHW060246230326
41458CB00094B/1468